The Biological Fractionation of Isotopes

The Biological Fractionation of Isotopes

ERIC M. GALIMOV

Institute of Geochemistry and Analytical Chemistry
Academy of Sciences of the USSR
Moscow, USSR

Translated by Dorothy B. Vitaliano

U.S. Geological Survey
Bloomington, Indiana

Translation Edited by W. G. Meinschein

Department of Geology
Indiana University
Bloomington, Indiana

1985

ACADEMIC PRESS, INC.

(Harcourt Brace Jovanovich, Publishers)
Orlando San Diego New York London
Toronto Montreal Sydney Tokyo

ACADEMIC PRESS, INC.
Orlando, Florida 32887

United Kingdom Edition published by
ACADEMIC PRESS, INC. (LONDON) LTD.
24/28 Oval Road, London NW1 7DX

Library of Congress Cataloging in Publication Data

Galimov, Erik Mikhaĭlovich.
The biological fractionation of isotopes.

Translation of: Priroda biologicheskogo fraktsionirovaniia izotopov.
1. Isotopes--Metabolism. 2. Biogeochemical cycles. 3. Compartmental analysis (Biology). I. Title. [DNLM: 1. Metabolism. 2. Isotopes. QU 130 G158p]
QP531.G3413 1984 574.19'2 84-456
ISBN 0-12-273970-1

PRINTED IN THE UNITED STATES OF AMERICA

85 86 87 88 9 8 7 6 5 4 3 2 1

Contents

4. Experimental Evidence of Thermodynamically Ordered Distribution of Isotopes in Biological Systems

5. Theoretical Model of Biological Fractionation of Isotopes

6. Explanation of the Regularities of the Biological Distribution of Isotopes in the Framework of the Proposed Theory

7. Interpretation of the Results of Some Biochemical Experiments

8. Fractionation of Isotopes in Biological Systems

9. Biological Fractionation of Oxygen Isotopes

Editor's Preface

Isotopes were discovered in 1913. A. Fleck, under the direction of F. Soddy, observed that certain members of different radioactive series have identical chemical properties but different atomic weights. In this same year J. J. Thompson established the presence of ions with mass-to-charge ratios of 20 and 22 while performing positive ray analyses in a neon-filled tube. These ions were specifically identified as ions of stable isotopes of neon in mass spectroscopic analyses run by F. W. Aston in 1919.

Interest in isotopic studies has increased with the growing realization of the unique potential of radioactive isotopes for dating geological and historical events and of isotopes in general for defining chemical and physical processes. Radioactive isotopes have received primary attention because radioactive disintegrations are conditionally invariant, secularly definitive, and easily assayed at high sensitivities. Most notably, radioisotopic "tracers" containing ^{14}C, ^{32}P, and ^{3}H (tritium) have provided the principal means by which biochemists and molecular biologists determine metabolic and genetic pathways. Since the 1940s, however, environmental concerns and improvements in analytical capabilities have expanded the use of stable isotopes in biological and geochemical investigations.

Although theoretical and experimental evidence for the fractionation of stable isotopes of carbon were presented by Urey and Greiff in 1935 and by Neir and Gulbransen in 1939, the major impetus for the scientific utilization of these isotopes was provided by the calculations of ^{12}C and ^{13}C partition functions in monocarbon compounds and diamond by Urey in 1947 and the empirical comparisons of carbon isotope abundances in natural carbonaceous substances by Craig in 1953. Craig's investigations revealed a bimodal distribution of carbon isotopes. Inorganic carbon compounds, such as carbonates and bicarbonates, were shown to be enriched in the heavy ^{13}C isotope relative to the carbon in biological compounds.

Geochemists were predominantly responsible for the initiation of stable isotope studies. Their recognition of the correspondence in the carbon isotope compositions of geological and biological organic materials led them to analyze plants and plant fractions. For example, in 1960 R. Park and S. Epstein investigated carbon isotope fractionation in tomato plants. Their results and subsequent analyses of plants that fixed carbon by the Hatch–Slack (C-4) as well as the Calvin (C-3) photosynthetic pathway supplied both an empirical basis for the interpretation of geochemical isotopic data and evidence of the complexity of isotopic differentiation in metabolic reactions. Isotopic measurements of the carboxyl, amino, and alkyl carbons by Abelson and Hoering in 1961 established that large variations in isotopic compositions occur within specific amino acids at the submolecular level.

In this book E. M. Galimov reviews the substantial amounts of isotopic data obtained in his laboratory and by numerous other investigators. He expands his earlier method of calculating the thermodynamic isotopic β factor for polyatomic carbon compounds.

He presents a detailed development of the assumptions and methods used for the calculation of β values. His assumptions require a microscale reversibility of chemical transformations at the active center of an enzyme. The equations he derives describe the possibility of the development of both kinetic and thermodynamic isotope effects and permit an assessment of the conditions under which each of these effects should prevail. He emphasizes that the equations derived for carbon are applicable for bonds of other elements, and he lists for the first time isotopic bond factors of nitrogen and oxygen compounds.

Galimov shows that the measured isotopic compositions of many biological compounds correlate closely with the values calculated by means of β isotopic bond factors. These correlations indicate that the distributions of isotopes in these compounds are thermodynamically ordered and that Galimov has recognized a significant characteristic of certain metabolic processes. He has made an important contribution to biogeochemistry.

W. G. Meinschein

Preface to the Russian Edition

According to dialectic-materialistic ideas, life is a form of the motion of matter which arises as a new form in the evolution of matter at a certain stage in its development.

Our terrestrial life resulted from the evolution of carbon compounds. The path of this evolution can be divided into two stages: chemical and prebiological. The individual links of the chemical stage can be found and studied not only on our planet but also in interstellar space, on the surface of the planets and the moon, and in the composition of comets and meteorites. In the stellar world, we can detect complex and biologically important organic compounds, but we find no signs of organisms. Here evolution stopped at the chemical stage; therefore, in contrast to the conditions on the earth's surface, we can study these objects in pure form, uncomplicated by the biological processes of the biosphere.

Today the paths of chemical evolution of carbon compounds are also being widely studied using model experiments which reveal the thermodynamic potential of organic substances and the possibility of synthesizing them under the conditions which can be presumed to have existed on the surface of the primitive earth.

Our accumulated knowledge gives us a solid basis for asserting that any organic substance, monomeric or polymeric, freely arose abiotically in some "subvital territory" of the earth, under unknown external conditions. It seemed that the main questions concerning the chemical stage of evolution of carbon compounds that resulted in the origin of life had been answered. However, the question of the original emergence of the finer details of the intramolecular structure of biological compounds, such as optical isomerism and the isotopic composition of carbon atoms (which are lacking in compounds obtained in model syntheses), still remains unanswered.

Weak optical activities can be abiotically induced in organic substances by use of quartz crystals that circularly polarize light. But today more and more facts are suggesting that the high purity of optical asymmetry of modern protoplasm did not arise until the biological stage of evolution, when a new regularity typical of life was superposed on the physiochemical regularities: natural selection of supermolecular phase-isolated systems based on their interaction with the surrounding external environment.

In addition to the optical isomerism typical of the world of living substances, there is the fact that the carbon of organisms is enriched in the light isotope ^{12}C compared with the carbon of inorganic sources of CO_2.

Until recently, it was thought that biological fractionation of isotopes was a process mainly localized in the stage of initial assimilation of CO_2 by autotrophs, and the physical meaning of isotopic fractionation was considered purely kinetically. In the ensemble of interacting particles, those of smaller mass were thought to have a faster reaction rate and to be more mobile. Therefore, the products of the reaction were enriched in the light isotope, and heavy carbon accumulated in the nonreacting residue.

In addition, an attempt was made to judge the biogenic or abiogenic origin of a natural organic substance solely on the overall ratio of ^{12}C and ^{13}C; but more recent studies of biogenic and abiogenic organic substances, either artificially synthesized or isolated from the earth's crust, meteorites, etc., showed that the process was much more complicated. It was discovered that isotopic effects arise not only in photosynthesis but also in several other biochemical processes, and that the overall ratio of carbon isotopes in an object does not always characterize its biogenic origin. Therefore, the systematic distribution of isotopes between and within the biomolecules inherent in organisms is much more important.

E.M. Galimov's book, "The Biological Fractionation of Isotopes," contains voluminous material on the carbon isotopic composition which has been recorded in scientific literature or was obtained experimentally by the author himself. It should by no means be considered only a systematic review of current information, although this is also an achievement of the book. The author has used all available facts on the solution of the fundamental and new-in-principle conclusions which he defends, in particular the experimental and theoretical evidence of the importance of thermodynamically ordered distribution of isotopes in biological systems, which according to the author correlates with a thermodynamic isotopic factor (the β factor). Based on this concept, the author developed his own method of isotopic investigation of biological systems and applied it in his experimental work on a large number of samples.

The author's general conclusions are in some cases still debatable, but they are of great interest for subsequent study of this complex problem.

Of special interest to me, as a biologist working on the origin of life, is the author's claim that the character of the intramolecular distribution of isotopes, which can result not merely from one individual reaction (or even an enzymatic one) but from a whole cycle of reactions accomplished in a phase-isolated open system, can serve as a better criterion for recognizing biogenic and abiogenic compounds. As is known, the thermodynamics of a system of that kind differ from classic thermodynamics, and this is reflected in the distribution of isotopes.

The production of polymolecular systems of that kind as a result of nonspecific self-assemblage of organic macromolecules characterizes the first stage of transition from chemical evolution to its biological stage. Thus (just as with respect to optical asymmetry) the paths of emergence of a systematic intramolecular distribution of isotopes in organic compounds should be sought not at the molecular level but at the level of phase-isolated polymolecular open systems interacting with the surrounding environment and evolving on the basis of primitive selection.

Of course, the extensive work of E. M. Galimov is not a book for light reading. It requires that the reader be open to new ideas. The book is rich in new facts and ideas and undoubtedly is a significant step forward in the study of the complex problem of biological fractionation of isotopes.

A.I. Oparin
Editor of the Russian Edition

Introduction

The properties of different isotopes of the same element are similar. The substitution of one isotope for another in a chemical compound has little effect on the course of the processes in which this compound participates. Therefore, isotopes can serve as nearly ideal indicators of the behavior of elements in physical and chemical processes. However, the properties of isotopes are not identical, and the differences among isotopes are of a different nature when compared with the differences among elements. Accordingly, the principles of the chemistry of isotopes are different from the principles describing the behavior of the chemical elements, i.e., ordinary chemistry. Therefore, by studying the fractionation of isotopes, i.e., the result of the dissimilar behavior of isotopes, one can obtain qualitatively new information on the nature and mechanism of processes compared with that offered by ordinary chemistry.

Historically, the first serious advances in the use of isotopic methods in biochemistry were related to the use of isotopic indicators. Biochemists immediately appreciated the advantages of using isotopes as "tracer atoms" for investigating complex processes of chemical transformation and transport of matter in organisms. The outstanding achievements of the past decades in understanding the chemistry of life are related to the use of isotopes, in particular carbon isotopes—first radioactive ^{14}C and then stable ^{13}C. In most cases the effects of isotopic fractionation could be neglected or taken into account by using corrections in the tracer-atom method. A number of fundamental investigations, especially C.Z. Roginskiy's (1956) book, "Theoretical Bases of Isotopic Methods of Studying Chemical Reactions," were devoted to analysis of the processes of isotopic fractionation mainly by estimating the values of those corrections for fractionation to be used in calculations when the method of isotopic tracers was used. Advances in the use of tracer atoms guided many biochemists to associate closely the concept of isotopy with the indicator properties of isotopes, including artificially introduced isotopic tracers.

Another sphere in which isotopes found extensive application was the geological sciences. Here the development of isotopic methods proceeded differently. Understandably, the method of isotopic tracers had limited prospects for studying geological processes which last millions of years. Therefore geochemists concentrated their attention on studying the fractionation of isotopes, trying to extract geologically useful information from the subtle isotopic effects that occur in natural processes. For this, it was necessary to obtain highly precise measurements of the isotopic compositions of elements. It should be mentioned that the remarkable progress recently achieved in the mass-spectrometric analysis of isotopes that now permits measurement of variations in isotopic content with a precision better than 0.01% was to a substantial extent initiated by the interests of geochemists and to a certain degree resulted from the efforts of experimental geochemists themselves. Systematic study of the isotopic composition of the elements of natural compounds led to voluminous factual data, which made it possible to develop the picture of the geochemical behavior of isotopes in great detail.

As this work developed it became increasingly obvious that many aspects of the distribution of isotopes in nature are controlled by processes which take place in the biosphere. Therefore, geochemical investigations more and more frequently included detailed studies of the isotopic compositions of biological substances. As a result, a curious situation arose: a substantial part of the experimental data concerning isotopic composition of biosystems was obtained in the course of geochemical investigations. Perhaps this explains the fact that many geochemists interested in the role of living matter in nature, including the author himself, turned or are turning to a consideration of biological fractionation of isotopes.

Originally, the concept of biological fractionation of isotopes was related exclusively to the fact that the isotopic composition of elements forming living matter (H, C, N, O, S) differs to some extent from the isotopic composition of those elements in compounds of an inorganic nature. An overwhelmingly large part of the known experimental data pertains to carbon, which is justified by the role this element plays in the chemistry of biological compounds. The carbon of organisms is enriched in the light isotope ^{12}C compared with the carbon of inorganic sources (CO_2, HCO_3^-). The $^{18}O/^{16}O$ isotopic ratio of the tissues of organisms also differs from that of carbon dioxide and water. Biological compounds contain less deuterium than intracellular water.

An established idea, still shared by many, is that biological fractionation of isotopes is a process which is localized mainly to initial assimilation of CO_2 by autotrophs. Epstein's experiments with enzymatic carboxylation of ribulose diphosphate *in vitro* showed that the carbon of phosphoglyceric acid—the initial product of fixation of CO_2 in the Calvin cycle—actually is enriched in the ^{12}C isotope by an amount sufficient to explain the enrichment of living organisms as a whole in that isotope. Therefore, interest in the theoretical examination of biological fractionation of isotopes was to a substantial extent limited to investigation of the kinetic isotopic effect associated with the initial carboxylation reactions. Other isotopic effects in organisms were considered to be of very limited importance, if they were accepted at all.

In the early 1970s it was found that plants which photosynthesize by the Hatch–Slack mechanism (C_4 plants) are much less enriched in the ^{12}C isotope than C_3 plants (M. Bender, S. Epstein, J. Lerman, B. Smith, and others). Still earlier it was established that individual biochemical fractions, individual compounds within these fractions, and even fragments of a compound have different isotopic compositions.

As a picture of more and more complex differentiation of the isotopic distribution in biosystems emerged, it was suggested that isotopic effects arise not only in the stage of photosynthesis but in several other biochemical processes, for instance, respiration and biosynthesis of lipids (S. Epstein), conversion of sugars (E. Degens), and biosynthesis of amino acids (P. Abelson). The difference in fractionation in C_3 and C_4 plants was explained by different isotopic effects in carboxylation of ribulose diphosphate and phosphoenolpyruvate (J. Cristeller, M. Estep, J. Lerman, J. Troughton, and T. Whelan). Regardless of how persuasive the proposed explanations were for individual cases, they shared the idea that a certain number of nodal points (barriers) existed where isotopic fractionation in living organisms was accomplished.

In the approach suggested by the author, biological fractionation was not related to specific paths or stages of biosynthesis but was considered a property inherent in all biochemical reactions taking place in organisms. It was suggested that a very similar mechanism of isotopic fractionation operates in all cases. The common feature of the mechanism and universality of the nature of the isotopic effect in various biochemical reactions were related to the fact that all biochemical reactions take place with the participation of enzymes. The theory proposed in this work is a development of that idea.

A newly discovered fact is the basis of the whole idea: a relationship exists between the isotopic composition of biomolecules and the value of their so-called thermodynamic isotopic factor. An explanation of the latter is as follows: if one of the atoms in any compound is replaced by its isotope, for instance, ^{12}C by ^{13}C, the free energy of the compound is changed. This change has a different value for different compounds. The thermodynamic isotopic factor (β factor) is a measure of the free energy change. At equilibrium this free energy of the system would be minimal. This condition is satisfied if the isotopic ratios characterizing the compound are distributed proportionally to their β factors: the higher the value of the β factor, the higher the content of the heavy isotope in the corresponding compound or structural position of the atom in the molecule.

As we shall see, a tendency toward such distributions of isotopes is a characteristic property of biological systems. Formally, it should be called equilibrium; but this does not completely correspond to the physical sense of the phenomenon. Therefore, we prefer to speak of the thermodynamically ordered distribution of isotopes in biosystems.

Before speaking of the biological aspect of the problem, it must be said that determination of the β factors of biomolecules is a complicated problem in itself.

The classic method of calculating the β factor requires knowledge of the vibrational frequencies of the isotopic forms of the molecule. Finding them necessitates cumbersome calculations but usually is attainable, with the introduction of a number of assumptions, for relatively uncomplicated molecules. For compounds such as biopolymers, a different approach was necessary. It was proposed on the basis of several principles of additivity and realized in the method of isotopic bond factors. Despite the known limitations inherent in the principle of additivity, the proposed method proved to be very fruitful for estimating the β factors of carbon compounds. Practically, its use was intended for interpretation of isotopic effects observed in natural hydrocarbon mixtures in the problems of the geochemistry of oil and gas. Therefore, it was first presented in the book "Carbon Isotopes in Oil and Gas Geology" (Galimov, 1973). However, in principle it was applicable to any complex compound including biomolecules. The additive method of estimating β factors is discussed in the third chapter of this volume. This is not a rehash of former publications. The basis of the method and the numerical values of the β factors themselves have changed little, but the ways of calculating the isotopic bond factors, their physical meaning, and the existing limitations of the principles of additivity are elucidated much more completely. Problems of the precision of method are examined. Additional corrections are introduced which allow broader usage. The isotopic bond factors of nitrogen and oxygen compounds are calculated for the first time. Some new isotopic factors of the bonds of carbon compounds are given. On the basis of the data of that chapter, the numerical value of the β factor of any carbon compound can be obtained.

In the fourth chapter, results of experimental investigations of intermolecular and intramolecular isotopic effects are given. Most of these data were obtained in the Laboratory of Carbon Geochemistry of the Vernadskiy Institute of Geochemistry and Analytical Chemistry of the Academy of Sciences of the USSR. Correlation of the isotopic composition of biological compounds with the values of the corresponding thermodynamic isotopic factors shows a significant relationship between them. The correlation proved to be characteristic of biological compounds which have different chemical compositions and biosynthetic pathways. Thus it is a general tendency for various types of biological compounds.

The purpose of this work is to formulate a theoretical model and an explanation of the observed phenomenon. It should be stressed that the proposed approach is not an application of thermodynamics for investigating biological fractionation of isotopes. The theoretical expression obtained in Chapter 5 describing the fractionation of isotopes in enzymatic reactions is sufficiently general and includes terms defining both the thermodynamic and the kinetic components of the isotopic effect. The fact that enzymatic reactions are characterized by thermodynamic isotopic effects is not the result of the special approach to their investigation but is an objective property of them. This property is determined by the mechanism of enzymatic catalysis. It is important to note that, as analysis shows, the thermodynamic character of the isotopic effects of individual enzymatic reactions alone is not sufficient for the production of a thermodynamically ordered distribution of isotopes within and between the molecules of a biosystem. In other words, even though all enzymatic reactions individually are characterized by thermodynamic isotopic effects, the observed relationship between isotopic compositions and β factors might not be manifested. That is a property of a system of higher order than an isolated enzymatic reaction. The essential properties of this system are the existence of direct and reverse paths of transformations and the steady-state nature of the occurrence of the processes, properties which to a certain extent characterize a metabolic system. The theoretical model of fractionation of isotopes in such a system is in agreement with the experimentally established regularities.

Thus, it is maintained that under certain conditions thermodynamic isotopic effects are inherent in enzymatic reactions, and under certain conditions a thermodynamically ordered distribution of isotopes is inherent in metabolic systems. Disruption of these conditions leads to disruption of the character of isotopic distribution—hence the practical possibility of diagnosing the state of the biosystems of an organism on the basis of the observed character of isotopic fractionation. Because a thermodynamically ordered distribution of isotopes in complex organic compounds is a specific property of living systems, criteria emerge for recognizing biogenic and abiogenic compounds. Moreover, it can be maintained that the manifestation of a thermodynamically ordered distribution of isotopes records the transition from nonliving to living in the course of evolution.

Although it is essentially biochemical, the proposed concept has important geochemical consequences. It then follows that by studying the distribution of isotopes in fossil forms of carbon one can establish the origin of organic compounds in ancient sediments, establish the genetic relationship of certain components (for instance, petroleum hydrocarbons) to their specific biological predecessors, and predict how the isotopic composition of natural products of the transformation of organic matter (for example, methane or carbon dioxide) will change as a function of the conditions and extent of transformation of the original organic matter. On this basis one can explain many aspects of the geochemical behavior of isotopes.

I am deeply grateful to Academician A.I. Oparin for his inspiring discourse and the honor he has shown me by writing the Preface to this work and reviewing its scientific content.

To my deep regret, the work was published after Alexander Ivanovich passed away. I dedicate this book to his long-enduring memory.

CHAPTER 1

Causes of Fractionation of Isotopes

As mentioned in the Foreword, this book is not a systematic compilation of information on the biochemistry of isotopes. Rather, it is devoted to a central idea: experimental and theoretical evidence of the phenomenon of thermodynamically ordered distribution of isotopes in biological systems, which, in our view, is a remarkable result of isotopic fractionation in enzymatic reactions.

However, it is obviously advisable to begin with very general topics. First, some applicable concepts of isotope chemistry will be recalled, because there is no established interpretation of some of them. Second, if the proposed thermodynamic concept is a new one for isotope biochemistry, the very concepts of isotope biochemistry and the natural fractionation of isotopes in general are new to the biological sciences in many respects. Therefore, an introduction to the range of corresponding ideas and information may prove useful.

I. FUNDAMENTAL CONCEPTS

A. Isotopic Composition

Isotopic composition is the relative distribution of isotopes of a given element, usually expressed in the form of the ratio of a less common isotope to the most common one: D/H, $^{13}C/^{12}C$, $^{18}O/^{16}O$, etc. The isotopic composition determines the atomic weight.

For a long time these values were considered to be constants characteristic of an element. "Precise" values of atomic weights and of the relative abundances of isotopes of the elements are usually given in chemical handbooks. For instance, the tabulated values for the atomic weight of carbon

(12.01115), for the $^{12}C/^{13}C$ ratio (88.69), and for its reciprocal $^{13}C/^{12}C$ (0.011275) are known. However, owing to nonequivalence in the properties of isotopes, their content in natural substances varies. The atomic weight of the same carbon ranges from 12.01144 to 12.01002 and the isotopic composition from 0.010225 to 0.011574. These variations are not large, but they exceed by several orders of magnitude the accuracy with which they can be measured. The precision of modern isotopic analysis (determination of atomic weight) using mass spectrometry is ± 0.000001.

B. Isotopic Effects

The configuration of the electron shells of different isotopes of an element is the same, and therefore the chemical behavior of those isotopes is essentially the same. However, the differences in isotopes, primarily in mass and also in nuclear spin, cause the isotopic forms of a compound to behave slightly differently. As a result, the isotopic composition of the product of a reaction may differ from the isotopic composition of the original compound. This difference is the isotopic effect.

Variation in the isotopic composition can also be caused by nuclear reactions, which result in the synthesis or transformation of isotopes. The intensity of nuclear reactions is not sufficient, under ordinary terrestrial conditions, to change the ratio of stable isotopes of elements such as H, C, O, N, and S, even if the effect is cumulative over hundreds of millions of years. The only practically significant nuclear isotopic effects occur in the interior of the stars or through the prolonged intense irradiation of cosmic bodies, for instance, the lunar surface and meteorites. Trace amounts of isotopes can be formed in the earth's atmosphere as a result of nuclear interactions with cosmic rays. As a result of radioactive decay and the accompanying nuclear reactions, several stable radiogenic isotopes are formed, including ^{3}He, ^{40}Ar, ^{87}Sr, lead isotopes, and others. These isotopic effects have no bearing on the problems examined in this book.

In contrast to the nuclear transformation of isotopes, isotopic fractionation is the consequence of the redistribution of isotopes between interacting compounds, which increases the relative concentration of an isotope in one compound while decreasing it in another. Isotopic fractionation results from the physicochemical inequality of isotopes, which can affect either the rate of a process or the energy state of a system. In the first case the so-called kinetic isotopic effect takes place; in the second case, the thermodynamic effect occurs.

C. Measured Values

The isotope ratios D/H, $^{13}C/^{12}C$, $^{18}O/^{16}O$, $^{15}N/^{14}N$, $^{34}S/^{32}S$, etc., can be measured directly in a mass spectrometer. However, the δ value, which is expressed either in percentages (%) or per mil (‰) and represents a ratio of the difference between isotopic composition of the investigated sample (R_x) and of a standard sample (R_{stand}) divided by R_{stand}, can be determined with far greater precision then the isotopic ratios. The values of δ are determined as follows:

$$\delta = [(R_x - R_{stand})/R_{stand}] \times 10^3\,‰$$

or

$$\delta = [(R_x - R_{stand})/R_{stand}] \times 10^2\,\%. \qquad (1.1)$$

In this book all δ values are given in per mil (‰).

Any substance can be used as a standard. Usually a substance convenient to prepare chemically for mass-spectrometric analysis is chosen. All laboratories customarily report their results with respect to a specific standard. At present there are several generally accepted standards. To designate a specific element, the symbol δ is written in combination with the symbol of the respective element.

In isotopic analysis of hydrogen and oxygen, the standard is a sample of water prepared so that its isotopic composition corresponds to that of standard mean ocean water (SMOW) (Craig, 1961):

$$\delta D = \frac{(D/H)_{sample} - (D/H)_{SMOW}}{(D/H)_{SMOW}} \times 10^3\,‰; \qquad (1.2)$$

$$\delta^{18}O = \frac{(^{18}O/^{16}O)_{sample} - (^{18}O/^{16}O)_{SMOW}}{(^{18}O/^{16}O)_{SMOW}} \times 10^3\,‰. \qquad (1.3)$$

Results of carbon and oxygen isotopic analysis are usually reported with respect to the PDB standard, which is a Cretaceous calcite fossil, *Belemnitella americana*. This standard was arbitrary (Craig, 1957), and none of it remains. However, results are traditionally recalculated on the basis of PDB standard:

$$\delta^{13}C = \frac{(^{13}C/^{12}C)_{sample} - (^{13}C/^{12}C)_{PDB}}{(^{13}C/^{12}C)_{PDB}} \times 10^3\,‰. \qquad (1.4)$$

In isotopic analysis of nitrogen, atmospheric nitrogen is the standard:

$$\delta^{15}N = \frac{(^{15}N/^{14}N)_{sample} - (^{15}N/^{14}N)_{atm}}{(^{15}N/^{14}N)_{atm}} \times 10^3\,‰, \qquad (1.5)$$

and in isotopic analysis of sulfur, the standard is troilite sulfur of the Canyon Diablo meteorite (MCD) (Grinenko and Grinenko, 1974):

$$\delta^{34}S = \frac{(^{34}S/^{32}S)_{\text{sample}} - (^{34}S/^{32}S)_{\text{MCD}}}{(^{34}S/^{32}S)_{\text{MCD}}} \times 10^3 ‰. \quad (1.6)$$

The precision of experimental determination of the value of δ on modern mass spectrometers is 0.01–0.02‰ (for the D/H ratio, 0.3–0.6‰).

The value and sign of δ are relative and depend on the standard chosen. For the standard, of course, the value of δ is zero. The minus sign indicates that the sample investigated is depleted in the less common isotope with respect to the standard. A positive value of δ characterizes a sample in which the content of the less common isotope is higher than the standard. For the light biogenic elements H, C, N, O, and S the heavy isotope is the less common one.

A measure of the difference in isotopic composition of two substances is the coefficient α, which is called the fractionation (separation) coefficient or coefficient of isotope distribution. It is the ratio of isotopic compositions of a given element in the substances being compared:

$$\alpha_{1-2} = R_{x_1}/R_{x_2}. \quad (1.7)$$

The coefficient α is expressed as follows in terms of the δ value:

$$\alpha_{1-2} = (1 + \delta_1 \times 10^{-3})/(1 + \delta_2 \times 10^{-3}). \quad (1.8)$$

Also a good approximation is the following relationship:

$$\alpha_{1-2} \cong 1 + (\delta_1 - \delta_2) \times 10^{-3}. \quad (1.9)$$

The difference in isotopic composition expressed in δ values is designated by the letter Δ:

$$\Delta_{1-2} = \delta_1 - \delta_2. \quad (1.10)$$

Sometimes a logarithmic expression is used, in the following form:

$$\Delta_{1-2} = \delta_1 - \delta_2 \cong 10^{-3} \log \alpha_{1-2}. \quad (1.11)$$

II. THE KINETIC ISOTOPIC EFFECT

The kinetic isotopic effect is produced by differences in the reaction rates of isotopic forms, i.e., of compounds completely analogous in composition and structure but differing in the presence, in one or more positions, of different isotopes of a given element.

A voluminous literature is devoted to the theory of kinetic isotopic effects. We shall discuss only the physical meaning. It is simple, although kinetic isotopic effects are much harder to calculate than other types of isotopic effects (Brodskiy, 1957; Roginskiy, 1956).

In an ensemble of interacting particles the smaller masses have the greater velocities, and molecules containing a light isotope are more mobile than molecules containing a heavy isotope.

Second factor is that the chemical bonds formed by a heavy isotope are stronger than the same bonds formed by a light isotope. The activation energy of the reaction in which an isotopically heavy form takes part is higher than for an isotopically light form ($E^* > E$). Here and henceforth the symbol $*$ will refer to the isotopically heavy form of a molecule. The kinetic isotopic effect of a chemical reaction is determined by the ratio of the constants of the reaction rates of the isotopic forms:

$$k/k^* = \sqrt{\mu^*/\mu}\,\exp[-(E - E^*)/RT], \tag{1.12}$$

where μ is the reduced mass of the molecule, T the absolute temperature, and R the gas constant.

The kinetic isotopic effect depends on the difference in activation energy, not on the absolute value of activation energy. In other words, the kinetic isotopic effect does not depend on the kinetics of the reaction but on the difference in the kinetics of the reaction of isotopic forms.

Because of the factors mentioned above, relationship (1.12) is greater than unity, i.e., the products of the reaction are enriched in the light isotope and the heavy isotope accumulates in the nonreacting residue. In rare cases there are exceptions to this rule, for instance, when free atoms take part in the reaction. As follows from expression (1.12), the kinetic isotopic effect depends on the difference in mass of the isotope molecules, temperature, and difference in activation energy of the isotopic forms.

As the temperature increases, the kinetic isotopic effect decreases. At $T \to \infty$ (practically, at $T > 800$–1000 K) the exponential term approaches unity and the value of the isotopic effect is determined only by the ratio of masses. The difference in activation energies of the isotopic forms is very difficult to calculate quantitatively. Therefore, theoretical calculations of kinetic isotopic effects are generally very approximate.

The calculation is usually accomplished in the framework of complex transient theory, which yields the following expression for the kinetic isotopic effect (Melander, 1964):

$$\frac{k}{k^*} = \frac{s^{*\#}s}{s^{\#}s^*}\left(\frac{v_L^{\#}}{v_L^{*\#}}\right)^{1/2} \prod_i^{3N-7} \frac{v_i^{\#}\exp(-hv_i^{\#}/2kT)[1 - \exp(-hv_i^{*\#}/kT)]}{v_i^{*\#}\exp(-hv_i^{*\#}/2kT)[1 - \exp(-hv^{\#}/kT)]}$$
$$\times \prod_i^{3N-6} \frac{v_i^*\exp(-hv_i^*/2kT)[1 - \exp(-hv_i/kT)]}{v_i\exp(-hv_i/2kT)[1 - \exp(-hv_i^*/kT)]}, \tag{1.13}$$

where ν_i is the frequency of vibrations, s the symmetry number, ν_L the imaginary frequency along the coordinate of expansion of the complex transient, N the number of atoms in the molecule, and h Planck's constant. The sign # refers to the parameters of the complex transient.

By introducing the symbols $u_i = h\nu_i/kT$ with the corresponding indices and using the Bigeleisen and Goeppert–Mayer functions, the last expression can be simplified considerably (Bigeleisen and Mayer, 1974):

$$\frac{k}{k^*} = \frac{s^{*\#}s}{s^*s^\#}\left(\frac{\nu_L^\#}{\nu_L^{*\#}}\right)^{1/2}\left(1 - \sum_i^{3N-7} G(u_i)^\# \,\Delta u_i^\# + \sum_i^{3N-6} G(u_i)\,\Delta u_i\right). \quad (1.14)$$

Sometimes as an approximation it is assumed that all the vibrational terms in Eq. (1.13) except one, which corresponds to the vibration of the bond broken in the reaction, are abbreviated. Then Eq. (1.13) is reduced to the Eyring–Keugel relationship, which can be used for rough estimation of the kinetic isotopic effect:

$$k/k^* = s^{*\#}s \,\mathrm{sh}(h\nu_a/kT)/s^*s^\# \,\mathrm{sh}(h\nu_a^*/kT), \quad (1.15)$$

where ν_a and ν_a^* are isotopic frequencies characterizing valency vibrations of the ruptured bond. In this case, it is assumed that $\nu_L = \nu_a$.

The vibrational frequency of molecules theoretically can be obtained experimentally from the Raman spectra and infrared spectra of the respective compounds. However, the isotopic shifts for elements heavier than hydrogen are small, and it is difficult to measure them with precision. Therefore the isotopic vibrational frequencies of a molecule are calculated theoretically.

The general method consists of solving the problem of vibrational motion of a molecule. The corresponding equation has the form (Wilson *et al.*, 1960) $|GF - I\lambda| = 0$, where G is the matrix of the kinematic elements, F the matrix of force constants, I the unit matrix, and λ the root of the secular equation which is related to the vibrational frequency relationship ν_i $(\mathrm{cm}^{-1}) = \sqrt{\lambda_i}$.

Calculation of the vibrational frequencies is the basis of estimating not only the kinetic but also the thermodynamic isotopic effect. Later we shall return to it in connection with the problem of the spectral method of estimating the values of the so-called β factors.

III. THE THERMODYNAMIC ISOTOPIC EFFECT

The inequality of the properties of isotopes, in addition to their inequality in kinetic respects, is also manifested in the fact that a compound which has a heavy isotope in its composition has a smaller reserve of free energy than the isotopically light form of the same compound. This inequality of isotopic forms is not the same for different compounds. Therefore the minimum free energy of a system is generally attained for dissimilar isotopic compositions of the compounds forming this system. The difference in isotopic compositions of the components produced by this is the thermodynamic isotopic effect.

The redistribution of isotope X* of element X between the components of a system can be expressed in the form of a reaction which is called an isotopic exchange reaction:

$$\mathrm{AX} + \mathrm{BX}^* = \mathrm{AX}^* + \mathrm{BX}, \tag{1.16}$$

where the asterisk designates the less common (heavy) isotope.

The equilibrium constant of this reaction, as of any other, can be expressed in terms of the concentration of the original reacting substances and their products:

$$K = [\mathrm{AX}^*][\mathrm{BX}]/[\mathrm{AX}][\mathrm{BX}^*]. \tag{1.17}$$

On the other hand, the value of the equilibrium constant is determined by the change in free energy in the reaction

$$-\Delta Z = RT \ln K. \tag{1.18}$$

Statistical thermodynamics makes it possible to express the free energy (isobaric-isothermal potential) of a compound in terms of the so-called partition function of the energy constants (distribution function) Q:

$$Z = -RT \ln Q, \tag{1.19}$$

therefore

$$K = (Q_{AX^*}/Q_{\mathrm{AX}})/(Q_{\mathrm{BX}^*}/Q_{\mathrm{BX}}). \tag{1.20}$$

The methods of quantum statistics make it possible to express the relationship of the partition functions in terms of the vibrational frequencies of the isotopic forms (Urey and Greiff, 1935):

$$f_{\mathrm{AX}} = \frac{Q_{\mathrm{AX}^*}}{Q_{\mathrm{AX}}} = \frac{s}{s^*} \prod^{3N-6} \frac{v_i^* \exp(-hv^*/2kT)[1 - \exp(-kv_i/kT)]}{v_i \exp(-hv_i/2kT)[1 - \exp(-hv_i^*/kT)]}, \tag{1.21}$$

where s and s^* are symmetry numbers.

Using Eq. (1.21), one can calculate the ratio of partition functions for the isotopic species if the vibrational frequencies v_i and v_i^* of the isotopic forms of the molecules are known, and thus the value of the constant K is also known.

On the other hand, relationship (1.17) for the constant is simply the ratio of isotopic compositions of element X in compounds AX and BX:

$$K = \frac{AX^*/AX}{BX^*/BX} = \frac{(X^*/X)_{AX}}{(X^*/X)_{BX}}, \tag{1.22}$$

i.e., the value α [see Eq. (1.7)].

Thus, proceeding from the vibrational spectrum of molecules, in the final analysis it is possible to determine the relationship of the isotopic components of compounds in the minimum free energy state of the corresponding system.

It should be mentioned that Eq. (1.16) describes the simplest case, where the redistribution of isotopes of element X, represented in compounds AX and BX by only one atom, is considered. Only in that case does

$$\alpha = K. \tag{1.23}$$

If isotopic redistribution occurs between compounds in which several atoms of X enter into the composition, then Eq. (1.16) takes a more general form:

$$m\mathrm{AX}_n + n\mathrm{BX}_m^* \rightleftarrows m\mathrm{AX}_n^* + n\mathrm{BX}_m, \tag{1.24}$$

and accordingly the constant

$$K = \frac{[\mathrm{AX}_n^*]^m[\mathrm{BX}_m]^n}{[\mathrm{AX}_n]^m[\mathrm{BX}_m^*]^n} = \left(\frac{Q_{\mathrm{AX}_n^*}}{Q_{\mathrm{AX}_n}}\right)^m \Big/ \left(\frac{Q_{\mathrm{BX}_m^*}}{Q_{\mathrm{BX}_m}}\right)^n. \tag{1.25}$$

Under the condition that all positions of X in compounds AX_n and BX_m are equivalent, for instance, of carbon in C_2H_6 or hydrogen in CH_4, the connection between α and K can be expressed by the relationship (Varshavskiy and Vaysberg, 1957)

$$\alpha = K^{1/mn}\left[\prod_{i=1}^{n}\prod_{j=1}^{m}\left(\frac{s}{s^*}\right)_i\left(\frac{s^*}{s}\right)_j\right]^{-1/mn}, \tag{1.26}$$

where i and j are the number of substitution of isotope X* for isotope X in all possible isotopic forms $AX_{n-i}X_i^*$ and $BX_{m-j}X_j^*$ of compounds AX_n and BX_m.

Expression (1.26) can be simplified if the following notation is used (Varshavskiy and Vaysberg, 1957):

$$Q_{\mathrm{AX}_n^*}/Q_{\mathrm{AX}_n} = (s/s^*)\beta^n_{\mathrm{AX}_n}. \tag{1.27}$$

Then the coefficient of distribution of isotopes α turns out to be

$$\alpha = \beta_{\mathrm{AX}_n}/\beta_{\mathrm{BX}_m}. \tag{1.28}$$

The connection between the values K and α is more complicated when atoms of X occupy nonequivalent positions in the molecules, as occurs for carbon and hydrogen in most organic compounds. But in this case, which was investigated previously (Galimov, 1971), the fractionation coefficient can also be expressed in the form of a relationship analogous to (1.28).

In the case mentioned above, the universal character of the value fs^*/s, which is the ratio of the partition functions of the isotopically *monosubstituted* (in the ith position) form to the nonsubstituted, shows up. We propose to call it the *thermodynamic isotopic factor* of the ith isotopic form and designate it β_i.

Calculating the β_i factors of all monosubstituted forms of a considered pair of compounds AX_n and BX_m, one can express their coefficient of fractionation of isotopes in the following fashion:

$$\alpha = \beta_{\Sigma AX_n}/\beta_{\Sigma BX_m}, \tag{1.29}$$

where

$$\beta_{\Sigma AX_n} = \frac{1}{n}\sum_i^n (\beta_i)_{AX_n} \qquad \text{and} \qquad \beta_{\Sigma BX_m} = \frac{1}{m}\sum_i^m (\beta_i)_{BX_m}.$$

The values of β_Σ characterizing the compound as a whole were called the thermodynamic isotopic factors of the compounds.

We considered it advisable to retain the symbol β to designate the thermodynamic isotopic factors. However, it should be mentioned that there is a certain difference between the use of this symbol in the work by Varshavskiy and Vaysberg (1957) and that adopted by us. The aforementioned authors use β to designate the value of the root n from the product in the right-hand part of expression (1.21), where n is the number of substitution. The concept of the thermodynamic isotopic factor refers only to the isotopically monosubstituted form (β_i) or to the arithmetic mean for a compound (β_Σ). If the atoms of X occupy equivalent positions in a molecule, both definitions coincide in meaning. However, if the structural positions of X atoms are not equivalent, the values of β_i characterizing the ith isotopic forms will not be the same in most cases, and in their definition (Varshavskiy and Vaysberg, 1957) β loses the physical meaning that we have given it, in particular the one related, as we shall show later, to the concept of intramolecular isotopic effects.

We shall return to a more detailed discussion of questions related to thermodynamic isotopic effects in multiatomic compounds in Chapter 3, Section I. Here it is only important to mention that the thermodynamic isotopic effect can be calculated theoretically by means of the value β through the relationships given above. Taking into account the form of representation of α via δ values (1.9) and the relationship of α to the values of β in the form (1.29), the formula for the thermodynamic isotopic effect can be written as

$$\delta_1 - \delta_2 = (\beta_1/\beta_2 - 1) \times 10^3 \text{‰}. \tag{1.30}$$

For cases in which it is necessary to indicate the element for which the isotopic distribution is being examined, it is convenient to use the symbol β in conjunction with the symbol of the respective isotope, for instance, $\beta^{13}C$ or $\beta^{18}O$.

The numerical values of the β factors are close to unity for all elements except hydrogen. The thermodynamic isotopics factors depend on the temperature: as the temperature increases, they decrease rapidly. The numerical values of β of many compounds of various elements are compiled in the work by Friedman and O'Niel (1977).

Two essential features of the thermodynamic isotope effects should be stressed. First, the relationship of thermodynamic isotopic factors characterizes the distribution of isotopes in the minimum free energy state of the system, i.e., it characterizes the equilibrium distribution. Second, like any other thermodynamic values, the thermodynamic isotopic factors define only the potential possibilities of a system for assuming a given state under given conditions. For the isotopic distribution corresponding to thermodynamic isotopic factors to arise, a mechanism of redistribution of isotopes between compounds is necessary. If the compounds do not interact, their isotopic composition remains unchanged, regardless of how long they are in physical contact with each other.

IV. THE NUCLEAR-SPIN ISOTOPIC EFFECT

Recently, still another type of isotopic effect was discovered—the nuclear-spin isotopic effect (Buchachenko *et al.*, 1976). Although a consideration of it is not related to the main idea of this book, we should discuss it briefly, as this effect may be important in some biological processes. In addition, it occupies a special place in the systematics of isotopic effects (Galimov, 1979). Whereas the kinetic, thermodynamic, and all other known types of isotopic effects are ultimately due to differences in the mass of isotopes, the nuclear-spin isotopic effect is the result of a difference in spins of the isotopic nuclei. For instance, the two stable isotopes of carbon, ^{13}C and ^{12}C, have half-integer and zero spins, respectively. Of the three stable isotopes of oxygen, the two isotopes ^{16}O and ^{18}O have zero spins but the ^{17}O isotope has a spin equal to $\frac{5}{2}$. Practically all biochemically important elements contain isotopes differing in nuclear spin.

The theoretical prerequisites for nuclear-spin isotopic effects are discussed below.

Some chemical processes, primarily radical processes, are related to a change in total electron spin of the reacting system or, as they say, by a change in spin multiplicity (for instance, by its change from the triplet to the singlet state). The expression for the energy characterizing this conversion contains a term which depends on the interaction of the spins of the electrons and the nucleus. As a result, the probability of the chemical process is dependent on the existence of nuclear spin and the magnitude of its interaction with the electron spin—the so-called hyperfine interaction. Whereas generally the isotopes of the same element may differ in magnitude of nuclear spin, the rate of the reaction may not be the same for isotopic compounds, which suggests the existence of an isotopic effect.

This mechanism can be realized in chemical radical reactions. Electrons are paired in molecules and electron spins are compensated, i.e., the total electron spin is equal to zero. That state is called the singlet state (S). When the molecule breaks down into radicals, the electrons are uncoupled. In a forming electron pair the spins of unpaired electrons may be either opposite (singlet) or additive (triplet) in orientation. In the triplet state the total spin is equal to unity, and in an external magnetic field it may be oriented, in correspondence with the laws of quantization, in three ways: $+1$, 0, and -1, respectively, forming the substates T_+, T_0, and T_-.

If a radical pair is in the triplet state, then before recombining they must return to the singlet state. The probability of the triplet–singlet conversion is different for nuclei having different spin. For a magnetic nucleus it is always greater than for a nonmagnetic one. The difference in probability determines the value of the nuclear-spin isotopic effect. The isotopic effect increases as the constant of hyperfine interaction a increases. In contrast to the usual kinetic isotopic effects, the nuclear-spin isotopic effect depends on the intensity of the external magnetic field.

As is known, there is a rule, which states that isotopic shifts for the isotopes of the same element are proportional to the difference in mass of the isotopes. For oxygen isotopes, for instance, $\Delta^{17}O \simeq 0.5\,\Delta^{18}O$. Therefore, analysis of the isotopic composition of oxygen is practically limited to measurements of $\delta^{18}O$, assuming that the ratio of the ^{17}O and ^{16}O isotopes is also determined. This rule, however, is valid only for isotopic effects that depend on mass. The existence of nuclear-spin effects, in the circumstance that ^{17}O has a different spin from the nuclei of ^{16}O and ^{18}O, causes a fractionation of ^{17}O and ^{16}O isotopes which is independent of $^{18}O/^{16}O$. It is noteworthy that this difference in fractionation of ^{17}O and ^{18}O is purely the result of the nuclear-spin isotopic effect. As a result, parallel study of $\delta^{18}O$ and $\delta^{17}O$ reveals the possibility of assessing the place and importance in natural processes, in particular in biological processes, of radical mechanisms of chemical reactions generating nuclear-spin isotopic effects. Such processes undoubtedly are inherent in biological systems, and apparently it is the biochemistry of oxygen isotopes which is most promising here.

V. SYSTEMATICS OF ISOTOPIC EFFECTS

Besides those considered above, there are a multitude of other types of isotopic effects related to diffusion, adsorption, evaporation, and phase transitions in general. Isotopic effects produced by electrolysis, thermodiffusion, and photoirradiation at the frequency of valency vibrations of an isotope (in particular, laser separation) are known. Other effects are isotopic fractionation in the gravitational field and fractionation caused by differences in the paths of isotopic ions in a magnetic field.

All these effects can be subdivided into two broad groups: those depending on the rate of the process and those depending on the state of the system, i.e., into kinetic and thermodynamic groups. For example, diffusional, tunnel, nuclear-spin, and other isotopic effects caused by a difference in the rate of processes for the isotopic forms should be assigned to the kinetic group.

The discovery of the nuclear-spin isotopic effect introduces a new logic into the systematics of isotopic effects (Galimov, 1979). The nuclear-spin isotopic effect pertains to a number of chemical isotopic effects, i.e., effects related to processes of a change in the configuration and energy of electron shells. On the other hand, it depends on a typically nuclear property, the spin of the nucleus. One can point out the opposite (and rare) case in which a nuclear process reveals a dependence on the structure of the electron shell, i.e., on a chemical property. This refers to β decay with preliminary trapping of an electron from the *K* shell; for instance, the decay of ^{40}K and its transformation into ^{40}Ar.

The analogy of these processes, nuclear-dependent chemical on the one hand and chemically dependent nuclear on the other, is rather profound and ultimately comes down to the quantum-mechanics phenomenon of the presence of electron density in the nucleus.

The constant of hyperfine interaction is determined by the relationship

$$a = (8\pi/3)\beta_e\beta_n|\psi_e(0)|^2, \tag{1.31}$$

where β_e and β_n are the magnetic moments of electron and nucleus and $|\psi_e(0)|^2$ is the probability density that an electron is present in the nucleus.

The probability of K capture P_K (thus the probability of β decay) also is proportional to the square of the modulus of the wave function of the K electron by the nucleus:

$$P_K \sim |\psi_e(0)|^2.$$

Thus both effects touch a boundary on both sides: one side is bordered by the field of chemical processes of isotopic *fractionation* and the other by the field of nuclear processes of isotopic *transformation.*

In a sense, the nuclear-spin isotopic effect is the link which makes it possible to connect chemical and nuclear isotopic effects into one system. As a result, the possibility exists for drawing up the systematics of isotopic effects based on the principle of their dependence on profoundness of properties of a substance involved in the process producing the isotopic effect.

In this classification, nuclear-spin isotopic effects occupy the position of the most profound chemical isotopic effects (Table 1.1). Second in this hierarchy comes tunnel effects, related to the quantum-mechanical escape of particles (dependent on mass) under the energy barrier of the reaction, and third is the usual chemical isotopic effects (thermodynamic and kinetic),

TABLE 1.1.

SYSTEMATICS OF ISOTOPIC EFFECTS

Factor of change of isotopic composition	*Property (the difference in which causes the isotopic effect)*	*Isotopic effect*	
		Type	*Variety (examples)*
Fractionation	Mass	Inertial (corpuscular)	Gravitational Diffusional Electromagnetic (fractionation of isotope ions in an electrical and magnetic field)
		Corpuscular–chemical	Phase Adsorptional
		Chemical	Thermodynamic Kinetic Tunnel
	Spin	Nuclear–chemical	Nuclear spin
Transformation	Nuclear properties		K capture
		Nuclear	Nuclear transformation

caused by processes localized in the electron shell. The most primitive isotopic effects (corpuscular) depend on the inert masses of isotopic particles.

The hierarchy of isotopic effects can be objectively related to the level of their quantum character. Inertial (corpuscular) effects are completely described by the means of classical physics. Chemical isotopic effects, which are quantum in their nature, can be partially calculated and described within the framework of classical approximations. The nature of the isotopic effects related to tunnel escape has not yet been subject to adequate classical interpretation.

Undoubtedly isotopic effects of a very different nature take part in biological fractionation of isotopes. However, their contribution is not the same. Up to now the point of view that biological fractionation is chiefly kinetic in nature has been common, not because that has been established, but because kinetic isotopic effects, for general considerations, have been more plausible for biological systems.

We hope, however, to show in this book that, because of the details of the mechanism of enzymatic fractionation of isotopes, thermodynamic isotopic effects play an essential role in biological systems, and the distribution of isotopes for a statistical set of biomolecules reveals a thermodynamically ordered character.

REFERENCES

Bigeleisen, J., and Mayer, M. G. (1947). Calculation on equilibrium constants for isotope exchange reactions. *J. Chem. Phys.* **15**, 261–267.

Brodskiy, A. I. (1957). "Isotope Chemistry." Izd-vo AN SSSR, Moscow.

Buchachenko, A. L., *et al.* (1976). Isotopic enrichment induced by magnetic interactions in chemical reactions. *Dokl. AN SSSR* **228** (2), 379–381.

Craig, H. (1957). Isotopic standards for carbon and oxygen and correction factors for mass spectrometric analysis of carbon dioxide. *Geochim. Cosmochim. Acta* **12**(N1/2), 133.

Craig, H. (1961). Standard for reporting concentrations of deuterium and oxygen-18 in natural waters. *Science* **133**, 1833–1834.

Friedman, I., and O'Neil, J. R. (1977). Compilation of stable isotope fractionation factors of geochemical interest. *In* "Data of Geochemistry," 6th ed., Chap. KK, Geol. Survey Prof. Pap. 440–K. U.S. Government Printing Office, Washington, D.C.

Galimov, E. M. (1971). On the relation of the coefficient of isotopic fractionation to the equilibrium constants of isotopic exchange reactions of carbon in hydrocarbon systems. *Zhurn. Fiz. Khimii* **5**, 1187–1191.

Galimov, E. M. (1979). The nuclear-spin isotopic effect—a new type of isotopic effect. *Geokhimiya* (2), 1274–1284.

Grinenko, V. A., and Grinenko, L. N. (1974). "Geochemistry of Stable Isotopes of Sulfur." Nauka, Moscow.

Melander, L. (1964). "Isotopic Effects in Reaction Rates" (Russian translation). Mir, Moscow.

Roginskiy, S. Z. (1956)."Theoretical Bases of Isotopic Methods of Studying Chemical Reactions." Izd-vo AN SSSR, Moscow.

Urey, H. C., and Greiff, L. J. (1935). Isotopic exchange equilibria. *J. Am. Chem. Soc.* **57**, 321.

Varshavskiy, Ya. M., and Vaysberg, S. E. (1957). Thermodynamic and kinetic particulars of isotope exchange reactions. *Uspekhi Khimii* **24**(12), 1434–1468.

Wilson, E., Decivs, J., and Cross, P. (1960). "Theory of Vibrational Spectra of Molecules." Izd-vo Inostr. Lit., Moscow.

CHAPTER 2

Isotopic Composition of the Carbon of Organisms

In biological fractionation of isotopes, two aspects can be distinguished. The first is separation of isotopes between the organism and the environment, and the second is separation of isotopes in the biosystems of the organism.

These aspects are interrelated. However, historically only the first of these attracted the attention of investigators. The carbon of living organisms is depleted in the ^{13}C isotope (and enriched in the ^{12}C isotope) compared with external CO_2, which is the carbon source for every living thing. This noteworthy fact was established in 1939 by Nier and Gulbransen (1939), later confirmed by Craig (1953) in his pioneering work, and subjected to thorough analysis in many subsequent works. The enrichment of living organisms in the light isotope of carbon was named the *biological isotopic effect*. Many assumed that this practically exhausted the concept of biological fractionation of isotopes. Although it was discovered later that differentiation of the isotopic composition takes place within the organism itself, this fact was not considered important. It was assumed that differentiation was produced by special causes. One of the cornerstones of this investigation is the claim that it is differentiation of isotopes in the biosystems of an organism that determines the main content of biological fractionation of isotopes. The fractionation of carbon isotopes between external CO_2 and primary biological acceptors of carbon is simply a particular manifestation of it.

I. FRACTIONATION OF ISOTOPES IN THE ORGANISM–ENVIRONMENT SYSTEM

As strange as it may seem, many natural (geochemical) regularities in the separation of isotopes in the organism–environment system can be understood without knowing the actual nature of biological fractionation of isotopes. This is explained by the existence of at least two external factors which affect the isotopic composition of organisms in natural environments to a great extent and which do not depend on how living organisms carry out the fractionation of isotopes. These factors are, first, variations in the isotopic composition of the carbon dioxide in the environment in which the organisms live and, second, the coefficient of utilization of CO_2.

Although it seems obvious that the isotopic composition of organisms (autotrophs) should depend on the isotopic composition of the original CO_2, this dependence is not always clear. In many cases it is mediated. Two examples are the isotopic compositions of organisms which is known to depend on the salinity of seas in near-shore regions, and of trees, which depends on the height of their growth. In the final analysis both are simply manifestations of the dependence of the isotopic composition of biological carbon on the isotopic composition of external CO_2.

For the coefficient of utilization of CO_2, it should be noted that, in any type of isotopic fractionation, the extent of transformation or coefficient of utilization of the original substance is critical in determining the ultimate distribution of isotopes between the original substance and the product. Resultant isotopic fractionation occurs only if the original substance is partially converted. If the original reagent is completely converted into the product, then the product acquires the isotopic composition of the original substance regardless of whether the corresponding process was accompanied by isotopic fractionation.

Other factors being equal, the maximum separation of isotopes between organisms and external CO_2 should occur when the rate is limited by the assimilation of CO_2 and not by its availability. Limiting the availability of CO_2 leads to a decrease in the apparent isotopic effect. If all the CO_2 that diffuses into cells from the surrounding environment is involved in the process of biosynthesis, then no separation of isotopes between organisms and environment occurs.

A. Geochemical Scale

Biological fractionation of isotopes is part of the general separation of isotopes in geological processes.

Today, the hundreds of works dealing with the investigation of the isotopic composition of the carbon of natural substances, taken together, give a fairly detailed picture of the geochemical behavior of carbon isotopes. A systematic discussion of the geochemistry of the stable carbon isotopes can be found in special monographs and reviews (Degens, 1969; Epstein, 1968; Galimov, 1968, 1973; Hoefs, 1973).

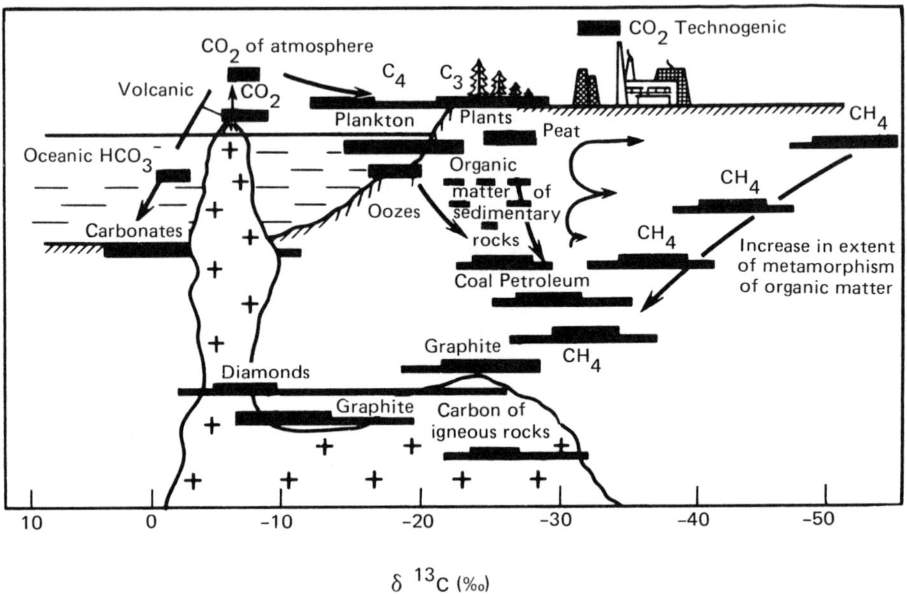

FIG. 2.1 Range of variations of the isotopic composition of carbon in some natural substances.

The sketch in Fig. 2.1 illustrates the position of biological materials in the general scale of distribution of carbon isotopes and of the scale of biological fractionation of isotopes compared with the overall range of variation of the isotopic composition of carbon in nature.

On the whole, the range of variation of the isotopic composition of carbon in nature amounts to 120‰. Biological fractionation is responsible for a substantial part of this variation.

Atmospheric CO_2 is the common link between the organic and inorganic cycles of carbon. Two very powerful processes of separation of carbon isotopes occurring in nature correspond to these two branches of the carbon cycle. The first, biological fractionation of isotopes, leads to enrichment of biospheric carbon in ^{12}C. The second process, isotope exchange in the atmosphere–hydrosphere system, leads to concentration of ^{13}C in sedimentary carbonates.

The isotopic composition of atmospheric carbon dioxides $\delta^{13}C = -7‰$. For carbon of the biosphere, $\delta^{13}C$ values from -15 to $-35‰$ are characteristic. Sedimentary carbonates have $\delta^{13}C$ from $+5$ to $-5‰$.

B. Higher and Lower Plants

No systematic differences in isotopic composition between the higher and lower plants have been observed. Although, marine plankton are usually isotopically lighter than the higher algae, the observed differences can generally be explained by ecological factors (in particular, the isotopic composition and availability of CO_2 in the environment inhabited). Blue–green algae are impoverished in the light isotope in some environments (Barghoorn *et al.*, 1977; Behrens and Frishman, 1971; Calder and Parker, 1973).

In higher plants, species belonging to the main classes of gymnosperms and angiosperms (and to the latter's subclasses of mono- and dicotyledons), do not differ in isotopic composition (Fig. 2.2) (Smith and Epstein, 1971). However, the isotopically heavy plants (-6 to -18‰) nearly always belong to a subclass of the angiosperms. This apparently is explained by the position they occupy in the phylogeny of C_4-type plants, since the isotopically heavy examples belong to this type. It has been noted that lichens and fungi are relatively impoverished in the ^{12}C isotope (Jacobson *et al.*, 1970).

C. Age of Plants

The age of plants has no appreciable effect on the isotopic composition of their carbon (Wickman, 1952). Young trees sometimes are somewhat enriched in the ^{12}C isotope, but this apparently is not related to the age but

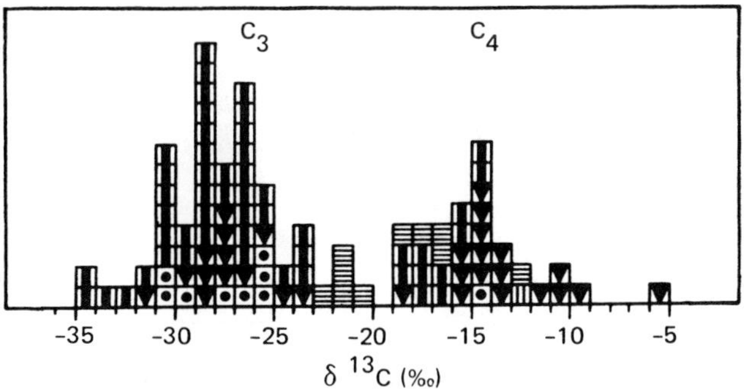

FIG. 2.2 Isotopic composition of the carbon of the higher plants. ◪, monocotyledons; ◧, dicotyledons; ⊡, gymnosperms; ▤, algae. (After Smith and Epstein, 1971.)

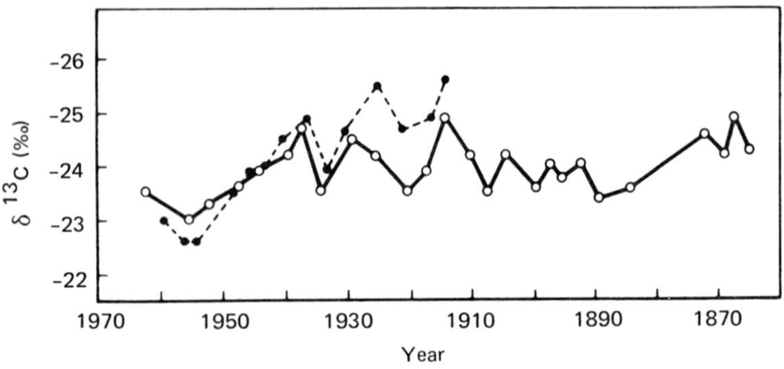

FIG. 2.3 Variation in carbon isotopic composition in annual rings of 100-year-old (—) and 50-year-old (---) spruce trees (high bank of Kama River, Perm district).

to the height of the plant (the concentration of isotopically light carbon dioxide formed as a result of decay and oxidation of organic remains is higher in the layer of air near the ground (Keeling, 1961a; Munnich and Vogel, 1959).

In one of our works we made a comparison of two trees of different age growing next to one another, tracing the variation in isotopic composition of the carbon of the wood on the annual rings (Galimov, 1976). Both trees showed synchronous variations in isotopic composition although one of them was 50 years older than the other (Fig. 2.3). Evidently these fluctuations in $\delta^{13}C$ were caused by a change in external conditions.

Craig (1954c) had a unique opportunity to study the variations in isotopic composition of carbon in plants over a span of 2500 years. He had at his disposal a section of a trunk of *Sequoia gigantea* which lived from 928 B.C. to A.D. 1649. The age variations in carbon isotopic composition proved to be insignificant—within 2‰. A similar range of variations in $\delta^{13}C$ was determined in a section of another old tree—a kauri from New Zealand—for the A.D. 1000 to 1800 time interval (Jansen, 1962).

D. C_3- AND C_4-TYPE PLANTS

Until recently the Calvin cycle was thought to be the common pathway for the assimilation of CO_2 by all photosynthetic organisms. Its primary product is three-carbon (C_3) phosphoglyceric acid, formed as a result of carboxylation of ribulose 1,5-diphosphate, and the subsequent reaction of cleavage of the six-carbon compound. In the middle of the 1960s another type of fixation of CO_2 carbon was recognized, in which the primary products are four-carbon compounds (C_4): malate and aspartate (Karpilov, 1960; Kortschak *et al.*, 1965). The corresponding reaction cycle was called the *Hatch–Slack cycle*, for the investigators who studied it and showed that the formation of four-carbon compounds is caused by condensation of exogenic CO_2 with phosphoenolpyruvate (Hatch and Slack, 1970).

Following this, it was established that plants having this type of photosynthesis are substantially less enriched in the ^{12}C isotope than ordinary C_3 plants (Bender, 1968; Smith and Epstein, 1971).

Terrestrial plants of the C_3 type are characterized by a $\delta^{13}C$ range of approximately -22 to -32‰. Plants of the C_4 type have a $\delta^{13}C$ range of variation from -10 to -18‰, almost continuous with the C_3 range. Members of the C_4 type, as a rule, are higher plants (angiosperms), that is, they stand rather high in the phylogenetic series. Maize, sorghum, and many plants of desert and arid regions are typical representatives of these C_4-type plants.

Even in early works, isolated examples of plants unusually depleted in the ^{12}C isotope were reported (Craig, 1954a; Wickman, 1952), although the overwhelming majority of investigated terrestrial plants were isotopically light. Craig (1954a) suggested that these anomalies of isotopic composition are explained by the special role of carbonate and bicarbonate carbon in these cases. Another explanation was that the plants in arid regions (the anomalous cases pertained to desert plants) utilize atmospheric carbon dioxide more fully, resulting in a decrease in the apparent separation of isotopes (Galimov, 1968). Afterward these plants were identified as plants of the C_4 type. A very economical use of carbon dioxide is typical of these plants (Hatch and Slack, 1970; Moss, 1962).

The causes of the unique isotopic composition of C_4 plants are complex. Many investigators (Lerman and Querioz, 1974; Troughton *et al.*, 1972) see these causes in the high coefficient of utilization of external CO_2, or in a difference in isotopic effect of enzymatic reactions of carboxylation of phosphoenolpyruvate in the Hatch–Slack cycle and of ribulose 1,5-diphosphate in the Calvin cycle (Deleens *et al.*, 1974; Whelan *et al.*, 1973). In Chapter VIII we will offer a somewhat different explanation.

It should be noted that, owing to intensified research in recent years on the isotopic composition of C_4-type plants, data pertaining to them now occupy an equally large place in different kinds of tables and histograms as data on the isotopic composition of C_3 plants (an example of this is Fig. 2.2). Actually, C_4-type plants are much less abundant than C_3-type plants, and their share of carbon in the total amount of carbon of the biosphere is relatively small.

E. Autotrophs and Heterotrophs

An appreciable enrichment in the light isotope of carbon compared with original carbon is inherent in almost all autotrophs. Heterotrophs differ little or not at all from their organic source in isotopic composition. This has been established both for organisms in the natural setting (Degens, 1969) and experimentally (Abelson and Hoering, 1961). Zooplankton, for example, do not differ in isotopic composition from the phytoplankton on which they feed. The ^{13}C content in the soft tissues of invertebrates falls approximately in the same limits as in marine algae (Craig, 1954a). Fish have the same isotopic composition (Parker, 1964). The isotopic composition of land animals coincides with that of terrestrial plants (Folinsbee *et al.*, 1970; Rafter, 1955). A correlation is observed between the isotopic composition of some animals and the proportions of C_3- and C_4-type plants in the feeding area (De Niro and Epstein, 1978).

F. Land and Water Plants

Land plants (except C_4-type plants) on the average are more enriched in the light isotope than plants growing in water. Most of the former fall in the $\delta^{13}C$ range of -22 to $-32‰$. A range of $\delta^{13}C$ from -10 to $-19‰$ is typical of algae. This fact was established some time ago (Baertschi, 1952; Craig, 1953). Apparently this was the next important observation after the discovery of the enrichment of biological carbon in the light isotope. Several explanations of this fact have been considered. The bicarbonate of seawater ($\delta^{13}C = -0.2‰$) is isotopically heavier than atmospheric CO_2 ($\delta^{13}C = -7‰$). This explanation is applicable if the original form of the carbon for photosynthesis is bicarbonate. But if it is dissolved carbon dioxide, then it must be remembered that dissolved CO_2 is lighter in isotopic composition than bicarbonate and in this respect is practically the same as atmospheric CO_2 (Thode *et al.*, 1965). A second explanation is that, in contrast to an atmospheric environment, carbon dioxide in an aquatic environment is in the same phase state as in the cytoplasm of the plant. Therefore, the isotopic effect accompanying the conversion of CO_2 from the gas phase into the cellular fluid possibly is absent here (Park and Epstein, 1960). Finally, the lesser enrichment of subaqueous plants in the light isotope may be related to the considerably lower rates of diffusion of CO_2 in water than in air. As a result, other factors being equal, the coefficient of utilization of CO_2 is higher in an aquatic than in an atmospheric environment. Hence a corresponding reduction of the isotopic effect is possible.

TABLE 2.1

DEPENDENCE OF THE ISOTOPIC COMPOSITION OF CARBON OF PLANKTON ON THE ABUNDANCE OF CO_2 IN THE ENVIRONMENT IN WHICH THEY LIVE[a]

Organism	*T* (°C)	pH	*Abundance of* CO_2	$\Delta^{13}C = \delta^{13}C_{C_{org}} - \delta^{13}C_{HCO_3^-}$
Cyclotella	20	8.6	Atmospheric concentration, normal aeration	−17.0
	20	8.2	Tenfold excess of HCO_3^-	−26.0
	20	5.8	5% Concentration of CO_2	−27.6
Skeletonema	18	8.6	Poor aeration	−12.3
	18	8.2	Energetic aeration	−19.1

[a] See Degens *et al.* (1968a).

The dependence of the isotopic composition of carbon of photosynthesizing organisms on CO_2 concentration actually has been observed in several experiments. For instance, the diatomaceous algae *Cyclotella* and *Skeletonema*, which grow in seawater under normal conditions, have the isotopic composition normal for marine plankton. In the case of a substantial excess of CO_2 and bicarbonate in solution, these same species were appreciably lighter isotopically (Table 2.1) (Degens *et al.*, 1968a). They virtually acquired the isotopic composition corresponding to that of terrestrial organisms.

Sensitivity to carbon dioxide concentration is also typical of plants which use carbon dioxide from the air (Table 2.2), although in view of the greater

TABLE 2.2

DEPENDENCE OF CARBON ISOTOPE IN TOMATOES ON CO_2 CONCENTRATIONS IN THE ATMOSPHERE[a]

CO_2 *concentration*	*Weight of plant* (g)	$\delta^{13}C$ (‰)
0.015	9.3	−24.7
0.05	35.1	−26.3
1.5	42.1	−28.5
1.5	50.1	−27.9

[a] $\delta^{13}C_{CO_2} = 0.0$‰. Data from Park and Epstein (1960).

mobility of atmospheric CO_2, it is manifested to a lesser extent than in aquatic plants.

Effects of analogous character have also been recorded in other experiments, all the way to the complete disappearance of the biological isotopic effect, as was observed, for instance, at certain concentrations of CO_2 in the experiments of Calder and Parker (1973). It must also be kept in mind that organisms can react physiologically when the concentration of carbon dioxide in the environment changes, and with this change isotopic fractionation is possible.

Under terrestrial conditions, depending on the local climatic and ecological setting, the concentration and isotopic composition of CO_2 may differ from those typical of the atmosphere on the average. In dense forests the percentage of carbon dioxide of organic origin increases in the ground-level layer of air. Therefore its isotopic composition is not $-7‰$ as for the atmosphere on the average, rather it is -8 to $-11‰$ (Galimov, 1966; Keeling, 1961a). This in turn leads to a shift in the carbon isotopic composition of the plants which assimilate this carbon dioxide.

In places with greater air circulation, i.e., in sectors exposed to the wind, carbon dioxide was a "normal" isotopic composition. Therefore plants on mountain slopes, for instance, are less enriched in the ^{12}C isotope by 1 to 2‰ than the same species in other settings (Keeling, 1961b).

As a result of human activity, a huge amount of carbon dioxide is discharged into the air, mainly from burning of various kinds of fuel, especially oil, gas, and coal. As the carbon of organic fuel is enriched in the ^{12}C isotope, technogenic carbon dioxide is isotopically light ($\delta^{13}C$ approximately $-30‰$). Therefore in cities and industrial areas atmospheric carbon dioxide is enriched by 2 to 8‰ in the light isotope (Friedman and Irsa, 1967), which affects the isotopic composition of plants in these areas.

In a number of works (Farmer and Baxter, 1972; Galimov, 1976; Rebello and Wagener, 1976) an attempt was made to establish from tree rings whether global accumulation of technogenic CO_2 in the atmosphere is occurring, which, as is known, could have certain consequences for the earth's climate. Our work cited above (Galimov, 1976) was undertaken for that purpose.

G. Freshwater and Marine Plants

As a rule, freshwater plants are enriched in the light isotope to a greater extent than marine plants. For marine plankton, $\delta^{13}C$ from -17 to $-22‰$ is typical, while freshwater plankton may have values of $\delta^{13}C$ from -20 to

$-30‰$. This is because a large amount of isotopically light carbon dioxide of organic origin reaches freshwater bodies with groundwater (Galimov, 1968). The $\delta^{13}C$ values of bicarbonate in river and lake waters ranges from -9 to $-14‰$ (Oana and Deevey, 1960; Vogel, 1959) in contrast to the $-0.2‰$ typical of ocean water. This is a property of small intracontinental water bodies rather than freshwater itself. The larger the body and the better developed the water circulation in it, the better the conditions for exchange in the system $HCO_3^-(aq)$–$CO_2(atm)$ and the closer the isotopic composition of the bicarbonate of the water to the equilibrium value, which in principle is the same for marine and freshwater bicarbonate.

In the transition zone from a continental to a marine environment, i.e., in shelf areas or estuaries of large rivers, a gradient of the carbon isotopic composition of organisms is observed (following the same trend as the change in salinity of the water), from $\delta^{13}C$ values of -24 to $-30‰$ typical of freshwater organisms to $\delta^{13}C$ values of -18 to $-22‰$ typical of marine organisms.

H. The Role of Temperature

The isotopic composition of the carbon of living organisms in some cases shows a relationship to the temperature of the environment. As a rule, the relationship consists of some enrichment of plants in the light isotope with a drop in temperature. Thus, oceanic plankton are isotopically lighter in colder high-latitude waters. This was established on the basis of the example of plankton collected in the Atlantic (Sackett *et al.*, 1965) and Indian oceans (Galimov *et al.*, 1975). In land plants the variation in isotopic composition of carbon in successive annual rings sometimes shows a correlation with variations in the regional temperature conditions during the life of the plant (Lerman *et al.*, 1967). In laboratory experiments the temperature dependence of the biological carbon composition of some species of plankton was ascertained (Degens *et al.*, 1968a). In some cases no systematic variation in isotopic composition of organic carbon with temperature was observed (Berry *et al.*, 1972; Calder and Parker, 1973; Troughton *et al.*, 1974).

Because the concentration of carbon dioxide is greater in cold water than in warm, the temperature dependence can be attributed in part to the change in the coefficient of CO_2 utilization with temperature (Degens, 1969). On the other hand, as will be shown later, the temperature dependence may be the result of the thermodynamically ordered distribution of isotopes in biological systems.

II. ISOTOPIC COMPOSITION OF BIOCHEMICAL COMPONENTS

Much of the experimental data on the isotopic composition of biochemical components and individual compounds entering into the composition of organisms is given in Chapters 4 and 6 in the discussion of the concept of thermodynamically ordered distribution of isotopes in biosystems. Only preliminary information is given here.

A. LIPIDS

Lipids differ from the other biochemical components in their greater enrichment in the light isotope. Table 2.3 gives the data of Park and Epstein (1961). Table 2.4 gives our data (Galimov and Shirinskiy, 1975), combined, they cover a fairly wide taxonomic range. In all cases the lipids invariably are enriched in the ^{12}C isotope relative to both the plant as a whole and to proteins (total amino acid fraction).

Differentiation of isotopic composition is found between individual components in the lipid fraction. The polar fraction of the lipids (extracted with an alcohol–benzene mixture) is less enriched in the ^{12}C isotope than the

TABLE 2.3

COMPARISON OF THE ISOTOPIC COMPOSITION OF THE CARBON OF PLANTS AND THEIR LIPID FRACTIONS[a]

	$\delta^{13}C$ (‰)			
Plant	*Plant as a whole*	*Lipids*	*Residue*	*Chloroform lipids*
Ulva sp.	−15.7	−23.5	−15.4	4.0
Gelidium sp.	−20.5	−24.3	−19.9	9.4
Macrocystis pyrifera	−12.6	−21.0	−12.7	1.7
Mixed phytoplankton (chiefly *Prorocentrum micans*)	−9.2	−14.2	−8.3	6.8
Asplenium bulbiferum	−29.2	−37.9	−30.7	3.8
Triticum vulgare	−28.5	−31.1	−27.6	11.6
Sequoia gigantea	−26.4	−28.3	−25.2	6.4
Solanum tuberosum	−25.6	−33.6	−25.6	0.2

[a] Data from Park and Epstein (1961).

TABLE 2.4

ISOTOPIC COMPOSITION OF THE CARBON OF LIPIDS, TOTAL AMINO ACID FRACTION, AND CARBON OF ORGANISMS AS A WHOLE[a]

	$\delta^{13}C$ (‰)			
Organism	*Organism as a whole*	*Lipids*	*Amino acids*	*Residue after extraction of lipids*
Laminaria saccharina	−18.4	−25.1	—	−17.0
Lupinus luteus	−27.6	−33.0	−26.0	−26.5
Euphasia superba	−25.2	−28.7	−24.0	−24.3
Anabaena variabilis	−27.5	−32.2	−26.4	−26.8

[a] Data from Galimov and Shirinskiy (1975).

fractions extracted with chloroform (Degens, 1968b; Park and Epstein, 1961). The nonsaponifiable fraction of the lipids (mainly isoprenoids) is 2‰ lighter isotopically than the saponifiable fraction (fatty acids) (Jacobson *et al.*, 1970).

Parker (1962) investigated the isotopic composition of individual fatty acids. The $\delta^{13}C$ values obtained turned out to be very similar for saturated and unsaturated fatty acids belonging to the lipid fraction of the same organism.

A thorough investigation of the lipid fraction, including study of the isotopic composition of the individual compounds, was later undertaken in our laboratory (Galimov and Shirinskiy, 1975). We shall discuss these results in Chapter 4.

B. PROTEINS

Park and Epstein (1961), who were the first to investigate several nonlipid components as well as lipids, found that the isotopic composition of proteins, starch, pectin, and lignin is nearly the same as the average isotopic composition of the carbon of the organism. Hence they came to the conclusion, shared by many investigators for some time, that the only mechanism of fractionation of isotopes existing in organisms (besides primary photosynthetic assimilation of carbon) is related to the biosynthesis of lipids.

From Table 2.4 it is seen that, in contrast with the lipids, proteins differ little in isotopic composition from the total carbon of the biomass of the

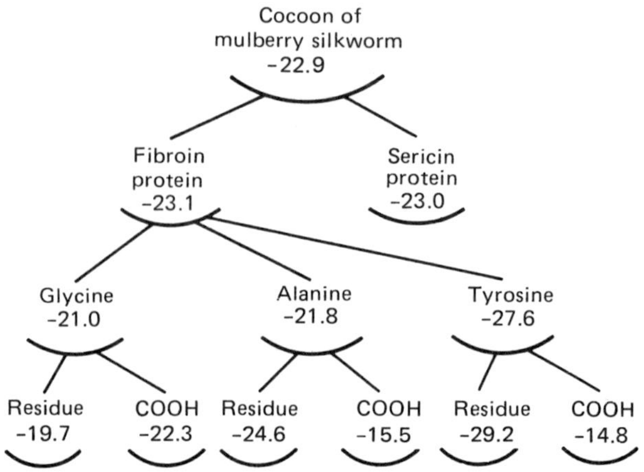

FIG. 2.4 Differentiation of isotopic composition of carbon in the protein of a mulberry silkworm cocoon.

organism. However, the impression that appreciable fractionation of isotopes is absent, which is created by the similarity of isotopic composition of proteins and the average carbon of the biomass, is deceptive. As illustrated in Fig. 2.4, which gives some data obtained in our laboratory (Galimov *et al.*, 1977), the fibroin and sericin proteins isolated from the cocoon of a mulberry silkworm almost coincide in isotopic composition with each other and with the carbon of the cocoon as a whole, whereas the isotopic composition of the amino acids forming the protein is markedly different.

The isotopic composition of amino acids was investigated in detail by Abelson and Hoering (1961). They established that the isotopic composition of amino acids isolated from the total hydrolysate of the protein fraction of an organism varies widely (Table 2.5), and certain consistent regularities are observed. Thus, leucine was relatively enriched in the light isotope in all the

TABLE 2.5

ISOTOPIC COMPOSITION (IN PER MIL) OF CARBON OF AMINO ACIDS IN PHOTOSYNTHESIZING MICROORGANISMS[a]

Amino acid	*Chlorella pyrenoides*	*Anacystis nidulans*	*Scenedesmus quadricauda*	*Chromatium* sp.	*Euglena gracilis*	*Gracilaria* sp.
Glutamic acid						
Total	−18.7	−11.1	−22.6	−15.6	−17.3	−17.2
Carboxyl	−8.8	11.7	−5.6	2.2	−0.2	−8.4
Residual	−21.1	−16.7	−26.6	−20.1	−21.5	−19.3

TABLE 2.5 *(continued)*

Amino acid	*Chlorella pyrenoides*	*Anacystis nidulans*	*Scenedesmus quadricauda*	*Chromatium* sp.	*Euglena gracilis*	*Gracilaria* sp.
Aspartic acid						
Total	−6.6	−9.7	−12.3	−21.8	−9.6	−14.4
Carboxyl	2.2	2.8	−0.8	−15.9	1.0	−5.3
Residual	−15.6	−22.1	−24.3	−27.6	−20.3	−18.4
Serine						
Total	−5.7	—	—	−6.6	−8.3	−14.1
Carboxyl	2.5	—	—	−2.9	−6.1	−8.8
Residual	−9.9	—	—	−8.4	−9.5	−16.8
Threonine						
Total	−14.5	—	—	—	−8.6	−14.5
Carboxyl	−0.5	—	—	—	2.1	−8.5
Residual	−19.2	—	—	—	−12.2	−16.4
Glycine						
Total	−14.3	−10.0	−14.7	—	−12.8	−10.2
Carboxyl	−5.3	−8.8	−11.8	—	−11.6	−12.8
Residual	−23.3	−11.1	−17.5	—	−14.1	−7.7
Alanine						
Total	−10.3	−9.8	−18.2	—	−14.3	−15.3
Carboxyl	−7.3	0.2	−13.5	—	−13.8	−11.5
Residual	−11.8	−14.8	−20.5	—	−14.6	−17.2
Isoleucine						
Total	−19.6	−10.5	−16.1	—	—	−18.8
Carboxyl	−7.4	−1.5	−13.1	—	—	−14.0
Residual	−22.0	−12.2	−17.1	—	—	−19.8
Leucine						
Total	−22.7	−17.3	−21.4	−23.5	−23.5	−22.5
Carboxyl	−16.3	−15.0	−17.8	−22.5	−13.6	−24.1
Residual	−26.0	−17.8	−22.2	−23.7	−25.6	−22.1
Lysine						
Total	−17.0	—	—	—	−22.8	—
Carboxyl	2.5	—	—	—	−24.8	—
Residual	−20.9	—	—	—	−22.4	—
Arginine						
Total	−19.2	—	—	—	−15.9	−13.8
Carboxyl	−2.6	—	—	—	3.6	−3.6
Residual	−22.6	—	—	—	−19.8	−15.9
Tyrosine						
Total	−19.8	—	—	—	−17.1	−23.0
Carboxyl	−4.1	—	—	—	−3.6	−8.3
Residual	−22.8	—	—	—	−18.8	−24.8
Phenylalanine						
Total	—	—	—	—	−18.6	−21.2
Carboxyl	—	—	—	—	−2.2	−9.4
Residual	—	—	—	—	−20.6	−22.7

[a] Value of original inorganic carbon is 0.0. Data from Abelson and Hoering (1961).

TABLE 2.6

ISOTOPIC COMPOSITION ($\delta^{13}C$‰) OF CARBOHYDRATES COMPARED WITH OTHER BIOCHEMICAL COMPONENTS

Biological object	*Lipids*	*Proteins*	*Starch*	*Glucose from starch*	*Published source*
Potato tubers	−34.8	−26.6	−25.5	—	Jacobson *et al.* (1970)
Potato tubers	—	—	−25.2	−24.9	Galimov *et al.* (1977)
Rice	—	−26.4	—	−26.1	Galimov *et al.* (1977)
Wheat	—	−25.1	−23.0	−23.1	Galimov *et al.* (1977)
Peas	—	—	−25.0	−24.9	Galimov *et al.* (1977)

investigations, and aspartic acid, as a rule, was isotopically the heaviest. The finding that the carboxyl carbon in amino acids is enriched in the heavy isotope was extremely important. This was the first evidence of intramolecular differentiation of isotopes in biological compounds.

C. CARBOHYDRATES

Like proteins, carbohydrates are a fraction in which the carbon differs little in isotopic composition from the carbon of the biomass on the average. Table 2.6 gives the data of Jacobson *et al.*, (1970) for starch, proteins, and lipids from potato tubers, supplemented by results of the investigations of proteins and carbohydrate fractions of rice and wheat carried out in our laboratory (Galimov *et al.*, 1977). The starch showed a small depletion (1 to 3‰) in the ^{12}C isotope compared with proteins.

Individual sugars were not investigated as thoroughly as the amino acids. As is seen from Table 2.6, the isotopic composition of glucose coincides with that of the starch from which it was obtained.

Degens *et al.* (1968b) found that the cellulose fractions in plankton contain somewhat more of the ^{12}C isotope ($\delta^{13}C_{av} = -22.4$‰) than hemicellulose ($\delta^{13}C = -18.7$‰). Because glucose is a monomer of cellulose and hemicelluloses contain pentoses in addition to hexoses (in a 1:2 ratio), the logical conclusion was drawn that there is differentiation of isotopes between individual sugars. In particular, the hexoses should be somewhat enriched in the ^{12}C isotope compared with the pentoses.

D. LIGNIN

In the work by Degens *et al.* (1968b) cited above, a broad range of biochemical constituents of marine plankton was investigated. The complete results of that work are given in Table 2.7. Among others, the lignin fraction was investigated. Lignin proved to be relatively enriched in the ^{12}C isotope.

The lignin fraction studied by Degens *et al.* (1968b) is not lignin in the strict sense of the word. It was aromatic metabolites, chiefly hydroxybenzoic acids, which in plankton can be regarded as arbitrary analogs of lignin in the higher plants. Incidentally, Manskaya and Kodina (1975) established that lignin may be present in several species of lower plants.

Later investigations of the isotopic composition of lignin of the higher plants, including intramolecular analysis, that were carried out in our laboratory (Galimov *et al.*, 1976) showed that this constituent of plants actually is relatively enriched in the light isotope. We shall return to these data later.

E. RESPIRED CARBON DIOXIDE

The carbon dioxide liberated in the process of respiration has been the object of experimental investigations in many works (Abelson and Hoering, 1961; Degens *et al.*, 1968a; Jacobson *et al.*, 1970; Park and Epstein, 1960). It was presumed that study of its isotopic composition might provide the key to an understanding of the mechanism of biological fractionation of isotopes.

It was noted that, at the beginning of a dark period, respired carbon dioxide is appreciably heavier than organic carbon. Then it gradually begins to be enriched in the light isotope (Table 2.8).

The organic carbon of cells also changes its isotopic composition during the dark period. In the absence of photosynthesis, liberation of respired CO_2, impoverished in the light isotope, leads to complementary enrichment of organic carbon in the light isotope (Degens *et al.*, 1968a). In an experiment with the green alga *Dunaliella*, the carbon of the cells was sharply depleted in the ^{13}C isotope during the first 5 days (Table 2.9), but on subsequent days the isotopic composition of the cells was practically unchanged. For the diatomaceous alga *Skeletonema* the effect was less marked but followed a similar trend. The experiments were carried out at different temperatures (see Table 2.9).

TABLE 2.7

ISOTOPIC COMPOSITION (IN PER MIL) OF ORGANIC CARBON OF THE MAIN BIOCHEMICAL CONSTITUENTS OF PLANKTON[a]

Sample number[b]	*Total carbon*	*Alcohol-extracted lipids*	*Chloroform-extracted lipids*	*Pectins*	*Hemicelluloses*	*Celluloses*	*Total sugars*	*Amino acids* (6N HCl)	*Amino acids* (2N H_2SO_4)	*Lignin*
1	−21.2	−21.2	−23.0	−16.3	−16.7	−20.4	−19.0	−17.9	−17.5	−23.8
2	−19.8	−20.6	−20.6	−37.4	−17.3	−20.2	−19.6	−16.1	−15.9	−25.2
9	−20.2	−21.7	−28.3	−17.0	−18.2	−21.6	−18.1	−16.5	−16.8	N.D.[c]
12	−19.9	−19.8	−23.5	−16.4	−18.3	−21.5	−18.4	−18.0	−17.5	−21.8
14	−20.2	−22.5	−31.5	−15.5	−19.7	−23.2	—	—	—	−20.2
16	−20.4	−21.0	−30.5	−17.2	−21.0	−23.6	−19.2	−18.8	−18.8	−22.6
18	−20.1	−20.9	−34.6	−17.1	−19.6	−26.7	−21.2	−17.7	−17.7	−25.3
Average	−20.2	−21.1	−29.8	−16.7	−18.7	−22.4	−19.2	−17.3	−17.3	−23.1

[a] Data from Degens *et al.* (1968b).
[b] In the work of Degens *et al.* there is a map of the sampling sites corresponding to the numbers given.
[c] N.D. stands for not determined.

TABLE 2.8

ISOTOPIC COMPOSITION OF RESPIRED CARBON DIOXIDE IN AN EXPERIMENT WITH TOMATOES[a]

Time elapsed from beginning of dark period (hr)	$\delta^{13}CO_2$ (‰)
0.16	−24.3
2.0	−25.8
4.0	−27.2
6.0	−27.6
16.0	−28.2
40.0	−30.0

[a] $\delta^{13}C = -32.4$‰. Data are from Park and Epstein (1960).

TABLE 2.9

VARIATION IN ISOTOPIC COMPOSITION OF PLANKTON KEPT IN THE DARK AS A FUNCTION OF DURATION OF EXPOSURE TO THE DARK[a]

Organism	*Respiration time (days)*	$\delta^{13}C$ (‰) *of bicarbonate substrate* (‰)	$\delta^{13}C$ *of cell carbon* (‰)	*Temperature* (°C)
Dunaliella	0	0.0	−16.2	18
	1	0.3	−18.8	18
	5	0.4	−20.9	18
	12	—	−21.8	18
	19	—	−21.7	18
Skeletonema	0	3.4	−13.4	8
	1	—	−10.9	8
	6	—	−11.8	8
	12	—	−11.8	8
Skeletonema	0	2.6	−9.7	18
	1	—	−9.2	18
	6	—	−10.5	18
	12	—	−10.4	18

[a] Data from Degens et al. (1968a).

III. ATTEMPTS TO EXPLAIN BIOLOGICAL FRACTIONATION OF ISOTOPES

For a long time biological fractionation of isotopes was identified with just one phenomenon, the enrichment of the light isotope in the whole organism.

In the first works an attempt was made to study the behavior of CO_2 to answer the question of why plants are enriched in the ^{12}C isotope. Baertschi (1952) and Craig (1953) pointed out the analogy between the process of solution of atmospheric CO_2 in the cell fluid and the process of bonding of carbon dioxide by inorganic absorbents. It was known that carbon dioxide absorbed from the air by a solution of barium hydroxide is enriched by 10 to 18‰ in the light isotope (Baertschi, 1952). Further enrichment of plants in the light isotope was related to cyclic mechanisms of a different kind. According to Wickman (1952), for instance, there is a gradual decrease in ^{12}C concentration in plants as a result of accumulation around plants of the products of their metabolism, which are reused in part in the process of photosynthesis. A modified version of this hypothesis was developed by Keeling (1961b) who suggested that periodic accumulation of $^{12}CO_2$ around a plant and removal of residual $^{13}CO_2$ occur as a result of the diurnal cycle of photosynthesis.

The first model of isotopic fractionation based on biological process per se was proposed by Park and Epstein (1960). According to their ideas, isotopic fractionation occurs in the course of photosynthesis, and is accomplished in two steps. In the first stage, preferential absorption of $^{12}CO_2$ from the air and its solution in the cytoplasm of plants occurs. Here fractionation of isotopes is produced by a kinetic isotopic effect like that proposed by Baertschi. In the second stage, the form preferentially enriched in ^{12}C is, in turn, extracted from the CO_2 dissolved in the cytoplasm and goes into the synthesis of organic compounds.

Carbon dioxide dissolved in cytoplasm and removed by sulfuric acid actually turned out to be 7‰ lighter than atmospheric carbon dioxide, on the average (leaves of tomato plants were investigated [Park and Epstein, 1960]). The variations in $\delta^{13}C$ value in the individual experiments were 3–5‰, which the experimenters explained by competition of the normal and reverse isotopic effects.

Park and Epstein modeled the second stage of separation of carbon isotopes by an *in vitro* reaction of enzymatic carboxylation of ribulose diphosphate—the initial reaction of the Calvin cycle. In this stage, the first stable product of photosynthetic fixation of carbon was obtained—two molecules of 3-phosphoglyceric acid.

As a result of the experiment it was shown that the carbon fixed in 3-phosphoglyceric acid is enriched in the light isotope by 17‰. It should be mentioned that this numerical value, which is widely quoted in literature, is very approximate, generally speaking. First, the measured isotopic effect actually was much less, as the isotopic composition of phosphoglyceric acid as a whole was analyzed, and only one carbon atom in two molecules comes from CO_2. Therefore the calculated value contains a sixfold experimental error. Second, this result proved to be difficult to reproduce. Even in the same work Park and Epstein obtained a different value (14‰) in a parallel experiment. In subsequent years this enzyme experiment was repeated by a number of investigators, who obtained different results. We shall return to a discussion of these data in later chapters (see Chapter 8).

In any case, by experimentally obtaining enrichment in the ^{12}C isotope in two stages and putting the results together ($7 + 17 = 24$‰) Park and Epstein were able to explain both the enrichment of organisms in the light isotope in general and the wide variations in the $\delta^{13}C$ values of different biogenic compounds.

Later, Epstein (1968) modified the model somewhat, pointing out that the two successive kinetic isotopic effects may not be additive. Depending on which process (penetration of exogenic CO_2 through the cell wall into the cytoplasm or photosynthetic fixation of CO_2 from the cytoplasm) is the rate-limiting one, only one of the isotopic effects appears. However, the magnitude of either is inadequate to explain the range of observed variations in isotopic composition of plants. Therefore Epstein suggested that there is a third way in which part of carbon dioxide is removed from cytoplasm, namely transport of cell fluid via the vascular system.

Fractionation of isotopes in the stage of photosynthesis (within the framework of the two-stage model) was considered almost entirely determinative of the whole extent of biological fractionation of isotopes. An exception was made only for the lipids. It was suggested that enrichment of the lipids in the ^{12}C isotope is a process complementary to impoverishment in ^{12}C of the carbon dioxide exhaled by the plant and occurs according to the following scheme (Epstein, 1968):

$$\begin{array}{lllll} & 2CO_2 & \text{exhaled } CO_2 & & \text{depleted in } ^{12}C \\ & \nearrow & \nearrow & & \\ \text{sugar } (C_6) \rightarrow & 2(C_3) \rightarrow & 2(C_2) \rightarrow & \text{lipids} & \text{enriched in } ^{12}C \end{array}$$

Thus the impoverishment of respired carbon dioxide in the ^{12}C isotope observed in the experiment was explained at the same time. However, this point of view was later abandoned. In a work by Jacobson *et al.* (1970) (with Epstein among the coauthors) it was acknowledged "that the rate of lipid synthesis on a carbon basis is less than 1% of the rate of CO_2 production and ... as much as 10% conversion of starch to lipid would cause but a 1 per mil increase in the δ-value of respiratory CO_2 over that of starch" (p. 13).

Following Park and Epstein, Abelson and Hoering suggested that "most of the isotope fractionation in photosynthesis occurs in the carbohydrate-forming process and specifically in the ribulose diphosphate step" (Abelson and Hoering, 1961, p. 630). They pointed to the Krebs cycle, in which CO_2 can be used directly from the environment, being condensed with pyruvate. The carbon of this CO_2 is also included later in the carboxyl groups of aspartic and glutamic acids. Because of this independent method of incorporating CO_2, which presumably is not connected with appreciable fractionation of isotopes, the carbon of the amino acid carboxyl groups is isotopically heavy. Enrichment of lipids in the light isotope compared with carbohydrates requires some additional process of isotope fractionation. The biological fractionation of isotopes is described by Abelson and Hoering (1961, p. 632) as follows:

> If the carbon of lipides is light, conservation laws require that some other portion of carbon is heavier. Or, stated another way, a change in isotope ratios implies the splitting of a population of molecules into two or more classes or streams. Something of this nature must occur after the fixation of CO_2 by the ribulose diphosphate. Part of the carbon goes to the citric acid cycle, part to carbohydrates, part to lipides, ... The isotopic fractionation in the process leading to lipides is probably related to a specific physical chemical step crucially related at the point where a flow of carbon splits into two or more fractions.

Similar considerations were expressed by (Degens *et al.*, 1968b; Degens, 1969). According to the results of experiments carried out by him and his coauthors, the list of possible barriers of isotopic fractionation in organisms was somewhat enlarged.

Like previous investigators, Degens proceeded from the assumption that the process predetermining the level of enrichment of organisms in the ^{12}C isotope is fractionation in the ribulose diphosphate stage. With no further fractionation, organic carbon formed at this stage is included in carbohydrates and proteins and, with a little fractionation, in lipids. However, in the course of interconversion of individual compounds of a given biochemical fraction, isotopic effects may take place which lead to variation in the isotopic composition of the compounds within that fraction, although its average isotopic composition remains unchanged.

Of course, much of the evidence supporting such a process is purely speculative. In his presumed isotopic fractionation between hexoses and pentoses (based on the established differences in isotopic composition of cellulose and hemicellulose), Degens and his coauthors limited themselves to pointing out that "the metabolic pathway leading to pentoses may involve a number of reactions. For instance, carboxylation (RuDP vs. PGA), decarboxylation (pentose shunt pathway) or a pronounced reassortment of carbon atoms (monosaccharide interconversions), may lead to carbon isotope fractionations" (Degens *et al.*, 1968b, p. 17).

The diversity of isotopic compositions of amino acids in the experiments of Abelson and Hoering was explained in the same spirit: "In view of the numerous ways in which carbon can be arranged in the TCA cycle, there is ample opportunity for carbon isotope fractionation," and later:

> The proteins in our plankton samples are ... similar in $\delta^{13}C$ to the carbohydrates. This ... can only imply that the carbon isotope labeling at the RuDP–PGA level is handed over without a significant fractionation to the carbohydrates and proteins. However, the $\delta^{13}C$ spread between the individual sugars and amino acids is rather pronounced as a result of metabolic carbon skeleton rearrangements. Consequently, under conditions of steady-state photosynthesis, the material balance of carbon isotopes will always remain a constant; in the present case the constant is principally determined at the RuDP–PGA barrier (Degens *et al.*, 1968b, p. 17).

Enrichment of the "lignin" fraction in the ^{12}C isotope was not considered an independent phenomenon but was explained by enrichment of its predecessors—the aromatic amino acids phenylalanine and thyrosine—in the light isotope.

Quoting the data of Parker (1962), who established small variations in the isotopic composition of fatty acids, Degens also postulated isotopic fractionation in the course of the interconversion of fatty acids.

Degens (1969; Degens *et al.*, 1968b) reduced the main stages of isotopic fractionation in an organism to seven barriers at which fractionation occurs: an effect related to arrival of external CO_2 in the cytoplasm, carboxylation at the RuDP–PGA level, interconversion of sugars, amino acid interconversion, respiration (decarboxylation reactions), fatty acid interconversion, and lipid–carbohydrate conversion.

Several other models have been proposed. Ivlev, for instance, attempted to explain the enrichment of an organism in the light isotope by the kinetic isotopic effect in the decarboxylation of pyruvate: the C_2 fragments enriched in the light isotope are involved in the biosynthesis of lipids, and the undecomposed pyruvate depleted in the ^{12}C isotope is removed from the organism via the Krebs cycle (Ivlev, 1976). In its formulation this model reiterates Epstein's idea of enrichment in the light isotope in the process of respiration. However, Ivlev's reasoning, concerning the biological fractionation of isotopes, is based on an error (Galimov, 1978). He incautiously transferred the results of experiments with an isotopic tracer to the interpretation of isotopic effects.

In summary, we can note, first, not only that biological fractionation of isotopes is characteristic of carbon migrating between an organism and its environment, but also that there are profound differences in isotopic composition between the biochemical fractions, between the individual compounds comprising these fractions, and even within these compounds. Second, the lack of systematic differences in isotopic composition between the higher and lower plants, between individual species of higher plants, and between autotrophs and heterotrophs—in other words, the lack of dependence of isotopic composition on the supercellular organization—indicates that biological fractionation of isotopes occurs at the cellular level, i.e., that the complex processes of transportation of substances and intercellular exchange characteristic of higher organisms play a much smaller role in biological fractionation of isotopes.

Therefore the attempts of investigators to find the causes of biological fractionation of isotopes in the physicochemical processes which accompany biosynthesis of organic compounds are justified. Their proposed hypotheses give a somewhat accurate interpretation of individual facts, but they share a series of defects.

First, they are qualitative and amount to surmises of enrichment or depletion of an organism (fraction, compound) in an isotope, but they contain no basis for quantitative estimation of biological isotopic effects.

Second, the existing hypotheses propose no general approach to an explanation of observed isotopic effects. As a result, they are adapted only to explaining established facts, but it is impossible to predict facts from them, i.e., to predict the distribution of isotopes in classes of compounds that have not been experimentally investigated as yet.

As we shall see, a new fact, the relationship of the isotopic composition of biomolecules to the values of their thermodynamic isotopic factors, lies completely outside the framework of established ideas. We shall now turn to the theory and experiments devoted to the study and substantiation of this phenomenon, and to an examination of the consequences stemming from them with respect to the mechanism of biological fractionation of isotopes.

REFERENCES

Abelson P. H., and Hoering T. C. (1961). Carbon isotope fractionation in formation of amino acids by photosynthetic organisms. *Proc. Nat. Acad. Sci. USA* **47**(5), 623–632.

Baertschi, P. (1952). The fractionation of carbon isotopes in the uptake of carbon dioxide. *Helv. Chim. Acta* **35**, 1030.

Baertschi, P. (1959). The fractionation of natural carbon isotopes in the carbon dioxide exchange of green plants. *Helv. Chim. Acta* **36,** 773–781.

Barghoorn, E. S., Knoll, A. N., Dembick, H., and Meinschein, W. G. (1977). Organic matter from the gunflint iron formation. *Geochim. Cosmochim. Acta* **41**, 425–430.

Bender, H. M. (1968). Mass spectrometric studies of carbon-13 variations in corn and other grasses. *Radiocarbon* **10**(2), 468.

Behrens, E. W., and Frishman, S. A. (1971). Stable carbon isotopes in blue–green algal mats. *J. Geol.* **79**, 94–100.

Berry J., Troughton, J. H., and Björkman, O. (1972). Effect of oxygen concentration during growth on carbon isotope discrimination in C_3 and C_4 species of atriples. *Carnegie Inst. Yearb.* **71**, 158–161.

Calder, J., and Parker, P. L. (1973). Geochemical implications of induced changes in ^{13}C fractionation by blue–green algae. *Geochim Cosmochim. Acta* **37**, 133.

Craig, H. (1953). The geochemistry of the stable carbon isotopes. *Geochim Cosmochim. Acta* **3**, 53–92.

Craig, H. (1954a). Carbon-13 in plants and relationships between carbon-13 and carbon-14 variations in nature. *J. Geol.* **62**, 115–149.

Craig, H. (1954b). Carbon-13 variations in sequoia rings and the atmosphere. *Science* **119**(3083), 141–143.

Deleens, E., Lerman, J. C., Nato, A., and Moyse, A. (1974). Carbon isotope discrimination by the carboxylating reactions in C_3, C_4 and CAM plants. *Proc. 3rd Intern. Congr. on Photosynthesis*, Rehovot, September 2–6.

Degens, E. T. (1969). Biogeochemistry of stable carbon isotopes. *In* "Organic Geochemistry" (G. Eglinton and M. T. Murphy, eds.), pp. 304–329. Springer, Heidelberg.

Degens, E. T., Guillard, R. R. L., Sackett, W. M., and Hellebust, J. A. (1968a). Metabolic fractionation of carbon isotopes in marine plankton. I. Temperature and respiration experiments. *Deep-Sea Res.* **15**, 1–10.

Degens, E. T., Behrend, M., Gotthardt, B., and Reppmann, B. (1968b). Metabolic fractionation of carbon isotopes in marine plankton. II. Data on samples collected off the coast of Peru and Ecuador. *Deep Sea Res.* **15**, 11–20.

De Niro, M., and Epstein, S. (1978). Influence of diet on the distribution of carbon isotopes in animals. *Geochim Cosmochim. Acta* **42**(N5), 495–506.

Epstein, S. (1968). Distribution of carbon isotopes and their biochemical and geochemical significance. *In* "Proc. Symp. on CO_2 Chemical, Biochemical and Physiological Aspects" (Forster *et al.*, eds.), pp. 5–14, NASA SP–188. Haverford PA.

Farmer, J. G., and Baxter, M. S. (1972). Short-term trends in natural radiocarbon dating. *In* "Proc. 8th Intern. Conf. on Radiocarbon Dating" (T. A. Rafter and T. Grant-Taylor, eds.), A58–71. Roy. Soc. N.Z., Wellington.

Friedman, L., and Irsa, A. P. (1967). Variations in the isotopic composition of carbon in urban atmospheric carbon dioxide. *Science* **158**, 263–264.

Folinsbee, R. E., Fritz, P., Krouse, H. R., and Robblee, A. R., (1970). Carbon-13 and oxygen-18 in dinosaur, crocodile and bird eggshells indicate environmental conditions. *Science* **168**, 1353–1356.

Galimov, E. M. (1966). Isotopic composition of the carbon of soil CO_2. *Geokhimiya* (9), 1110–1118.

Galimov, E. M. (1968). "Geochemistry of the Stable Carbon Isotopes." Nedra, Moscow, 224 p.

Galimov, E. M. (1973) "Carbon Isotopes in Oil and Gas Geology." Nedra, Moscow (English translation: NASA TT F-682, Washington, D.C., 1975, 395 pp.

Galimov, E. M. (1976). Variations of Carbon Cycle at the Present and in the Geological Past. *In* "Environmental Biogeochemistry" (J. O. Nriagu, ed.), Vol. 1, pp. 1–14. Science Publ., Ann Arbor.

Galimov, E. M. (1978). On the concept of thermodynamically ordered distribution of isotopes in biological systems and errors related to its comprehension. *Geokhimiya* (10), 186–200.

Galimov, E. M., and Shirinskiy, V. G. (1975). Ordered distribution of isotopes of carbon in individual compounds and components of the lipid fraction of various organisms. *Geokhimiya* (4), 503–528.

Galimov, E. M., Shirinskiy, V. G., Borodoviskiy, O. K., and Zaikin, V. G. (1975). Isotopic composition of carbon of lipids of Indian Ocean plankton. *Geokhimiya* (6), 895–904.

Galimov, E. M., Kodina, L. A., and Generalova, V. N. (1976). Experimental investigation of intra- and intermolecular isotopic effects in aromatic compounds of biological nature. *Geokhimiya* (1), 11–16.

Galimov, E. M. *et al.* (1977). Investigation of the Distribution of Carbon Isotopes in Biogenic Compounds. *In* "Proc. VIII International Congress on Organic Geochemistry" (A. V. Sidorenko, ed.), Vol. 3. Nauka, Moscow.

Hatch, M. D., and Slack, C. R. (1970). Protosynthetic CO_2-fixation pathways. *Ann. Rev. Plant Physiol.* **21**, 141–162.

Hoefs, J. "Stable isotope Geochemistry." Springer, Heidelberg.

Ivlev, A. A. (1976). Problems of the theory of carbon isotope fractionation in photosynthetic organisms. *Uspekhi Sovrem. Biologii* **81**(1), 84–105.

Jacobson, B. S., Smith, B. N., Epstein, S., and Laties, G. G. (1970). The prevalence of carbon-13 in Respiratory carbon dioxide as an indicator of the type of endogenous substrate. *J. Gen. Physiol.* **35**, 1–17.

Jansen, H. S. (1962). Depletion of carbon-13 in a young kauri tree. *Nature* **196**, 84–85.

Karpilov. Yu. S. (1960). Distribution of radioactive carbon ^{14}C among the products of photosynthesis of maize. *Trudy Kazakh. S.-Kh. Inst.* **41**(1).

Keeling, Ch. D. (1961a). The concentration and isotopic abundances of carbon dioxide in rural and marine air. *Geochim. Cosmochim. Acta.* **24**, 277–299.

Keeling, Ch. D. (1961b). A mechanism for cyclic enrichment of carbon-12 by terrestrial plants. *Geochim. Cosmochim. Acta* **24**, 299–313.

Kortschak, H. P., Hartt, C. E., and Burr, G. O. (1965). Carbon dioxide fixation in sugarcane leaves. *Plant Physiol.* **40**, 209–213.

Lerman, J. C. (1972). Soil CO_2 and Groundwater: Carbon isotope compositions. *In* "Proc. 8th Intern. Conf. on Radiocarbon Dating" (T. A. Rafter, and T. Grant-Taylor, eds.), D–93–105. Roy Soc. N.Z., Wellington.

Lerman, J. C., and Queiroz, O. (1974). Carbon fixation and isotope discrimination by a crassulacean plant. Dependence on the photoperiod. *Science* **183**, 1207–1209.

Lerman, J. C., Mook, W. C., and Vogel, J. C. (1967). Effect of the Tunguska meteor and sunspots on radiocarbon in three rings. *Nature* **216**, 990–991.

Manskaya, S. M., and Kodina, L. A. (1975). "Geochemistry of Lignin." Nauka, Moscow, 229 pp.

Moss., D. M. (1962). The limiting carbon dioxide concentration for photosynthesis. *Nature* **193**, 587.

Münnich, K. O., and Vogel, J. C. (1959). ^{14}C age determination of freshwater limestone deposits. *Naturwissenschaften* **46**(5), 168–169.

Nier, A. O., and Gulbransen, E. A. (1939). Variations in the relative abundance of the carbon isotopes. *J. Am. Chem. Soc.* **61**, 697.

Oana, S., and Deevey, E. S. (1960). Carbon-13 in lake waters and its possible bearing on paleolimnology. *Am. J. Sci.* **258,** 253–272.

Park, R., and Epstein, S. (1960). Carbon isotope fractionation during photosynthesis. *Geochim. Cosmochim. Acta* **27**(1/2), 110–126.

Park, R., and Epstein, S. (1961). Metabolic fractionation of ^{12}C and ^{13}C in plants. *Plant Physiol.* **36**, 133.

Parker, P. L. (1962). The isotopic composition of the carbon of fatty acids. *Carnegie Inst. Yearbook* **61**, 187–190.

Parker, P. L. (1964). The biogeochemistry of the stable isotopes of carbon in a marine bay. *Geochim. Cosmochim. Acta* **28**, 1155–1164.

Rafter, T. A. (1955). C14-variations in nature and effect on radiocarbon dating. *N.Z. J. Sci. Tech.* **37**(13), 20.

Rebello, A., and Wagener, K. (1976). Evaluation of ^{12}C and ^{13}C data on atmospheric CO_2 on the basis of a diffusion model for oceanic mixing. *In* "Environmental Biogeochemistry" (J. Nriagu, ed.), Vol. 1, pp. 13–23. Ann Arbor Science, Ann Arbor, Michigan.

Sackett, W. M., Eckelmann, W. R., Bender, M. L., and Be, A. W. H. (1965). Temperature dependence of carbon isotope composition in marine plankton and sediments. *Science* **148**, 235–237.

Smith, B. N., and Epstein, S. (1971). Two categories of $^{13}C/^{12}C$ ratios for higher plants. *Plant Physiol.* **47**, 380–384.

Thode, H. G., Shima, M., Rees, C. E., and Krishnamurty, K. V. (1965). Carbon-13 isotope effects in systems containing carbon dioxide, biocarbonate, carbonate, and metal ions. *Can. J. Chem.* **43**, 582–595.

Troughton, J. H. (1972). Carbon isotope fractionation by plants. *In* "Proc. 8th Internat. Conf. on Radiocarbon Dating" (T. A. Rafter and T. Grant–Taylor, eds.), E39–57. Roy. Soc. N.Z., Wellington.

Troughton, J. H., Wells, P. V., and Mooney, U. A. (1974). Photosynthetic mechanisms and paleoecology from carbon isotope rations in ancient specimens of C_4 and CAM plants. *Science* **185**(4151), 610.

Vogel, J. C. (1959). On the isotopic composition of carbon in freshwater limestone deposits. *Geochim. Cosmochim. Acta* **16**(4) 236–242.

Whelan, T. W., Sackett, M., and Benedict, C. R. (1973). Enzymatic fractionation of carbon isotopes by phosphoenolpyruvate carboxylase from C_4-Plants. *Plant. Physiol.* **51**, 1051–1054.

Wickman, F. E. (1952). Variations in the relative abundance of the carbon isotopes in plants. *Geochim. Cosmochim. Acta* **2**, 243.

CHAPTER 3

Method of Calculating the Thermodynamic Isotopic Factors of Polyatomic Compounds

As a rule, biomolecules are polyatomic compounds containing at least a few atoms of the same element in different structural positions. Thermodynamic isotopic analysis of systems containing such compounds is a complicated task. Some time ago, the application of isotopic thermodynamics to biopolymers, which are the basis of life, seemed to be unrealistic. However, that problem was generally avoided because two considerations apparently ruled out any connection between isotopic effects in biomolecules and their thermodynamic parameters. First, the thermodynamic distribution presumes isotopic exchanges to be at equilibrium. Second, as a rule isotopic exchange takes place only between simple compounds. Polyatomic compounds exchange isotopes very slowly, especially isotopes of carbon, which forms the framework of organic molecules, and from isotope geochemistry it is known that organic compounds can retain their original isotopic composition for many millions of years.

Nevertheless, it resulted that, in many cases, the isotopic composition of biomolecules shows a dependence on thermodynamic isotopic factors, and as we shall attempt to demonstrate, this is not inconsistent with either the biochemistry or the chemistry of isotopes. The prerequisite and essential link in establishing the phenomenon of thermodynamic ordering of isotopes in biosystems was the development of a method of calculating the thermodynamic isotopic factors of polyatomic compounds which could be extended to biomolecules (Galimov, 1972, 1973a,b, 1974). Therefore, this method should be considered a component part of the thermodynamic concept of biological fractionation of isotopes discussed in this work.

Essentially only Tables 3.13 and 3.19, which contain the numerical values of the isotopic bond numbers, and formulas (3.29) and (3.44) are required for use in the succeeding chapters. However, we thought it advisable to give an extensive exposition of the thermodynamic isotopic factor method here.

This will allow the reader to form his own opinion concerning the validity and limitations of the premises which are the basis for the calculation of the thermodynamic isotopic factors of biomolecules. We shall base our quantitative interpretation of biological isotopic effects on values computed from the numerical calculation. In addition, many problems (for example, correction for the distant surroundings, assessment of precision, calculation of the nitrogen and oxygen bond numbers) are being considered for the first time in this work. For completeness and comprehension of subsequent discussion, it was also necessary to review questions which have been discussed in earlier publications but which are closely related to the thermodynamic isotopic factor method, for instance, the relationship between the α and β values of polyatomic compounds and the concept of intramolecular thermodynamic effects.

I. INTRA- AND INTERMOLECULAR THERMODYNAMIC ISOTOPIC EFFECTS: RELATIONSHIP BETWEEN β_i AND β_Σ VALUES

If some compound AX containing one atom of a given element (for instance, carbon in the methane molecule) is considered, the thermodynamic isotopic effect of the compound is determined by the partition functions and symmetry numbers of its isotopic forms:

$$\beta_{AX} = s^* Q^*_{AX} / s Q_{AX}. \tag{3.1}$$

The isotopic effect, i.e., the difference between the isotopic compositions of two similar compounds AX and BX, is defined as the ratio of isotopic abundances of the two compounds:

$$\alpha = (AX^*/AX)/(BX^*/BX). \tag{3.2}$$

On the other hand, this ratio is determined by the relationship of the β factors of compounds AX and BX:

$$\alpha = \beta_{AX}/\beta_{BX}. \tag{3.3}$$

However, when the compound contains several atoms of a given element (AX_n), it forms a multitude of isotopic forms in which the main isotope is substituted by the heavy one in various positions: $AX_1^*X_2 \cdots X_n$, $AX_1X_2^* \cdots X_n$, etc., and also isotopic forms with two or more isotopic substitutions. In this case, the determination of the isotopic fractionation factors must be defined more precisely. Exactly which isotopic isomers are used to evaluate the isotopic fraction factor for such a compound?

In a compound containing several atoms of element X, the isotopes X* are distributed so that the various isotopic isomers occur in proportion to the number of atoms substituted in that isomer. Therefore, the isotopic composition of the compound is determined by the relationship

$$\left(\frac{X^*}{X}\right)_{AX_n} = \frac{n[AX_n^*] + \cdots + (n-i)[AX_iX_{n-i}^*]}{n[AX_n] + \cdots + (n-i)[AX_i^*X_{n-i}]}. \tag{3.4}$$

Therefore, instead of the simple relationship (3.2), the fractionation factor between compounds AX_n and BX_m is expressed by the relationship

$$a = \frac{n[AX_n^*] + \cdots + (n-i)[AX_iX_{n-i}^*]}{n[AX_n] + \cdots + (n-i)[AX_i^*X_{n-i}]} \bigg/ \frac{m[BX_m^*] + \cdots + (m-j)[BX_jX_{m-j}^*]}{m[BX_m] + \cdots + (m-j)[BX_j^*X_{m-j}]}, \tag{3.5}$$

where i and j are the numbers of the substituted atoms in the isotopic forms of both compounds; the substitution should be made for all nonequivalent atoms in each isotopic isomer.

It is impossible to determine the relationships among all the isotopic forms of a polyatomic compound experimentally. Therefore the isotopic composition is determined by another method: the investigated polyatomic compound is converted to simpler compounds containing one or two equivalent atoms of a given element (CO_2, in investigations of $^{13}C/^{12}C$ and $^{18}O/^{16}O$; SO_2 in investigations of $^{34}S/^{32}S$; and H_2 in investigations of D/H, etc.). Then the isotopic composition of these samples is analyzed on a mass spectrometer. The value obtained when the investigated sample was quantitatively converted to a mass-spectrometer sample corresponds precisely to the value $(X^*/X)AX_n$, which characterizes the ratio of isotopes of the investigated element in the original polyatomic compound. Of course, information about the intramolecular distribution of isotopes is lost in this process.

What are the numerical parameters and how can they be compared with the spectrometer-determined isotopic composition and ratio of isotopic compositions α for the compounds forming an isotope-exchange system?

For polyatomic compounds the isotope-exchange reaction takes the following form:

$$mAX_n + nBX_m^* \rightleftharpoons mAX_n^* + nBX_m. \tag{3.6}$$

Actually, within the framework of this reaction a set of reactions of the individual isotopic forms takes place:

$$AX_{n-i+1}X_{i-1}^* + BX_{m-j}X_j^* \rightleftharpoons AX_{n-i}X_i^* + BX_{m-j+1}X_{j-1}^*. \tag{3.7}$$

Each set of reactions is characterized by an equilibrium constant

$$K_{ij} = [AX_{n-i}X_i^*][BX_{m-j+1}X_{j-1}^*]/[AX_{n-i+1}X_{i-1}^*][BX_{m-j}X_j^*]. \quad (3.8)$$

Solving Eq. (3.8) for concentration and considering the dependence of the constant on the value of β, one can find [via relationship (3.5)] the connection between the α and β factors of the isotopic forms (Galimov, 1971).

Deriving a general expression is rather cumbersome. Therefore we shall use the simple propane–methane system to demonstrate these relationships.

In this system the propane molecule is polyatomic for carbon, and it contains two groups of nonequivalent atoms: a central and two peripheral carbon atoms. The latter are equivalent; they convert into one another in the symmetry operations of the point group to which the molecule belongs.

The isotopic fractionation factor in the system C_3H_8–CH_4 by definition has the form

$$\alpha = \frac{\begin{array}{c}3[^{13}C_3H_8] + 2[^{13}C^{12}C^{13}CH_8] + 2[^{13}C^{13}C^{12}CH_8] \\ + [^{12}C^{13}C^{12}CH_8] + [^{13}C^{12}C^{12}CH_8]\end{array}}{\begin{array}{c}3[^{12}C_3H_8] + 2[^{12}C^{13}C^{12}CH_8] + 2[^{13}C^{12}C^{12}CH_8] \\ + [^{13}C^{12}C^{13}CH_8] + [^{13}C^{13}C^{12}CH_8]\end{array}} \div \frac{[^{13}CH_4]}{[^{12}CH_4]}. \quad (3.9)$$

The full isotope-exchange reaction

$$^{12}C_3H_8 + 3\,^{13}CH_4 \rightleftharpoons \,^{13}C_3H_8 + 3\,^{12}CH_4 \quad (3.10)$$

is characterized by the constant

$$K = \frac{[^{13}C_3H_8][^{12}CH_4]^3}{[^{12}C_3H_8][^{13}CH_4]^3} = (Q^*/Q)_{(^{13}C_3H_8/^{12}C_3H_8)}/(Q^*/Q)_{(^{13}CH_4/^{12}CH_4)}. \quad (3.11)$$

The value β_n is the nth root of the product of the ratio of partition functions times the ratio of their symmetry numbers s^* and s, where n is the number of substituted atoms in the isotopically substituted form.

For the completely isotopically substituted form of propane, $n = 3$ and

$$\beta_{1,2,3} = (Q^*_{^{13}C_3H_8} s^*_{^{13}C_3H_8}/Q_{^{12}C_3H_8} s_{^{12}C_3H_8})^{1/3}. \quad (3.12)$$

The subscripts of β show the number of atoms of propane carbon substituted by the heavy isotope. Because the symmetry numbers of the isotopically completely substituted and unsubstituted forms are the same, from (3.12) and (3.11) we have

$$K = K_{1,2,3} = \beta^3_{1,2,3}/\beta^3_{CH_4}. \quad (3.13)$$

To avoid repeating the expressions for the symmetry numbers in each case below, we immediately establish that, from an analysis of the geometry of propane and methane, the isotopic molecules ${}^{13}C_3H_8$, ${}^{12}C_3H_8$, ${}^{12}C^{13}C^{12}CH_8$, ${}^{13}C^{12}C^{13}CH_8$, ${}^{13}C^{13}C^{12}CH_8$, ${}^{12}C^{12}C^{13}CH_8$, ${}^{12}CH_4$, and ${}^{13}CH_4$ have the respective symmetry numbers 2, 2, 2, 2, 1, 1, 12, and 12.

From the reaction of the isotopic forms we find the following relationships:

$$ {}^{12}C_3H_8 + {}^{13}CH_4 \rightleftharpoons {}^{13}C^{12}C^{12}CH_8 + {}^{12}CH_4, $$

$$ K_1 = \frac{[{}^{13}C^{12}C^{12}CH_8][{}^{12}CH_4]}{[{}^{12}C_3H_8][{}^{13}CH_4]} = 2\frac{\beta_1}{\beta_{CH_4}}, \tag{3.14} $$

$$ {}^{12}C_3H_8 + {}^{13}CH_4 \rightleftharpoons {}^{12}C^{13}C^{12}CH_8 + {}^{12}CH_4, $$

$$ K_2 = \frac{[{}^{12}C^{13}C^{12}CH_8][{}^{12}CH_4]}{[{}^{12}C_3H_8][{}^{13}CH_4]} = \frac{\beta_2}{\beta_{CH_4}}; \tag{3.15} $$

$$ {}^{12}C_3H_8 + 2\,{}^{13}CH_4 \rightleftharpoons {}^{13}C^{12}C^{13}CH_8 + 2\,{}^{12}CH_4, $$

$$ K_{1,3} = \frac{[{}^{13}C^{12}C^{13}CH_8][{}^{12}CH_4]^2}{[{}^{12}C_3H_8][{}^{13}CH_4]^2} = \frac{\beta_{1,3}^2}{\beta_{CH_4}^2}; \tag{3.16} $$

$$ {}^{12}C_3H_8 + 2\,{}^{13}CH_4 \rightleftharpoons {}^{13}C^{13}C^{12}CH_8 + 2\,{}^{12}CH_4, $$

$$ K_{1,2} = \frac{[{}^{13}C^{13}C^{12}CH_8][{}^{12}CH_4]^2}{[{}^{12}C_3H_8][{}^{13}CH_4]^2} = 2\frac{\beta_{1,2}^2}{\beta_{CH_4}^2}. \tag{3.17} $$

Because both terminal atoms of propane are equivalent, the distribution of isotopes between them is equally probable. Consequently, $\beta_1 = \beta_3$ and $\beta_{1,2} = \beta_{2,3}$. In addition, there are the relationships $\beta_{1,2}^2 = \beta_1\beta_2$; $\beta_{1,3}^2 = \beta_1\beta_3$; $\beta_{1,2,3}^3 = \beta_1\beta_2\beta_3$, etc. Hence

$$ K_{1,3} = \frac{\beta_1^2}{\beta_{CH_4}^2}; \tag{3.18} $$

$$ K_{1,2} = 2\frac{\beta_{1,2,3}^3}{\beta_3\beta_{CH_4}^2}. \tag{3.19} $$

Now let us divide the numerator and denominator of expression (3.9) by ${}^{12}C_3H_8$ and introduce the notation $[{}^{13}CH_4]/[{}^{12}CH_4] = R$. Substituting in (3.9) the values of concentrations of isotopic forms of the molecules, expressed using relationships (3.14)–(3.19) via the respective β factors, we obtain

$$ \alpha = \frac{\beta_{1,2,3}}{\beta_{CH_4}} \frac{\dfrac{3\beta_{1,2,3}^2}{\beta_{CH_4}^2}R^2 + \dfrac{2\beta_1^3 + 4\beta_{1,2,3}^3}{\beta_{1,2,3}\beta_1}\dfrac{R}{\beta_{CH_4}} + \dfrac{1}{\beta_{1,2,3}}(\beta_2 + 2\beta_1)}{\dfrac{2\beta_{1,2,3}^3 + \beta_1^3}{\beta_1}\dfrac{R^2}{\beta_{CH_4}^2} + \dfrac{2\beta_2 + 4\beta_1}{\beta_{CH_4}}R + 3}. \tag{3.20} $$

First, let us note that, if all carbon atoms in the propane molecule were equivalent, i.e., $\beta_1 = \beta_2$ and thus $\beta_1 = \beta_2 = \beta_{1,2,3}$, then this would lead to abbreviation of the term in brackets in the numerator and denominator of expression (3.20). As a result the expression $\alpha = \beta/\beta_{CH_4}$ would be obtained, analogous to relationship (3.3), for compounds containing one atom of the given element. Hence for compounds containing several equivalent atoms, the fractionation factor α is determined simply by the ratio of their β factors. The latter can be calculated from any isotopic form (in view of the condition $\beta_1 = \beta_2 = \cdots = \beta_{1,2,\ldots,n}$).

The value R in expression (3.20) characterizes the isotopic ratio of carbon. The $^{13}C/^{12}C$ ratio is 0.011. For other biogenic elements, $R \ll 1$ also. Therefore, by neglecting the terms containing R, expression (3.20) can be reduced to the simple relationship

$$\alpha \simeq \frac{2\beta_1 + \beta_2}{3}\frac{1}{\beta_{CH_4}}. \tag{3.21}$$

From the general properties of expression (3.20), formula (3.21) for molecules of arbitrary composition can be generalized:

$$\alpha = m\sum_{i}^{q_A} p_A(\beta_{AX_n})_i \bigg/ n\sum_{i}^{q_B} p_B(\beta_{BX_m})_i. \tag{3.22}$$

The summation is carried out for the number of q_A and q_B groups of equivalent atoms in the molecules, where p_A and p_B are the numbers of equivalent atoms in each q_A and q_B group. These numbers are equal to the ratio of symmetry numbers of the unsubstituted molecule and of the molecule with the isotope substituted for one of its atoms of the corresponding group.

Thus, to describe fractionation of isotopes in a system containing polyatomic molecules, it is necessary to know the β factors of their isotopic forms. The number of possible isotopic forms of the compound is equal to

$$\sum_{i=1}^{n} \frac{n!}{i!(n-1)!}, \tag{3.23}$$

where n is the number of atoms of a given element, and i the extent of isotopic substitution.

Expression (3.22) does not require the β factors of all possible isotopic forms. For pentane, for instance, the number of these forms is 31. However, the values of three β factors are enough for an isotopic description of pentane according to (3.22). These β factors correspond to isotopically monosubstituted forms characterizing independent symmetry types of the given compounds isotopic forms. We designate them β_i.

The given ratios of the partition functions of the constants of any other isotopic forms are not independent. These can be expressed as combinations of thermodynamic factors β_i. The β value of any multiply substituted isotopic form is the geometric average of the β factors characterizing all isotopic monosubstitutions in the composition of a given isotopic form, i.e.,

$$\beta_{AX_{n-k}X_k} = \left(\prod_i^k \beta_i\right)^{1/k}, \tag{3.24}$$

where k is the multiplicity of substitution. For the completely substituted isotopic form,

$$\beta_{AX_n^*} = \left(\prod_i^n \beta_i\right)^{1/n}. \tag{3.25}$$

Nonequivalent atoms of a given element in the molecules are characterized by different β_i values. Consequently the distribution of isotopes between them is not equally probable. In other words, polyatomic compounds which contain nonequivalent atoms will be characterized by *intramolecular thermodynamic isotopic effects*. The difference in isotopic composition of an element in any two nonequivalent positions is determined by the relationship of the respective intramolecular thermodynamic factors:

$$\delta_{i(1)} - \delta_{i(2)} = \left(\frac{\beta_{i(1)}}{\beta_{i(2)}} - 1\right) \times 10^3\,‰. \tag{3.26}$$

The separation of isotopes between two different compounds is determined by relationship (3.22). Its numerator and denominator contain the mean arithmetic β_i factors of the respective compounds

$$\alpha = \frac{1}{n}\sum_i^n \beta_{i(AX_n)} \Big/ \frac{1}{m}\sum_i^m \beta_{i(BX_m)}. \tag{3.27}$$

In this notation β_i designates the β factors of all isotopically monosubstituted forms of a given element in the molecule (from 1 to n). Some of the β_i values, corresponding to equivalent atoms, turn out to be the same (singular).

Obviously, the values in the numerator and denominator of (3.27) characterize the isotopic thermodynamic properties of the compared compounds as a whole. Designating them β_Σ, we obtain

$$\alpha = \beta_{\Sigma(AX_n)}/\beta_{\Sigma(BX_m)}. \tag{3.28}$$

Then α again stands for the ratio of β factors, as in relationship (3.3) for simple compounds. We call the β_Σ value characterizing the compound as a whole the *thermodynamic isotopic factor of the compound.*

As follows from (3.27) and (3.28), between the thermodynamic isotopic factors of the compound and the isotopically monosubstituted forms, there exists the relationship

$$\beta_\Sigma = \frac{1}{n} \sum_i \beta_i, \tag{3.29}$$

which links molecular and intramolecular isotopic effects and plays an essential role in our future discussion.

Let us note that, owing to the additive character of this important relationship, we prefer not to use the logarithmic form of representation of factors widespread in literature (more precisely, the reduced partition functions in the form $\ln f(s^*/s)$. In using the value $\ln \beta \simeq \beta - 1$, more convenient additive forms of representation are obtained for the essentially multiplicative relationships of the values α, β, and δ (see Chapter 1). However, for complex polyatomic compounds, this does not compensate for the inconveniences related to the use of logarithmic expressions in the additive relationship (3.29).

II. SPECTRAL METHOD OF CALCULATING β_i FACTORS

Thus, as has been shown above, if the β_i values are known, it is possible to describe the intramolecular thermodynamic isotopic effects and the isotopic properties of a compound. As mentioned in Chapter 1, there is a theoretical method for calculating β_i factors (ratios of partition functions of the constants of isotopic forms). It is based on the quantum-statistical expression

$$\begin{aligned}\beta_i &= (Q^*s^*)_{AX_{n-1}X^*}/(Qs)_{AX_n} \\ &= \prod_i \nu_i^* \exp(-h\nu^*/2kT)(1 - \exp(-h\nu_i/kT))/[\nu_i \exp(-h\nu_i/2kT) \\ &\quad \times (1 - \exp(-h\nu^*/kT))],\end{aligned} \tag{3.30}$$

which was first obtained by Urey (1935, 1947). A modification of it which simplified the calculation procedures was developed by Bigeleisen (Bigeleisen and Mayer, 1947).

This is sometimes called the "quantum-statistical" method, however, "spectral method" is more precise, as it is related to calculation of the vibrational frequencies of the isotopic forms of a compound and is ultimately based on spectroscopic data. Also, the fundamental errors and limitations of the method are related to the procedure of determining the vibrational frequencies.

The vibrational frequencies of the main and of the isotopically substituted forms of the molecules studied that are needed to calculate the β factors can be obtained directly from the Raman and infrared spectra of the respective compounds. However, spectroscopic data on frequencies, especially of polyatomic and highly symmetrical molecules, are incomplete, and the isotopic shifts of frequencies for elements heavier than hydrogen are small and difficult to measure with precision. Therefore, in calculating β factors, theoretically established isotopic frequencies are used.

The methods of vibration theory on which the calculation of isotopic frequencies is based are well developed. In a number of handbooks and monographs (Gribov, 1965; Nakamoto, 1973; Sverdlov *et al.*, 1970; Vol'kenshteyn *et al.*, 1949; Wilson, 1960) vibration theory is expounded in connection with its application in the fields of spectroscopy and physical chemistry. The procedure for finding isotopic frequencies with the calculation of thermodynamic isotopic factors of a number of carbon compounds has been discussed fairly thoroughly in our work (Galimov, 1973b). A voluminous literature deals with calculation of the values of the thermodynamic isotopic factors of individual compounds on the basis of the spectral method.†

Vibrational frequencies are found by solving the equation of vibrational motion of molecules. In matrix form it is written as

$$|\mathbf{GF} - \mathbf{I}\lambda| = 0, \tag{3.31}$$

where **F** is the matrix of force constants; **G** the kinetic matrix; and **I** the so-called unit matrix; $\lambda_i = \sqrt{2\pi\nu_i}$.

Finding the root λ of the secular equation is a very laborious procedure, although if a computer is used, it becomes a routine mathematical problem, solved by means of standard programs for finding matrix values per se.

The calculation involves finding the best fit of calculated and experimental frequencies. If the agreement is unsatisfactory, the force constants are corrected and a new set of frequencies is obtained. When the convergence of the calculated spectrum with the experimental spectrum is acknowledged to be satisfactory, one must calculate the frequencies of the isotopically substituted form. This is achieved by substituting the mass of

† See, among others, Bachinski, 1969; Bottinga, 1968, 1969; Craig, 1953; Galimov and Ivlev, 1973–1974; Hartshorn and Shiner, 1972; Ivlev *et al.*, 1974a, b; Johansen, 1964; Ohmoto, 1972; Sakai, 1968; Singh and Wolfsberg, 1975; Urey, 1947; Vojta, 1960a, b, c.

the other isotope and correspondingly changing the **G** matrix using the equation

$$|\mathbf{FG}^* - \mathbf{I}\lambda^*| = 0. \tag{3.32}$$

The ν_i and ν_i^* values found are substituted in (3.30) and the value of the β_i factor of the corresponding isotopic form is determined.

By this method the β_i factors of many compounds of various elements have been calculated, the temperature dependences of isotopic effects have been found, the range of thermodynamic isotopic effects has been appraised, and the fractionation coefficients have been determined in various isotope-exchange systems.

However, the method has several limitations, which appear noticeably when polyatomic compounds are analyzed. The principal limitation is related to the uncertainty of the force field of a molecule, with the force constants found by solving the so-called inverse spectral problem, i.e., solving Eq. (3.31) with respect to elements $\mathbf{F}_{ij}$. However, because the number of force constants $\frac{1}{2}N(N + 1)$ is always greater than the number of original equations, which is $3N - 6$, it is nearly impossible to determine all the force constants. Therefore, one has to assume that some force constants are negligibly small, others equal to each other, and some analogous to related force constants from other molecules. The uncertainty can be somewhat reduced and the number of equations for finding the constants increased by introducing experimental data on isotopic frequencies. By substituting isotopic frequency in the secular equation and correspondingly recalculating the kinematic elements $\mathbf{G}_{ij}$, an additional initial equation can be obtained. That method is especially effective for compounds containing hydrogen, as the deuterium-substituted molecules give a substantial shift of isotopic frequency which can be measured with fairly high precision. Therefore (and this should be stressed), data on the force constants and correspondingly on β_{spectr} values of compounds containing hydrogen are very reliable. However, in the general case the uncertainty in calculations of the force constants remains substantial. As mentioned above, the originally chosen system of force constants is corrected at each step by comparing the calculated spectrum with the experimental one until a satisfactory fit is obtained. What is considered satisfactory agreement is rather subjective. The spectroscopic data themselves which are the basis of the comparison are of limited precision. Different authors use different data. In many cases the observed spectra are incomplete, i.e., there is no information on all the frequencies characterizing the molecule, especially in the case of high-molecular-weight compounds. Certain errors may be related to incorrect assignment of frequencies. Uncertainty in calculating the geometry of the molecule may be a source of error in calculating the elements of the **G** matrix. This is generally a problem with complex compounds. Another source of error is the model of a harmonic approximation of the vibrational motion of the molecule.

As a rule the importance of these limitations increases as the number of atoms and the complexity of the molecular structure of the compound increase. Therefore, precise calculation of the β factors of polyatomic complex compounds, which include most biologically interesting compounds, is practically impossible.

It should be mentioned that the spectral method of calculating β factors is very laborious. The determination of the β factors of even such a relatively simple compound as hexane requires operations with matrices of order 54. Modern computer techniques make it possible to cope with the mathematical difficulties. But with increasing complexity of molecules, the awkwardness of the computer problem becomes an obstacle to the calculation of the β factors.

Isotopic thermodynamic analysis of biological systems necessitates a more general approach to the calculation of the isotopic factors of complex compounds. Such an approach and, correspondingly, such a method of calculation were found on the basis of the "additivity principle" formulated below.

III. THE PRINCIPLE OF ADDITIVE CALCULATION OF THE β FACTOR

The procedure of determining the value of the thermodynamic isotopic factor is such that in calculating the β factor of a compound we cannot use knowledge of the value of the β factor of another compound even if the latter is similar to the former in structure and composition. In each case, one must completely solve the equation of vibrational motion of the molecule and find the frequencies of the isotopic forms of the compound for which the β factor has to be determined. However, it must be mentioned that, in solving the vibration equation, "transfer" of the force constants is widely practiced: the force constants characterizing the interaction of atoms in some molecules are used in the **F** matrices of other molecules containing analogous chemical bonds. This has raised the question of whether the β factor value itself cannot be represented as a combination of numerical values, each of which characterizes a certain type of bond and enters into the β factors of those compounds, which includes this bond (Galimov, 1973b).

Starting from very general considerations which do not rest on experience with the valency–force field theory, one concludes that the β factor may represent a value that is additive with respect to increments characterizing the individual bonds in a molecule.

First we shall take note of the facts that the thermodynamic isotopic factors of a free atom and of molecules have values of unity and greater than unity, respectively. This means that the β factor is different from unity only when a chemical bond between atoms is present. Therefore, the superiority of a heavy isotope in terms of energy is manifested only in molecules.

It follows that, in principle, the thermodynamic isotopic factor can be represented in the form

$$\beta = 1 + x, \tag{3.33}$$

where x is a term dependent on the presence and character of chemical bonds.

If any parameter of a given atom in a molecule depends on the character of the surrounding atoms, it will depend most strongly on the immediate surroundings, or, stated another way, it is determined primarily by the chemical bonds which bind the given atom. This is a manifestation of what is known in chemistry as the principle of localization. On the other hand, the widespread property of additivity of thermodynamic values makes a linear combination of increments characterizing individual bonds likely. In other words, one can attempt to represent x in the form $\sum_{j=1}^{n} L_j$, where L_j is a value characterizing the jth bond of n bonds formed by the ith atom, i.e.,

$$\beta_i = 1 + \sum_{j=1}^{n} L_j. \tag{3.34}$$

Finally, the parameters characterizing a chemical bond, to some extent, have the property of transferability, i.e., they retain their intrinsic values in different chemical compounds. To some degree L_j should have this property.

To what extent the aforementioned approximations are valid and how satisfactory the method of calculating β factors based on them is depend on how applicable the general ideas of localization, additivity, and transferability are to isotopic thermodynamics.

These principles involve known limitations that may or may not be important. Expression (3.34), because of these general considerations, could prove to be completely unrealizable in isotopic thermodynamics. Actually, as we shall see, it is satisfactorily workable. This fact is primarily of value for understanding isotopic thermodynamics in that the aforementioned generalities have proved to be thoroughly characteristic. Consequently, a method of calculating thermodynamic isotopic factors on a very general basis arises.

Although postulation of the additivity principle does not depend on the theory of the valency-force field on which the calculations of β factors via vibrational frequencies are based, it can be shown that the structure of Eq. (3.34) is derivable from these ideas and, in a certain sense, is included in quantum-statistical relationship (3.30) (Galimov, 1974).

Therefore, the principle of additivity presumes the possibility of representing the β factor by means of increments, the numerical values of which depend on the chemical bond type. We propose that the value of the corresponding increment be called the *isotopic bond number*.

As is known, there are several approaches to the description of a chemical bond. In essence, the principle of additivity is logically tied to the understanding of a chemical bond in the theory of localization of electron pairs (the method of valency bonds).

According to this idea, the bond between atoms arises as a result of the overlapping of directional atomic orbitals which are in a valency state. The overlap of s orbitals or sp hybrid orbitals leads to the formation of a bond in which the electron cloud is axially elongated along the bond. Such a bond is called a σ bond. Mutual overlap of nonhybrid p orbitals leads to the production of a shared electron cloud with two sides parallel to the axis of the bond. Such a bond is called a π bond.

In the ethane molecule, for instance, the carbon atoms are in a state of sp_3 hybridization. Each carbon atom forms a bond with hydrogen, the atomic orbital of which is in the s state. The electron cloud is elongated axially along the bond. From this point, we use the following designation of the bond: in parentheses we indicate first the symbol of the atom in whose bond we are interested, then the symbol of the atom with which the bond is formed; next to the parentheses we indicate the type of bond, and in a subscript, the state of hybridization of the atom in question and of the atom with which it forms the bond. Thus we designate the bond of carbon to hydrogen bond in ethane $(\mathrm{C{-}H})\sigma_{sp_3-s}$. The carbon to carbon bond in ethane is designated $(\mathrm{C{-}C})\sigma_{sp_3-sp_3}$ forms. In the ethylene molecule the orbits of the carbon atoms are in a state of sp_2 hybridization. With hydrogen the carbon forms a σ bond: $(\mathrm{C{-}H})\sigma_{sp_2-s}$. Between the carbon atoms, one σ bond due to overlap of sp_2 orbitals and an additional π bond due to overlap of $2p_z$ nonhybridized orbitals form. Thus in ethylene the carbon atoms form a $(\mathrm{C{=}C})\sigma\pi_{sp_2-sp_2}$ bond. Because four, rather than two, electrons participate in this bond, it is a double bond.

The interaction of atomic orbitals imposes the geometric structure on the molecule. For example, the tetrahedral configuration of the methyl groups of ethane, the angular structure of water, as well as the pyramidal structure of NH_3, and the linear structure of acetylene are direct consequences of the configuration of the atomic orbitals in the valency state of the atoms forming the respective molecules.

A bond can be achieved both by overlap of atomic orbitals, each of which contains one electron, and by overlap of a filled atomic orbital of one atom with a vacant atomic orbital of another. In the latter case the bond bears the special name of donor–acceptor. Its electron structure coincides with that of an ordinary σ bond. They are indistinguishable chemically and, as we shall see, are characterized by the same isotopic bond number.

It should be mentioned that, traditionally, in the structural formulas of chemical compounds the donor–acceptor bond is often marked by two dashes (similar to a double bond), although it is an ordinary bond. For instance, in illustrating the structural formula of SO_4^{2-}, to convey the hexavalent character of the sulfur, its bond with the oxygen atom is sometimes depicted as two double and two ordinary bonds. Actually, in the sulfate ion, the sulfur atom (which is in a state of sp_3 hybridization) forms four bonds of the same electron structure with the oxygen atoms, two ordinary σ bonds and the two donor–acceptor bonds. Because we characterize the bond by the isotopic number assigned in the theory of localized electron pairs, the β factor of hexavalent sulfur in the SO_4^{2-} molecule is determined by the sum of four equivalent $L_{(S-O)\sigma_{sp3-sp3}}$ bonds.

On the other hand, bonds that are actually equivalent, in some cases, should be differentiated (specialized). For instance, in the carbonate ion the carbon atom, in a state of sp_2 hybridization, forms three equivalent bonds with the oxygen atoms. However, in the localized electron pairs method, two different types of bonds are present: a $\sigma\pi$ bond with an sp_2 hybridized oxygen atom and two ordinary σ bonds with sp_3 hybridized oxygen atoms. Accordingly, the $\beta^{13}C$ factor of CO_3^{2-} would be determined by the sum $1 + L_{(C=O)\sigma\pi_{sp2-sp2}} + 2L_{(C-O)\sigma_{sp2-sp3}}$.

By these examples, we hope to emphasize that, in calculating the value of the β factor, the set of isotopic bond numbers would be determined by the electron structure of the chemical bonds in the method of localized electron pairs.

It must be remembered that, despite the fruitfulness of this concept, which has been proven by many applications, it is based on certain approximations which may be unsuitable for some cases. The theory of localized electron pairs does not enable us to explain some properties of compounds, for instance, color and magnetic properties. It is based on the idea of localization of electrons on bonds, although the idea of the completeness of delocalized molecular orbits is physically rigorous. The coefficients of hybridization (the ratios of the p state to the s state) are considered to be whole numbers: three in sp_3 hybridization, two in sp_2 hybridization, and one in sp_1 hybridization. Meanwhile, the coefficients of hybridization actually differ from integral values; for instance, in the ethane molecule (sp_3) it is 3.2; in benzene (sp_2) it is 2.18; in acetylene (sp_1) it is 1.30. The idea of localized chemical bonds becomes inadequate for describing the bonds in molecules with an odd number of electrons, for instance, NO and NO_2. To some degree these limitations affect the applicability of the principle of additivity to the calculation of the β factors of compounds.

IV. ISOTOPIC BOND NUMBERS OF CARBON

The isotopic bond numbers L_{C-X} can be calculated from relationship (3.34) if the appropriate $\beta_i{}^{13}C$ values are known. At the present the values of the thermodynamic isotopic factors have been established, by the spectral method, for a substantial number of carbon compounds. By determining L_{C-X} on the basis of several values of β_{spectr} and then using the bond number to calculate the β factors of other compounds, it is possible to check the validity of the principle of additivity in the approximation adopted.

The β factors for the gaseous state should be used as the starting data. In condensed mediums intermolecular interactions contribute to the values of the β factors. Therefore the bond numbers L_j and the β_{addit} values obtained will characterize the substance in the gaseous form. However, the thermodynamic isotopic effects in the vapor–liquid system have an appreciable value only for hydrogen isotopes and to a lesser extent for oxygen isotopes in simple compounds, for instance water. For organic carbon compounds, the interphase isotopic effects are insignificant (Galimov, 1973b) and the $\beta^{13}C$ values calculated for the gaseous state can be used without large error when examining the condensed phases.

Because of the uncertainties inherent in the spectral method, the β_{spectr} values are calculated with an error which may, in some cases, be considerable. This forces us to choose cautiously the β_{spectr} values which serve as the basis for determining L_{C-X}. Similarity of β_{spectr} and β_{addit} can also serve as the basis for assessing the validity of the principle of additivity. Nevertheless, the determination of L_{C-X} from known values of β_{spectr} is the simplest way of realizing the additive method.

All the isotopic bond numbers calculated below are referred to a temperature of 300 K.

A. The L_{C-H} and L_{C-C} Isotopic Bond Numbers

In the methane molecule, carbon forms four equivalent bonds of $(C—H)\sigma_{sp_3-s}$ type. Thus we have $\beta^{13}C = 1 + 4L_{(C-H)\sigma_{sp_3-s}}$. The $\beta^{13}C$ value we calculated by the spectral method is 1.1136 at 300 K (Galimov and Ivlev, 1973–1974). From this calculation, a value of 0.028 can be obtained for $L_{(C-H)\sigma_{sp_3-sp_3}}$. In the ethane molecule, each carbon atom forms three $(C—H)\sigma_{sp_3-s}$ bonds and one of the $(C—C)\sigma_{sp_3-sp_3}$ type. Using the $L_{(C-H)\sigma_{sp_3-s}}$ value obtained for the methane molecule and the value of the thermodynamic isotopic factor calculated for ethane on the basis of spectral data ($\beta^{13}C = 1.1316$; Galimov and Ivlev, 1973–1974), we find that $L_{(C-C)\sigma_{sp_3-sp_3}} = 0.046$. Combining the values found, the $\beta^{13}C$ values for other hydrocarbons can be obtained. Comparing the β_{spectr} (Table 3.1, column 2) and β_{addit} (column 4) values shows that there is satisfactory agreement between them. However, the CH_2 groups of different hydrocarbons are characterized by the same $\beta_i{}^{13}C$ values, whereas the solution of the vibrational problem for the respective isotopic forms yields values of $\beta_i{}^{13}C$ which differ by about $\pm 0.2\%$ for the CH_2 groups in different molecules. In this case the difference may be explained by the influence of "distant surroundings." This influence can be accounted for by introducing corrective isotopic factors with subscripts corresponding to the types of bonds formed by the partner atoms of a given carbon atom (column 6 in Table 3.1). The values of the β_i factor then is determined using the expression

$$\beta_i = 1 + \sum_j L_j + \sum_k l_k,$$

where k designates the bond formed by atoms adjacent to a given carbon atom.

Using the known β_i values given in Table 3.1, the values of l_{C-H} and l_{C-C} can be determined. One of the values inevitably turns out to be dependent. Therefore it is convenient to take $l_{C-H} \equiv 0$. Then we obtain a value of 0.0013 for l_{C-C}.

As seen from Table 3.1, correction for the distant surroundings leads to agreement of the results obtained by the additive and by the spectral methods for homologs of methane, with a precision to the fourth decimal place. At the same time the value of the number $l_{C-C} = 0.0013$ is small. It constitutes about 3% of the $L_{C-C} = 0.0464$ value, which gives an idea of the contribution of the distant surroundings to the value of the β factor.

The values of the isotopic bond numbers L_{C-H}, L_{C-C}, and l_{C-C} are found by the least squares method using all nine known values of β_i given in Table 3.1. Determinations of $\beta^{13}C$ values for these molecules made by

TABLE 3.1

VALUES USED FOR CALCULATING ISOTOPIC BOND NUMBERS $L_{C-H} = 0.0284$, $L_{C-C} = 0.0464$, $l_{C-H} = 0$, AND $l_{C-C} = 0.0013$

Compound	$\beta_{spectr}\,{}^{13}C$[a]	$\sum L_{C-X}$	$\beta_{addit}\,{}^{13}C$ (*neglecting* l_{C-X})	$\Delta = \beta_{addit}\,{}^{13}C - \beta_{spectr}\,{}^{13}C$	$\sum l_{C-X}$	$\beta_{addit}\,{}^{13}C$	$\Delta = \beta_{addit}\,{}^{13}C - \beta_{spectr}\,{}^{13}C$[b]
$*CH_4$	1.1136	$4L_{C-H}$	1.1136	0.0000	—	1.1136	0.0000
$*CH_3CH_3$	1.1317	$3L_{C-H} + L_{C-C}$	1.1316	−0.0001	$3l_{C-H}$	1.1316	−0.0001
$*CH_2CH_2CH_3$	1.1331	$3L_{C-H} + L_{C-C}$	1.1316	−0.0015	$l_{C-C} + 2l_{C-H}$	1.1329	−0.0002
$CH_3{*}CH_2CH_3$	1.1493	$2L_{C-H} + 2L_{C-C}$	1.1496	0.0003	$6l_{C-H}$	1.1496	0.0003
$*CH_3CH_2CH_2CH_3$	1.1331	$3L_{C-H} + 2L_{C-C}$	1.1316	−0.0015	$l_{C-C} + 2l_{C-H}$	1.1329	−0.0002
$CH_3{*}CH_2CH_2CH_3$	1.1507	$2L_{C-H} + 2L_{C-C}$	1.1496	−0.0011	$l_{C-C} + 5l_{C-H}$	1.1509	0.0002
$*CH_3CH_2CH_2CH_2CH_3$	1.1330	$3L_{C-H} + L_{C-C}$	1.1316	−0.0014	$l_{C-C} + 2l_{C-H}$	1.1329	−0.0001
$CH_3{*}CH_2CH_2CH_2CH_3$	1.1506	$2L_{C-H} + 2L_{C-C}$	1.1496	−0.0010	$l_{C-C} + 5l_{C-H}$	1.1509	0.0003
$CH_3CH_2{*}CH_2CH_2CH_3$	1.1519	$2L_{C-H} + 2L_{C-C}$	1.1496	−0.0023	$2l_{C-C} + 4l_{C-H}$	1.1522	0.0003

[a] Data are from Galimov and Ivlev (1973–1974)

[b] Standard deviation is ± 0.0002.

different authors differ somewhat from each other. We prefer to use only the results of our own calculations as base values of the β factors (Galimov and Ivlev, 1973–1974) because they were obtained using a single system of force constants.

Three of the nine values of β_{addit} given in Table 3.1 are not independent. However, the other six degrees of freedom guarantee a fairly high level of significance in the calculations. The standard deviation between the $\beta_{\mathrm{addit}}{}^{13}C$ and $\beta_{\mathrm{spectr}}{}^{13}C$ values is ± 0.0002, or $\pm 0.5\%$ of the range of the $\beta_{\mathrm{spectr}}{}^{13}C$ values used (1.1136 to 1.1519; see Table 3.1).

B. The L_{C-O} and $L_{C=O}$ Isotopic Bond Numbers

In the CO_2 molecule, carbon forms two bonds of $(C{=}O)\sigma\pi_{sp_1-sp_2}$ type. The $\beta^{13}C_{\mathrm{spectr}}$ value for CO_2 at 300 K is 1.1909 (Urey, 1947). The value obtained for the corresponding isotopic number is 0.0955. In the CO molecule there also is a bond of $(C{=}O)\sigma\pi_{sp_1-sp_2}$ type. The $\beta^{13}C$ value calculated by the spectral method is 1.0970. Thus the isotopic bond number is 0.0970. Despite the essential difference in $\beta^{13}C$ values of CO_2 and CO, the analogous bond numbers prove to be similar, that is, satisfactory transferability is shown. It was mentioned above that, according to the theory of valency bonds, one $(C{=}O)\sigma\pi_{sp_2-sp_2}$ bond and two $(C{-}O)\sigma_{sp_2-sp_3}$ bonds enter the structure of the $CO_3{}^{2-}$ molecule. In this case the actual equivalence of the carbon–oxygen bonds in the carbonate ion can be considered the result of delocalization of π electrons and formally represented as the result of the mixture of three forms, in each of which a $\sigma\pi$ bond occupies one of the directions designated.

By identifying the isotopic numbers for the analogous bonds in all three molecules, we obtain $L_{C-O} = 0.0550$ and $L_{C=O} = 0.0958$ (Table 3.2). In this case we ignore the fact that the $\sigma\pi$ bonds in the CO_2 and CO molecules

TABLE 3.2

Values Used for Calculating Isotopic Bond Numbers $L_{C-O} = 0.0550$, $L_{C=O} = 0.0958$, $l_{C-O} = 0.0019$, and $l_{C=O} = 0.0028$

Compound	$\beta^{13}C_{\mathrm{spectr}}$[a]	$\sum L_{C-X}$	$\sum l_{C-X}$	$\beta^{13}C_{\mathrm{addit}}$	Δ[b]
CO_2	1.1909	$2L_{C=O}$	—	1.1915	0.0006
$CO_3{}^{2-}$	1.2057	$2L_{C-O}+L_{C=O}$	—	1.2058	0.0001
CO	1.0970	$L_{C=O}$	—	1.0958	−0.0012

[a] Data from Urey (1947).
[b] Standard deviation = ±0.0009.

TABLE 3.3

β^{13}C VALUES DETERMINED BY THE SPECTRAL AND ADDITIVE METHODS

Compound	$\beta^{13}C_{spectr}$[a]	$\sum L_{C-X}$	$\sum l_{C-X}$	$\beta^{13}C_{addit}$	Δ[b]
CH_3OH	1.1328	$3L_{C-H} + L_{C-O}$	—	1.1402	0.0074
$*CH_3COOH$	1.1381	$3L_{C-H} + L_{C-C}$	$l_{C=O} + l_{C-O}$	1.1363	−0.0018
CH_3*COOH	1.1978	$L_{C-C} + L_{C-O} + L_{C=O}$	—	1.1974	−0.0004
$*CH_3CHO$	1.1293	$3L_{C-H} + L_{C-C}$	$l_{C=O}$	1.1344	0.0051
CH_3*CHO	1.1731	$L_{C-H} + L_{C-C} + L_{C=O}$		1.1708	−0.0023

[a] Data from Ivlev *et al.* (1974a).
[b] Standard deviation = ±0.0036.

and those in CO_3^{2-} are formed by atoms in different states of hybridization. This omission also proves to be valid in the consideration of other types of bonds discussed later.

Table 3.3 gives the values of the β_i factors of five more compounds containing C—O and C=O bonds. The calculations of $\beta^{13}C_{spectr}$ for these are less reliable than for CO, CO_2 and CO_3^{2-}. Also, in addition to C—O and C=O bonds, they contain C—H, C—C, and O—H bonds. Consequently, in this case there are other sources of divergence of β_{spectr} and β_{addit} besides the uncertainty introduced by error in the determination of L_{C-O} and $L_{C=O}$ and the limitation of the principle of additivity. Nevertheless, the amount of divergence is rather modest: $\Delta = \pm 0.0036$, which is about 4% of the range of variation of the β^{13}C values in question.

A correction for the distant surroundings must be taken into account for the isotopic forms $*CH_3COOH$ and $*CH_3CHO$. Following the logic of the principle of additivity it can be presumed that the l_{C-X} values have the Same relationships to each other as the L_{C-X} values. Then the l_{C-X} number for other types of bonds can be found using the following expression:

$$l_{C-X} = (L_{C-X} - mL_{C-H}) \times 0.0722, \tag{3.35}$$

where m is the multiplicity of the C—X bond. Therefore, $l_{C-O} = 0.0019$ and $l_{C=O} = 0.0028$.

B. THE $L_{C=C}$ ISOTOPIC BOND NUMBER AND AROMATIC COMPOUNDS

Ethylene is typical of compounds containing a bond of the $(C{=}C)\sigma\pi_{sp_2-sp_2}$ type. The value of the $\beta_i{}^{13}$C factor of $*CH_2{=}CH_2$ was determined by Hartshorn and Shiner (1972). More precisely, they calculated the coefficients of isotope fractionation with respect to acetylene, rather than β values. With their coefficient of fractionation in the system CO_2—C_2H_2, one can take the value of $\beta_i{}^{13}C = 1.1909$ for CO_2 and calculate the $\beta_i{}^{13}C$ factor of the CH_2 group of ethylene, which turns out to be 1.1354. Neglecting the effect of the extent of hybridization of carbon (the bond of carbon with hydrogen in ethylene, $(C—H)\sigma_{sp_2-s}$, is somewhat different from the same bond in methane hydrocarbons, $(C—H)\sigma_{sp_3-s}$), and taking a value of $L_{C-H} = 0.0284$ for ethylene, we obtain a value of 0.0785 for the isotopic number $L_{C=C}$ (Table 3.4).

For the aromatic compounds it originally proved necessary to introduce the specialized number $L_{C\overset{...}{=}C} = 0.062$ (Galimov, 1974), which was calculated starting with the β_i value of 1.1533 for benzene that we had determined by the spectral method (Galimov and Ivlev, 1973–1974). However, it was ascertained that the value $L_{C\overset{...}{=}C}$ is precisely equal to half the sum of L_{C-C} and $L_{C=C}$. In other words, for purposes of calculating the isotopic bond numbers, the equivalence of the carbon—carbon bonds in the benzene molecule can be considered to be the result of alternating ordinary and double bonds.

An analogous situation has already been mentioned in the case of the carbonate-ion molecule. Such specialization of equivalent bonds, according to the method of localized electron pairs, proves to be justified also in the calculation of the isotopic bond numbers of nitrogen and oxygen, as will be shown later.

In view of the π coupling in the benzene molecule, the effect of the distant surroundings evidently should be corrected for differently. To retain the general nature of the correction method, it is necessary to add the correction numbers l_{C-C}, $l_{C=C}$ and an additional correction number l_0 for the aromatic ring. In first approximation the total correction should come to zero, i.e., one should take $l_0 = -0.0029$ $(-l_0 = l_{C-C} + l_{C=C})$. Generally, the value of this correction would depend on the factors which determine delocalization of the π electrons, i.e., the number of substituents, the character of the substituents, and the position of the carbon atom in the ring relative to the substituent.

More study is needed to consider the problem of second-order corrections in aromatic compounds.

C. The $L_{C\equiv C}$ Isotopic Bond Number

The isotopic bond number $(C\equiv C)\sigma\pi\pi_{sp-sp}$ originally was determined from the β_i values of the isotopic form of propynal *CH≡C—CHO—the only acetylenic compound for which a β_{spectr} value was available. The value of

TABLE 3.4

VALUES USED FOR CALCULATING ISOTOPIC BOND NUMBERS $L_{C=C} = 0.0785$, $l_{C=C} = 0.0016$, AND $l_0 = -0.0029$

Compound		$\beta^{13}C_{spectr}$ Value	$\beta^{13}C_{spectr}$ Reference	$\sum L_{C-X}$	$\sum l_{C-X}$	$\beta^{13}C_{addit}$	Δ[a]
$CH_2{=}CH_2$		1.1354	Hartshorn and Shiner (1972)	$2L_{C-H} + L_{C=C}$	—	1.1354	0.0000
		1.1533	Galimov and Ivlev (1973–1974)	$L_{C-H} + L_{C-C} + L_{C=C}$	$l_{C-C} + l_{C=C} + l_0$	1.1533	0.0000
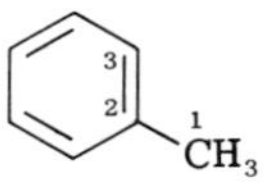	(1)	1.1293	Galimov and Ivlev (1973–1974)	$3L_{C-H} + L_{C-C}$		1.1316	0.0023
	(2)	1.1677	Galimov and Ivlev (1973–1974)	$2L_{C-C} + L_{C=C}$	$l_{C-C} + l_{C=C} + l_0$	1.1712	0.0035
	(3)	1.1545	Galimov and Ivlev (1973–1974)	$L_{C-H} + L_{C-C} + L_{C=C}$	$2l_{C-C} + l_{C=C} + l_0$	1.1546	0.0001

[a] Standard deviation is ± 0.0020.

TABLE 3.5

VALUES USED FOR CALCULATING ISOTOPIC BOND NUMBERS $L_{C\equiv C} = 0.0881$ AND $l_{C\equiv C} = 0.0003$

	β_{spectr} ^{13}C					
Compound	*Value*	*Reference*	$\sum L_{C-X}$	$\sum l_{C-X}$	β_{addit} ^{13}C	Δ
$CH\equiv CH$	1.1173	Hartshorn and Shiner (1972)	$L_{C-H} + L_{C\equiv C}$	—	1.1165	−0.0008
*$CH\equiv C-CHO$	1.117	Johansen (1964)	$L_{C-H} + L_{C\equiv C}$	$l_{C=O}$	1.1178	0.0008

$L_{C\equiv C}$ was found to be 0.088 (Galimov, 1973b). It should be noted that a relatively large spread was obtained between the calculations of the $\beta^{13}C$ value by the spectral and additive methods for another isotopic form of propynal—CH≡*C—CHO (see Table 3.15). However, the β_{spectr} value for this isotopic form seemed questionable to us, and the value 0.088 was selected for $L_{C\equiv C}$. Later the work of Hartshorn and Shiner (1972) became available to us, in which the value $\alpha_{(CO_2-C_2H_2)} = 1.0659$ is given. Since this value is equal to the ratio $\beta_{CO_2}/\beta_{C_2H_2}$ and the value 1.1909 is known for β_{CO_2}, $\beta_{C_2H_2}$ turns out to be 1.1173. Therefore, the value $L_{C\equiv C}$ is 0.0889, which is in good agreement with the value adopted previously. As in the case of ethylene, the insignificance of an effect of the character of hybridization of carbon on the L_{C-H} value is confirmed.

The isotopic bond numbers $(C—H)_{sp_3-s}$, $(C—H)_{sp_2-s}$, and $(C—H)_{sp-s}$ are within the limits of the uncertainties in calculations of β factors.

Table 3.5 gives values of the divergence of β_{addit} and β_{spectr} for a value of $L_{C\equiv C} = 0.0881$. From expression (3.35) a value of $l_{C\equiv C} = 0.0003$ is determined.

D. The L_{C-N} and $L_{C\equiv N}$ Isotopic Bond Numbers

The $(C—N)\sigma_{sp_3-sp_3}$ isotopic bond number was determined from two estimates of the β_{spectr} for the compound $*CH_3NH_2$. Its value was found to be 0.050.

The $(C\equiv N)\sigma\pi\pi_{sp-sp}$ isotopic bond number was determined from data on β_i values for the HCN molecule and CN^- ion (Table 3.6). For a value of $L_{C\equiv N} = 0.090$ the divergence between β_{spectr} and β_{addit} is ± 0.0025.

E. The Isotopic Bond Numbers L_{C-F}, L_{C-Cl}, L_{C-Br}, L_{C-I}

The isotopic bond number C—F was originally determined using the value $\beta_{CF_4} = 1.2241$, calculated via vibrational frequencies in the work (Galimov, 1973b). The value L_{C-F} was found to be 0.056.

At the present, values of β_{spectr} are known for several compounds containing the C—F bond (Table 3.7). The determination of the isotopic number L_{C-F} by the least squares method on the basis of three compounds yields a value of 0.0564, which coincides closely with the value established earlier. The standard deviation between β_{addit} and β_{spectr} in Table 3.7 is ± 0.0050, which corresponds to approximately 6% of the maximum variation between the β values. Considering that the β_{spectr} of the compounds were calculated by different authors, the observed similarity of β values

TABLE 3.6

VALUES USED FOR CALCULATING ISOTOPIC BOND NUMBERS $L_{C-N} = 0.050$, $L_{C\equiv N} = 0.090$, $l_{C\equiv N} = 0.0003$, AND $l_{C-N} = 0.0016$

	$\beta_{spectr}\,^{13}C$					
Compound	*Value*	*Reference*	$\sum L_{C-X}$	$\sum l_{C-X}$	$\beta_{addit}\,^{13}C$	Δ
H*CN	1.1206	Urey (1947)	$L_{C-H} + L_{C\equiv N}$	—	1.1184	−0.0022
*CN⁻	1.0875	Urey (1947)	$L_{C\equiv N}$	—	1.0900	0.0025
$^*CH_3NH_2$	1.1361	Galimov (1973b)	$3L_{C-H} + L_{C-N}$	—	1.1352	−0.0009
$^*CH_3NH_2$	1.1344	Ivlev *et al.* (1974a)	$3L_{C-H} + L_{C-N}$	—	1.1352	0.0008

determined by the spectral and additive methods can be considered quite satisfactory.

The β_{spectr} and β_{addit} values for compounds containing the C—Cl bond show approximately the same level of convergence (Table 3.8). The isotopic number of this bond $L_{C-Cl} = 0.030$.

In the work by Hartshorn and Shiner (1972) cited previously, values are given for the isotopic fractionation factors between halogen substitutions of methane: CH_3F, CH_3Cl, CH_3Br, CH_3I, and acetylene. It is advisable to convert to values of the β factors of these compounds via β_{CH_4} values, using for the latter the value $\beta_{CH_4} = 1.1136$ which we have calculated (Galimov and Ivlev, 1973–1974) and used above to determine L_{C-H}. The conversion via β_{CH_4} (rather than via β_{CO_2}) provides the possibility of estimating the relationship between values of the isotopic numbers of the halogen—carbon bonds and L_{C-H} more precisely.

The values of β_{CH_3F} and β_{CH_3Cl} are given in Tables 3.7 and 3.8, respectively. Tables 3.9 and 3.10 give values of β_{spectr} for CH_3I and CH_3Br, respectively, which, combined with the data of Vojta (1960b) for ICN and BrCN, allows the calculation of the isotopic bond numbers of carbon—bromine and carbon—iodine. In Vojta's work (1960b) the β_{spectr} values were determined at 273.15 K.

We converted to β_{spectr} values for 300 K by multiplying $\beta_{273.15}$ by 0.9766. The precision of that extrapolation falls within ± 0.003. Although this conversion may be an additional source of uncertainty of the starting values of β_{spectr}, the convergence of values of β_{spectr} and β_{addit} for compounds containing C—Br and C—I bonds is satisfactory. The isotopic numbers of these bonds are $L_{C-Br} = 0.027$ and $L_{C-I} = 0.018$. The two compounds used to calculate each of these bond numbers have similar β-factor values, for example, $\beta_{ICN} = 1.107$ and $\beta_{CH_3I} = 1.1031$. However, they are characterized by different sets of bond numbers; in one case, there is $L_{C\equiv N}$ and in the other $3L_{C-H}$, in addition to $L_{(C-I)}$. Therefore, the fact that the values of β for these compounds, determined by the additive method, agree with the

TABLE 3.7

VALUES USED FOR CALCULATING ISOTOPIC BOND NUMBERS $L_{C-F} = 0.0564$ AND $l_{C-F} = 0.0020$

	$\beta_{spectr}\,^{13}C$					
Compound	*Value*	*Reference*	$\sum L_{C-X}$	$\sum l_{C-X}$	$\beta_{addit}\,^{13}C$	Δ^a
CF_4	1.2241	Galimov (1973b)	$4L_{C-F}$	—	1.2256	0.0015
CH_3F	1.1368	Hartshorn and Shiner (1972)	$3L_{C-H} + L_{C-F}$	—	1.1416	0.0048
HCOF	1.186	Johansen (1964)	$L_{C-H} + L_{C=O} + L_{C-F}$	—	1.1808	−0.0052

[a] Standard deviation is ± 0.0050.

TABLE 3.8

VALUES USED FOR CALCULATING ISOTOPIC BOND NUMBERS $L_{C-Cl} = 0.030$ AND $l_{C-Cl} = 0.0001$

	$\beta_{spectr}\,^{13}C$					
Compound	*Value*	*Reference*	$\sum L_{C-X}$	$\sum l_{C-X}$	$\beta_{addit}\,^{13}C$	Δ
CH_3Cl	1.1149	Hartshorn and Shiner (1972)	$3L_{C-H} + L_{C-Cl}$	—	1.1152	0.0003
$^*CH_2ClCOOH$	1.133	Johansen (1964)	$2L_{C-H} + L_{C-C} + L_{C-Cl}$	$l_{C-O} + l_{C=O}$	1.1379	0.005
$CH_2{=}^*CCl_2$	1.1436	Hartshorn and Shiner (1972)	$2L_{C-Cl} + L_{C=C}$	—	1.1385	−0.0051

TABLE 3.9

VALUES USED FOR CALCULATING ISOTOPIC BOND NUMBERS $L_{C-I} = 0.018$

Compound	β_{spectr} ^{13}C Value	β_{spectr} ^{13}C Reference	$\sum L_{C-X}$	$\sum l_{C-X}$	β_{addit} ^{13}C	Δ
ICN	1.107	Vojta (1960b)	$L_{C\equiv N} + L_{C-J}$	—	1.106	−0.001
CH_3I	1.1031	Hartshorn and Shiner (1972)	$3L_{C-H} + L_{C-J}$	—	1.1032	0.000

TABLE 3.10

VALUES USED FOR CALCULATING ISOTOPIC BOND NUMBERS $L_{C-Br} = 0.027$

Compound	β_{spectr} ^{13}C Value	β_{spectr} ^{13}C Reference	$\sum L_{C-X}$	$\sum l_{C-X}$	β_{addit} ^{13}C	Δ
BrCN	1.119	Vojta (1960b)	$L_{C\equiv N} + L_{C-Br}$	—	1.117	−0.002
CH_3Br	1.111	Hartshorn and Shiner (1972)	$3L_{C-H} + L_{C-Br}$	—	1.1122	0.001

values calculated via vibrational frequencies illustrates the workability of the principle of additivity.

A systematic decrease in L_{C-X} values is observed in the series C—F, C—Cl, C—Br, and C—I.

E. THE $L_{C=S}$ ISOTOPIC BOND NUMBER

Determination of the $L_{C=S}$ value and calculation of its additive properties is possible on the basis of the β_{spectr} values for the two compounds COS and CS_2 (Table 3.11) calculated by Vojta (1960b). The β_{spectr} values for

TABLE 3.11

VALUES USED FOR CALCULATING ISOTOPIC NUMBER $L_{C=S} = 0.0495$

Compound	β_{spectr} $^{13}C^a$	$\sum L_{C-X}$	$\sum l_{C-X}$	β_{addit} ^{13}C	Δ
COS	1.140	$L_{C=O} + L_{C=S}$	—	1.145	0.005
CS_2	1.104	$2L_{C=S}$	—	1.099	−0.005

[a] Data are from Vojta (1960b)

300 K were obtained by extrapolation of the data for $\beta_{273.15}$, as mentioned above. The divergence of β_{spectr} and β_{addit} is ± 0.005.

F. THE $L_{C=N}$ ISOTOPIC BOND NUMBER

The value of the $L_{C=N}$ isotopic number can be determined from β_{spectr} values for the isotopic form CH_3NC^* (Vojta, 1960c). The $\beta_{CH_3NC^*}$ value for 300 K is 1.079. Hence $L_{C=N} = 0.079$. The reliability of the value obtained is difficult to estimate directly, as the compound in question is the only compound containing a bond of C=N type for which the β_{spectr} value is known. However, it can be concluded indirectly that the $L_{C=N}$ value of 0.079 is plausible. Analysis of the isotopic bond numbers shows that their values increase systematically as the multiplicity of the bond increases (Table 3.12). The value of $L_{C=N} = 0.079$ shows the expected relationship to $L_{C\equiv N} = 0.090$.

TABLE 3.12

CARBON—CARBON AND CARBON—NITROGEN ISOTOPIC NUMBERS FOR BONDS OF DIFFERENT MULTIPLICITY

Bond	Type of bond $\sigma_{sp3-sp3}$	$\sigma\pi_{sp2-sp3}$	$\sigma\pi\pi_{sp-sp}$
Carbon—carbon	0.0464	0.0785	0.0881
Carbon—nitrogen	0.050	0.079	0.090

V. PRECISION OF THE ADDITIVE METHOD

A. ERROR OF CALCULATION OF ISOTOPIC BOND NUMBERS

Our values obtained for the isotopic bond numbers of carbon are summarized in Table 3.13. The values of Δ recorded can be regarded as the error in the calculation of L_{C-X}. Because Δ is the difference between β_{spectr} and β_{addit}, it is an accumulative error resulting not only from the calculation of the bond number being determined but also from calculations of other L_{C-X} values, which are increments of β_{addit}. The nature of the error of L_{C-X} is dual. Owing to the errors inherent in the spectral method of calculating β factors, some uncertainty is inherent in the information on which

TABLE 3.13

ISOTOPIC BOND NUMBERS OF CARBON

	L_{C-X}				l_{C-X}	
N	*Symbol*	*Value*	*Degrees of freedom*	Δ	*Symbol*	*Value*
1	L_{C-H}	0.0284	6	±0.0002	l_{C-H}	0
2	L_{C-C}	0.0464	6	±0.0002	l_{C-C}	0.0013
3	$L_{C=C}$	0.0785	4	±0.002	$l_{C=C}$	0.0016
4	$L_{C\equiv C}$	0.088	1	±0.001	$l_{C\equiv C}$	0.0002
5	L_{C-N}	0.050	1	±0.001	l_{C-N}	0.0016
6	$L_{C=N}$	0.079	0		$l_{C=N}$	0.0016
7	$L_{C\equiv N}$	0.090	1	±0.002	$l_{C\equiv N}$	0.0003
8	L_{C-O}	0.055	2	±0.001	l_{C-O}	0.0019
9	$L_{C=O}$	0.096	2	±0.001	$l_{C=O}$	0.0028
10	L_{C-F}	0.056	2	±0.005	l_{C-F}	0.0020
11	L_{C-Cl}	0.030	2	±0.005	l_{C-C}	0.0001
12	L_{C-Br}	0.027	1	±0.002	l_{C-Br}	−0.0001
13	L_{C-I}	0.018	1	±0.001	l_{C-J}	−0.0007
14	$L_{C=S}$	0.050	1	±0.005	$l_{C=S}$	−0.0005

the calculation of L_{C-X} is based. The difference Δ in part reflects this uncertainty, caused partially by the limitation of the postulated principle of additivity.

To achieve the minimum uncertainty in the calculation of isotopic bond numbers caused by errors in the determination of β_{spectr}, a limited number of very reliable determinations of β_{spectr} is used. The number of degrees of freedom in Table 3.13 shows the excess values of β_{spectr} taken for determining the isotopic number. In other words, it corresponds to the number of independent tests of the additive method by comparing β_{addit} with fairly reliable values of β_{spectr}. A more complete analysis, including comparison of β_{addit} with determinations of β_{spectr}, obtained from literature, is given below.

VI. STATISTICAL ANALYSIS OF THE DIVERGENCE BETWEEN $\beta_i\,{}^{13}C_{spectr}$ AND $\beta_i\,{}^{13}C_{addit}$

The questions of the precision of the additive method of determining β factors and the limits of applicability of the additivity principle are rather complex. The usual method of checking a theory—experimentation—is practically inapplicable in this case. The present experimental determination of isotopic fractionation factors is not precise enough, and it is possible to achieve isotopic equilibrium only for a very limited range of systems. Therefore, the only method of estimating the correctness of the additive method is to compare it with another theoretical method, the spectral method of calculating β factors. If the spectral method were rigorous and devoid of errors, comparison with it could provide an estimation of the precision of the additive method. However, experience shows that calculation of β factors using the spectral method is accompanied by errors (the sources of which were mentioned previously), which are sometimes substantial.

For these reasons, marked discrepancies have been obtained in calculations of β factors made by different authors (Table 3.14) for the same compounds, including some fairly simple ones. They reflect the uncertainty of the spectral method for calculating β factors. The average value of the standard deviation between calculations of β_{spectr} for the same compound is ± 0.0055. Thus the observed disagreement between β_{spectr} and β_{addit} may be caused both by the error inherent in the additive method and by errors in the spectral method.

Figure 3.1 compares the values of β_{spectr} and β_{addit}. The standard deviation between them is ± 0.006. The values of the β_i factors vary from $\beta_i = 1.0875$ to 1.225. The uncertainty in calculating β_i factors amounts to approximately 5% of their variability, which includes the uncertainty caused by the error of the additive method. Note that in Fig. 3.1 the β values from which the bond numbers were determined are not given; only independent calculations of β_{spectr} and β_{addit} are compared. Because the bond numbers are practically determined by the least squares method from β_{spectr} values for several compounds, one of them is excluded for each bond number—the one for which the agreement between β_{spectr} and β_{addit} was closest. An exception is made for only two compounds. Instead of the values of β_{CH_3Cl} and β_{CF_4}, characterizing the minimum values of $\Delta = \beta_{spectr} - \beta_{addit}$, the values of $\beta_{CH_2CCl_2}$ ($\Delta = -0.004$) and β_{HCOF} ($\Delta = -0.005$) were eliminated in the first case to retain the parallel values (joined by lines in Fig. 3.1), in the second to leave a value characterizing the upper limit of the range of β-factor values.

Table 3.15 gives the 76 known values of $\beta_i{}^{13}C_{spectr}$. The corresponding values of β_{addit} were obtained by summing the isotopic bond numbers. Thirteen types of L_{C-X} were used. Because the L_{C-X} themselves were determined from β_{spectr} values, the investigated set is characterized by $75 - 13 = 62$ degrees of freedom, which constitutes a sampling of sufficient volume. The L_{C-H} isotopic number was used as increment 128 times in determining the value of β_{addit}, 41 times for L_{C-C}, 20 for $L_{C=O}$, 12 for L_{C-O},

TABLE 3.14

THERMODYNAMIC ISOTOPIC FACTORS CALCULATED USING THE SPECTRAL METHOD (β_{spectr}) BY DIFFERENT AUTHORS FOR THE SAME COMPOUND

	Reference								
Compound	*Singh and Wolfsberg* (1975)	*Hartshorn and Shiner* (1972)[a]	*Galimov and Ivlev* (1973–1974) *Ivlev et al.* (1974*a*)	*Bottinga* (1968)	*Johansen* (1964)	*Vojta* (1960*c*)	*Craig* (1953)	*Urey* (1947)	*Standard deviation*
CH_4	1.1192	1.1224	1.1136	1.112	—	—	1.1220	—	±0.005
CH_3CH_3	1.1366	1.1363	1.1317	—	—	—	—	—	±0.003
$*CH_3CH_2CH_3$	1.1411	1.1413	1.1331	—	—	—	—	—	±0.004
$CH_3{*}CH_2CH_3$	1.1483	1.1488	1.1493	—	—	—	—	—	±0.001
CO_2	1.1904	1.1904	—	—	—	1.1948	—	1.1909	±0.002
$*CH_3CHO$	—	—	1.1293	—	1.141	—	—	—	±0.008
$CH_3{*}CHO$	—	—	1.1731	—	1.160	—	—	—	±0.009
CO_3^{2-}	—	1.1982	—	1.202	—	—	—	1.2057	±0.004
$*CH_3CN$	—	1.1353	—	—	—	1.1424	—	—	±0.005
									±0.0055[b]

[a] Data of these authors recalculated to the value $\beta_{CO_2} = 1.1904$.

[b] This average value of the standard deviation equals the sum of the standard deviations of the calculated values as determined from the average value divided by the number of standard deviations summed.

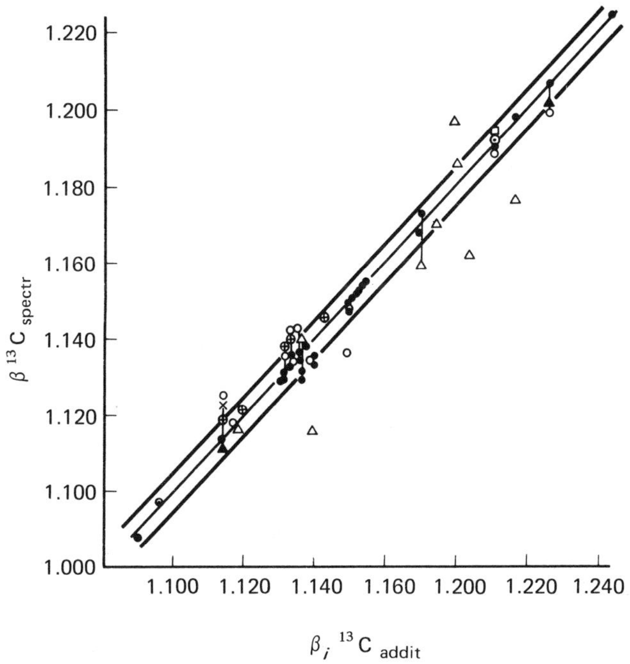

FIG. 3.1 Comparison of values of thermodynamic isotopic factors of carbon calculated using the spectral (β_{spectr}) and the additive method (β_{addit}). The vertical lines connect β_{spectr} values of one compound obtained by different authors. The graph shows the corridor of errors corresponding to the mean square divergence between β_{spectr} and β_{addit} values; $\sigma = \pm 0.006$. Symbols refer to the particular references: ⊙, Urey, 1947; ×, Craig, 1953; □, Vojta, 1960b, c; △, Johansen, 1964; ▲, Bottinga, 1969; ●, Galimov and Ivlev, 1973–1974; ○, Hartshorn and Shiner, 1972; ⊕, Singh and Wolfsberg, 1975.

8 for $L_{C=C}$, 8 for $L_{C=N}$, 3 for $L_{C=C}$, 2 for L_{C-N}, 7 for L_{C-F}, 6 for L_{C-Cl}, and 3 for L_{C-Br}, L_{C-I}, and $L_{C=S}$. On the whole, the statistics were considerable, but the various bond numbers are unequally represented in the sampling, which decreases the generality of the conclusions which can be drawn with respect to the limits of applicability of the additive method.† On the other hand, bonds typical of biological molecules are well represented in the sampling. This is especially important for the purposes of this work.

† Since there is only one compound containing the C=N bond for which the β_{spectr} value is known, the $L_{C=N}$ isotopic factor and corresponding $\beta_{CH_3NC^*}$ value are clearly related and therefore are left out of consideration.

TABLE 3.15

β_i ^{13}C VALUES CALCULATED BY THE SPECTRAL AND ADDITIVE METHODS

		β_{spectr}			
Number	*Compound*	*Value*	*Reference*	β_{addit}	Δ
1	CH_4	1.1136	Galimov and Ivlev (1973–1974)	1.1136	0.0000
2	CH_4	1.112	Bottinga (1969)	1.1136	0.0016
3	CH_4	1.122	Craig (1953)	1.1136	−0.0084
4	CH_4	1.1224	Hartshorn and Shiner (1972)	1.1136	−0.0088
5	CH_4	1.1112	Singh and Wolfsberg (1975)	1.1136	−0.0056
6	CH_3CH_3	1.1317	Galimov and Ivlev (1973–1974)	1.1316	−0.0001
7	CH_3CH_3	1.1368	Hartshorn and Shiner (1972)	1.1316	−0.0052
8	CH_3CH_3	1.1366	Singh and Wolfsberg (1975)	1.1316	−0.0050
9	$^*CH_3CH_2CH_3$	1.1331	Galimov and Ivlev (1973–1974)	1.1329	−0.0002
10	$^*CH_3CH_2CH_3$	1.1418	Hartshorn and Shiner (1972)	1.1329	−0.0089
11	$^*CH_3CH_2CH_3$	1.1411	Singh and Wolfsberg (1975)	1.1329	−0.0082
12	$CH_3{}^*CH_2CH_3$	1.1493	Galimov and Ivlev (1973–1974)	1.1496	0.0003
13	$CH_3{}^*CH_2CH_3$	1.1492	Hartshorn and Shiner (1972)	1.1496	0.0004
14	$CH_3{}^*CH_2CH_3$	1.1483	Singh and Wolfsberg (1975)	1.1496	0.0013
15	$^*CH_3CH_2CH_2CH_3$	1.1331	Galimov and Ivlev (1973–1974)	1.1329	−0.0002
16	$CH_3{}^*CH_2CH_2CH_3$	1.1507	Galimov and Ivlev (1973–1974)	1.1509	0.0002
17	$^*CH_3CH_2CH_2CH_2CH_3$	1.1330	Galimov and Ivlev (1973–1974)	1.1329	−0.0001
18	$CH_3{}^*CH_2CH_2CH_2CH_3$	1.1506	Galimov and Ivlev (1973–1974)	1.1509	0.0003
19	$CH_3CH_2{}^*CH_2CH_2CH_3$	1.1519	Galimov and Ivlev (1973–1974)	1.1522	0.0003
20	Cyclo-C_3H_8	1.1355	Hartshorn and Shiner (1972)	1.1522	0.0167
21	Cyclo-C_3H_8	1.1353	Singh and Wolfsberg (1975)	1.1522	0.0169
22	Cyclo-C_6H_6	1.1544	Galimov and Ivlev (1973–1974)	1.1522	−0.0022
23	CO_2	1.1909	Urey (1947)	1.1915	0.0006
24	CO_2	1.1904	Singh and Wolfsberg (1975)	1.1915	0.0011
25	CO_2	1.1948	Vojta (1960c)	1.1915	−0.0033
26	$CO_3{}^{2-}$	1.2057	Urey (1947)	1.2058	0.0001
27	$CO_3{}^{2-}$	1.2020	Bottinga (1969)	1.2057	0.0037
28	$CO_3{}^{2-}$	1.1988	Hartshorn and Shiner (1972)	1.2057	0.0069
29	CO	1.0970	Urey (1947)	1.0958	−0.0012
30	$^*CH_3COOH$	1.1381	Ivlev *et al.* (1974a)	1.1363	−0.0018
31	$CH_3{}^*COOH$	1.1978	Ivlev *et al.* (1974a)	1.1974	−0.0004
32	$^*CH_3CHO$	1.1293	Ivlev *et al.* (1974a)	1.1344	0.0051
33	$^*CH_3CHO$	1.140	Johansen (1964)	1.1344	−0.0056
34	$CH_3{}^*CHO$	1.1731	Ivlev *et al.* (1974a)	1.1708	−0.0023
35	$CH_3{}^*CHO$	1.160	Johansen (1964)	1.1708	0.0108
36	CHC*CHO	1.170	Johansen (1964)	1.1708	0.001
37	$HCOO^-$	1.162	Johansen (1964)	1.1791	−0.017
38	HCOOH	1.197	Johansen (1964)	1.1791	−0.017
39	$CClH_2{}^*COOH$	1.177	Johansen (1964)	1.1974	0.020
40	CH_3OH	1.1328	Ivlev *et al.* (1974a)	1.1402	0.0074

TABLE 3.15 (*continued*)

		β_{spectr}			
Number	*Compound*	*Value*	*Reference*	β_{addit}	Δ
41	CHCH	1.1173	Hartshorn and Shiner (1972)	1.1165	−0.0008
42	*CHCCHO	1.117	Johansen (1964)	1.1178	0.0008
43	CH*CCHO	1.116	Johansen (1964)	1.1373	0.0213
44	CH_2CH_2	1.1354	Hartshorn and Shiner (1972)	1.1353	−0.0001
45	CH_2CH_2	1.1351	Singh and Wolfsberg (1975)	1.1353	0.0002
46	$*CH_2CBr_2$	1.1293	Hartshorn and Shiner (1972)	1.1351	0.0058
47	$*CH_2CF_2$	1.1334	Ivlev *et al.* (1974a)	1.1393	0.0059
48		1.1533	Galimov and Ivlev (1973–1974)	1.1532	−0.0001
49	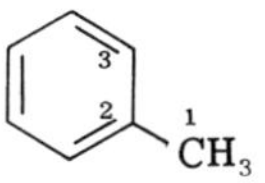 (1)	1.1293	Galimov and Ivlev (1973–1974)	1.1316	0.0023
50	(2)	1.1677	Galimov and Ivlev (1973–1974)	1.1713	0.0036
51	(3)	1.1545	Galimov and Ivlev (1973–1974)	1.1546	0.0001
52	HCN	1.1206	Urey (1947)	1.1184	−0.0022
53	CN^-	1.0875	Galimov and Ivlev (1973–1974)	1.0900	0.0025
54	CH_3*CN	1.1325	Vojta (1960c)	1.1364	0.0039
55	$*CH_3CN$	1.1424	Vojta (1960c)	1.1319	−0.0105
56	$*CH_3CN$	1.1358	Hartshorn and Shiner (1972)	1.1319	−0.0039
57	$*CH_3NH_2$	1.1344	Ivlev *et al.* (1974a)	1.1352	0.0008
58	$*CH_3NH_2$	1.1361	Galimov (1973b)	1.1352	−0.0009
59	CF_4	1.2241	Galimov (1973b)	1.2256	0.0015
60	CH_3F	1.1368	Hartshorn and Shiner (1972)	1.1416	0.0048
61	HCOF	1.186	Johansen (1964)	1.1808	−0.0052
62	CH_3F	1.1457	Singh and Wolfsberg (1975)	1.1416	−0.0041
63	$*CH_3Cl$	1.1149	Hartshorn and Shiner (1972)	1.1162	0.0013
64	$*CH_2ClCOOH$	1.133	Johansen (1964)	1.1389	0.0059
65	CH_2*CCl_2	1.1436	Hartshorn and Shiner (1972)	1.1395	−0.0041
66	ClCN	1.126	Vojta (1960b)	1.1210	−0.0050
67	CH_3Cl	1.1240	Singh and Wolfsberg (1975)	1.1162	−0.0078
68	BrCN	1.119	Vojta (1960b)	1.117	−0.002
69	CH_3Br	1.111	Hartshorn and Shiner (1972)	1.112	0.001
70	CH_3Br	1.1177	Johansen (1964)	1.1122	−0.0055
71	ICN	1.107	Vojta (1960b)	1.108	0.001
72	CH_3I	1.103	Hartshorn and Shiner (1972)	1.103	0.000
73	CH_3I	1.1101	Johansen (1964)	1.1032	−0.0069
74	COS	1.140	Vojta (1960b)	1.146	0.006
75	CS_2	1.104	Vojta (1960b)	1.100	−0.004
76	CH_3N*C	1.0789	Vojta (1960c)	1.079	—

The mathematical expectancy of values of the difference Δ being close to zero is

$$M\Delta = \frac{1}{n}\sum_{i}^{76} \Delta_i = +0.0003.$$

This means that the set of increments (bond factors) taken is adequate for determining the corresponding additive value ($\beta_i{}^{13}C$). Otherwise a systematic deviation would have appeared.

The dispersion of Δ values amounts to

$$D\Delta = M(\Delta_i - M)^2 = \frac{1}{n}\sum_{i}^{76} (\Delta_i - M\Delta)^2 = 4372 \times 10^{-8}$$

and the standard deviation is

$$\sigma = \pm\sqrt{D\Delta} = \pm 0.0066.$$

This last characterizes the total error of determinations of β_{spectr} and β_{addit}. Because we are interested in assessing the precision of the additive method, the differences Δ that are probably due to errors in the determination of β_{spectr} should be excluded from the sample.

First of all, from the β_{spectr} values for which there are several determinations made by different authors, it is advisable to exclude those in poor agreement.

Then certain extreme values, for which the corresponding determinations of β_{spectr} may have been made with substantial error, are excluded even though there are no parallel determinations. We shall consider them individually. Let us begin with $\beta_{HCOO^-} = 1.162$ and $\beta_{HCOOH} = 1.197$. For both compounds the set of L_{C-X} bond numbers is the same. In the structure of these two compounds there is no characteristic distinguishing them from other compounds for which the principle of additivity is valid. The properties of additivity do not depend essentially on the ionic or molecular form of the compound, as is seen from the example of CN^- and $CO_3{}^{2-}$. Bonding of hydrogen to the oxygen atom can only affect the level of the value l_{C-X}. Therefore, the considerable difference between β_{HCOO^-} and β_{HCOOH} values raises doubt as to their reliability.

The value of β_{spectr} for $CClH_2{}^*COOH$, equal to 1.177, is also doubtful. For the carbon of the carboxyl group of acetic acid a value of $\beta_{CH_2{}^*COOH} = 1.1978$ has been obtained (Ivlev *et al.*, 1974a), which is in good agreement with the value obtained by the additive method ($\beta_{addit} = 1.1974$). Therefore there is no reason to expect that the carboxyl carbon of monochloracetic acid, which forms the same chemical bonds, will have a sharply different β_i factor, especially because the value of L_{C-Cl} is close to L_{C-H}. For the bonding of carbon with chlorine the additivity rule is well obeyed for all investigated compounds, as is seen from Table 3.15. A fourth compound, for which an anomalously large divergence between β_{addit} and β_{spectr} has been obtained, is the isotopic form of propynal: CH≡C*—CHO. For the second carbon atom of this compound a value of $\beta_i = 1.116$ has been obtained, which practically coincides with the value of the β_i factor of the third atom (1.117). Because these carbon atoms occupy a nonequivalent position, such coincidence seems doubtful. In contrast to the carbon atom bound to hydrogen, the second carbon atom bound to aldehyde carbon should have a higher value of the β_i factor, as is observed in all other analogous cases.

Let us note that these four determinations of β_{spectr} are by the same author (Johansen, 1964). His calculated β_i-factor values of carbon in the methyl and aldehyde groups of acetaldehyde also differ substantially from the corresponding values of β_i obtained by the additive method. But these values of β_{spectr} can be ruled out unconditionally because in the case of acetaldehyde there are parallel determinations made by other authors which show good agreement with the additive calculations.

Finally, there is one other value of β_{spectr} for which the difference of β_{spectr} and β_{addit} is so large that it requires special consideration. This is cyclopropane, for which a value of $\beta_{spectr} = 1.135$ has been obtained, whereas the additive method yields a value of 1.152. The structure of cyclopropane has no analogs among other compounds for which $\beta^{13}C$ values are known. The tetrahedral configuration of the sp_3-hybridized carbon in the cyclopropane molecule is highly deformed. Owing to the contiguity of indirectly interacting σ orbits, additional overlap is possible. Possibly for this reason, the β_{addit} value obtained without correcting for the effects of conjugation turned out to be too high. Of course, the possibility exists that it is simply an error in the determination of β_{spectr}. This question is in need of investigation. Apparently $\beta^{13}C_{cyclo\text{-}C_3H_8}$ should be excluded from the sampling, as the large value of the difference of β_{spectr} and β_{addit} related to this compound distorts the dispersion characterizing the sampling.

Application of the criteria of coincidence to the distribution, improved as a result of eliminating large differences, shows that it is not Gaussian. Rather, the distribution obtained is characterized by an excess of large values. In such cases it can be presumed that the results obtained with a substantial deviation are superposed on the results of experiments containing a small error; with respect to the latter, they are "coarse" errors (Pustyl'nik, 1968).

The analysis shows that, if all terms with deviations greater than 0.004 are eliminated from the sampling, the remaining sample containing 28 elements does obey a Gaussian distribution. The standard deviation characterizing this distribution is $\sigma_{\beta_i} = \pm 0.0016$.

Analysis of the sample shows that it includes compounds containing all types of bonds for which L_{C-X} have been determined. In other words, the considerable differences of Δ are are not related to any particular type of bond. Therefore, we are inclined to assume that the distribution, with a standard deviation of $\sigma_{\beta_i} = \pm 0.0016$, reflects the limits of validity of the additivity principle for compounds containing chemical bonds of the type in question. If the divergence between β_{spectr} and β_{addit} exceeds 0.004, then it is probably related to an error in the determination of β_{spectr} if, of course, the chemical structure of the compound has no features which could not be taken into account by the additive method in its present form.

B. Error in the Determination of β_Σ Factors

The spread of the sampling established above, $\beta_{addit} - \beta_{spectr}$, should be regarded as a dispersion characterizing the calculation of β_i values which depend on the uncertainty σ_{LX} of calculation of the corresponding bond factors:

$$D\beta_i = \sigma^2_{\beta_i} = \sum^m \sigma^2_{LX}. \tag{3.36}$$

The thermodynamic isotopic factor of a compound as a whole is determined according to (3.34) as the arithmetic mean of β_i values of the isotopic forms. Consequently,

$$D_{\beta_\Sigma} = D\frac{1}{m}\sum^m \beta_i. \tag{3.37}$$

According to the known properties of dispersion, if all values of a random quantity are multiplied by some factor, the dispersion is multiplied by the square of this factor:

$$D\beta_\Sigma = \frac{1}{m^2}\sum^m \beta_i. \tag{3.38}$$

The dispersion of the independent random quantities' sum is equal to the dispersions sum:

$$D\beta_\Sigma = \frac{1}{m^2}\sum^m D\beta_i \tag{3.39}$$

and

$$\sigma_{\beta_\Sigma} = \pm\sqrt{\sum^{m} D\beta_i}\Big/ m = \sqrt{\sum^{m} \sigma^2_{\beta_i}}\Big/ m. \tag{3.40}$$

If the determination errors of the values of $\beta^{13}C$ are assumed approximately equal, i.e., $\sigma_{\beta_i(1)} = \sigma_{\beta_i(2)} = \sigma_{\beta_i(m)}$, then

$$\sigma_{\beta_\Sigma} = \pm\sigma_{\beta_i}/\sqrt{m}. \tag{3.41}$$

Thus the calculation of the thermodynamic isotopic factor of a compound containing many carbon atoms is derived with a smaller uncertainty than calculation of the β_i factor. If, for instance, the error in calculating β_i is taken as ± 0.004, then for molecules containing 20–25 carbon atoms $|\sigma_{\beta_\Sigma}| < 0.001$, that is, the error in determination of the β_Σ factor is in the fourth decimal place.

What has been stated is valid under the condition that the value of $\beta_{i(\text{addit})}$, due to the imperfection of the additivity principle, randomly deviates from the true value. The symmetry of the distribution of values of the deviation $\beta_{i(\text{addit})} - \beta_{i(\text{spectr})}$ would appear to be evidence of this. However, it must be remembered that we do not know the actual form of the distribution function of $\beta_{i(\text{addit})} - \beta_{i(\text{true})}$. Therefore relationships (3.39) and (3.41) reflect the tendency toward the decrease in σ_{β_Σ} as m increases rather than precisely conveying the relationship between the errors in determining the β_Σ and β_i factors.

Most experimental data on biological isotopic effects have been obtained by study of intermolecular isotopic effects, i.e., the differences between isotopic compositions of individual compounds. Because these isotopic effects sometimes are very small, it is important to be able to obtain β_Σ factors for a compound with relatively high precision.

Calculation of β_i factors is less precise. However, the value of intramolecular isotopic effects is usually much greater than that of intermolecular. Therefore, a larger error is less perceptible in this case.

VII. ISOTOPIC BOND NUMBERS OF NITROGEN

The number of compounds for which the values of $\beta^{15}N_{\text{spectr}}$ has been calculated is insufficient for a reliable estimation of the isotopic bond numbers of nitrogen. Nevertheless, from a preliminary analysis one can obtain an idea of the theoretical applicability of the concept of additivity to the isotopic thermodynamics of nitrogen.

TABLE 3.16

THERMODYNAMIC ISOTOPIC FACTORS OF NITROGEN COMPOUNDS IN CHEMICAL BONDS FORMED BY ATOMS OF NITROGEN IN THE RESPECTIVE ISOTOPIC FORMS

	$\beta_i{}^{15}N_{spectr}$		
Compound	*Value*	*Reference*	*Chemical bond*
NH_3	1.0683	Ivlev *et al.* (1974b)	$3L_{(N-H)\sigma sp_3-s}$
$NH_4{}^+$	1.0947	Ivlev *et al.* (1974b)	$4L_{(N-H)\sigma sp_3-s}$
CH_3NH_2	1.0853	Ivlev *et al.* (1974b)	$2L_{[N-H)\sigma sp_3-s} + L_{(N-C)\sigma sp_3-sp_3}$
C_4H_4NH	1.1074	Ivlev *et al.* (1974b)	$L_{(N-H)\sigma sp_3-s} + 2L_{(N-C)\sigma sp_3-sp_2}$
$C_6H_5-{}^*N=N-C_6H_5$	1.0970	Ivlev *et al.* (1974b)	$L_{(N-C)\sigma sp_3-sp_3} + L_{(N=N)\sigma\pi sp_2-sp_2}$
CN^-	1.0641	Urey (1947)	$L_{(N\equiv C)\sigma\pi\pi sp_1-sp_1}$
HCN^-	1.0655	Urey (1947)	$L_{(N\equiv C)\sigma\pi\pi sp_1-sp_1}$
N_2	1.0814	Urey (1947)	$L_{(N\equiv N)\sigma\pi\pi sp_1-sp_1}$

Table 3.16 gives published values of $\beta^{15}N_{spectr}$ and the set of additive increments (bond numbers) determining the corresponding thermodynamic isotopic factors.

From these data the values of some isotopic bond numbers of nitrogen can be calculated. They are given in Table 3.17. It is not possible to verify the transferability for each bond number. However, in cases where the respective value of L_{N-X} was determined from several compounds, the variations of the values fell within relatively narrow limits.

The nitrogen—oxygen isotopic bond numbers can be calculated only provisionally. A value of $\beta^{15}N = 1.1534$ is known for the $NO_3{}^-$ ion (Ivlev *et al.*, 1974b). Physically, all three N—O bonds in this molecule are indistinguishable (similar to the C—O bonds in the $CO_3{}^{2-}$ ion). However, following the valency bond method it can be shown that there are simultaneously present in the $NO_3{}^-$ ion two $(N-O)\sigma\pi_{sp_2\text{-}sp_3}$ bonds and one

TABLE 3.17

ISOTOPIC BOND FACTORS OF NITROGEN

Symbol	*Range of variation*	*Average value*
L_{N-H}	0.023–0.025	0.023
L_{N-C}	0.039–0.042	0.040
$L_{N=N}$	0.058	0.058
$L_{N\equiv C}$	0.064–0.065	0.065
$L_{N\equiv N}$	0.081	0.081

$(N{=}O)\sigma\pi_{sp_2-sp_2}$ bond. The $\beta^{15}N$ values for the NO and NO_2 molecules are equal to 1.0659 (Urey, 1947) and 1.1079 (Ivlev *et al.*, 1974b), respectively. Assuming that nitrogen in these compounds is in sp_2 hybridization (although the method of valency bonds does not justify describing a mole cule with an odd number of electrons) and correcting for the aforementioned character of the bonds in the NO_3^- ion, we obtain the numbers $L_{N-O} = 0.044$ and $L_{N=O} = 0.065$ for the nitrogen—oxygen bond.

The value of the isotopic number of the σ bond formed by interaction of unfilled orbits of atoms is the same (within the precision determinable) as that of the donor–acceptor type of bond. In particular, it follows from this that the thermodynamic isotopic factors of nitrogen do not depend on its valency state, i.e., $\beta^{15}N$ values in both trivalent and pentavalent nitrogen are determined by the same increments for bonds of the same type.

Inherent in the isotopic bond numbers of nitrogen are the same regularities as in the isotopic bond numbers of carbon. The L_{N-H} isotopic number has the minimum values; it characterizes the bond of nitrogen with hydrogen. The values of L_{N-X} increase as the electronegativity of the partner atom increases. For nitrogen, as for carbon, there is a change in value of isotopic number which is a function of multiplicity of the bond. Evidently other regularities established on the basis of the more voluminous material on carbon compounds are transferable to the $\beta^{15}N$ value.

Thus analysis of available data provides grounds for presuming that the principle of additivity can be extended to the isotopic thermodynamics of nitrogen. The limited amount of reference data precludes computing the error of calculation of $\beta^{15}N_{addit}$.

VIII. ISOTOPIC BOND NUMBERS OF OXYGEN

Oxygen is a divalent element. It forms two ordinary σ bonds or one double $\sigma\pi$ bond. The value of $\beta_i{}^{18}O$ is determined by only one or two additive increments. Therefore the idea of additivity, as applicable to the calculation of the β_i factors of oxygen via the bond numbers, loses its meaning to a substantial extent, and only the realizability of the principle of transferability is of interest.

Table 3.18 presents the reference information. The value of the L_{O-H} bond number was determined using the value of $\beta^{18}O_{spectr}$ for the water molecule. Using available data for the $\beta^{18}O$ factors of CO, CO_2, and CO_3^{2-} one can determine the values of the $L_{O=C}$ and L_{O-C} isotopic numbers. The results are presented in Table 3.19.

TABLE 3.18

THERMODYNAMIC ISOTOPIC FACTORS OF OXYGEN COMPOUNDS AND CHEMICAL BONDS FORMED BY OXYGEN ATOMS

	$\beta_i{}^{18}O_{spectr}$		
Compound	*Value*	*Reference*	*Chemical bonds*
CO	1.1053	Urey (1947)	$L_{(O=C)\sigma\pi sp_2-sp_1}$
CO_2	1.1172	Bottinga (1968)	$L_{(O=C)\sigma\pi sp_2-sp_1}$
$CO_3{}^{2-}$	1.0945[a]	Urey (1947)	$\frac{1}{3}L_{(O=C)\sigma\pi sp_2-sp_2} + \frac{2}{3}L_{(O-C)\sigma sp_3-sp_2}$
O_2	1.0818	Urey (1947)	$L_{(C=O)\sigma\pi sp_2-sp_2}$
H_2O	1.0636[b]	Bottinga (1968)	$2L_{(O-H)\sigma sp_3-s}$
$SO_4{}^{2-}$	1.0924	Urey (1947)	$L_{(O-S)\sigma sp_3-sp_3}$
SO_2	1.0888	Urey (1947)	$\frac{1}{2}L_{(O=S)\sigma\pi sp_2-sp_2} + \frac{1}{2}L_{(O-S)\sigma sp_3-sp_2}$

[a] Value is given for the completely isotopically replaced form, i.e., $1.0945^3 = (1 + L_{O-C} + 2l_{O-C})(1 + L_{O-C} + l_{O=C} + l_{O-C})^2$.

[b] $\beta^{18}O$ value for water vapor is given.

If it is assumed that the difference in $\beta^{18}O$ factors of the oxygen of CO and CO_2 is explained by the presence of the second oxygen atom in the CO_2 molecule, the values $l_{O=C}$ and l_{O-C} can be found. They are 0.012 and 0.007, respectively.

In the SO_2 molecule, oxygen forms four indistinguishable $(O-S)\sigma_{sp_3-sp_3}$ bonds with sulfur, which is in a state of sp_3 hybridization. In the SO_2 molecule, sulfur is in a state of sp_2 hybridization. Thus it forms two equivalent bonds with oxygen. However, the latter can be considered a combination of two bonds, $(O=S)\sigma\pi_{sp_2-sp_2}$ and $(O-S)\sigma_{sp_3-sp_2}$. Therefore, the values of the isotopic numbers L_{S-O} and $L_{S=O}$ can be determined.

TABLE 3.19

ISOTOPIC BOND NUMBERS OF OXYGEN

Symbol	*Value*
L_{O-H}	0.032
L_{O-C}	0.065
$L_{O=C}$	0.105
L_{O-S}	0.092
$L_{O=S}$	0.085

IX. SOME REGULARITIES CHARACTERIZING ISOTOPIC BOND NUMBERS

From the preceding section, several general conclusions can be drawn.

(1) The principles of localization, additivity, and transferability, on which the additive method of calculating β factors is based, prove to be intrinsic to isotopic thermodynamics to a substantial degree.

(2) The property of additivity of increments (bond numbers), in principle, is valid for different types of carbon bonds, for carbon bonds with various elements (including H, N, O, F, S, Cl, Br, I), and also for different elements (C, N, O).

(3) The value of the β factor is determined mainly by the immediate chemical surroundings or by the atoms directly bound to a given one. The contribution of the distant surroundings (bonds joining atoms to atoms that are attached to the given one) constitutes about 3% of β.

(4) The value of the isotopic number characterizing the bond of a given type depends little on the state of hybridization of the atom forming the bond. For instance, the bond of carbon with hydrogen in methane (sp_3 hybridization), in ethylene (sp_2), and in acetylene (sp_1) is characterized by the same number $L_{C-H} = 0.0284$.

(5) The molecular structure has an insignificant effect on the value of the β_i factor. This is in agreement with the preceding point, because the molecular structure of carbon compounds (tetrahedral, planar, or linear) is intimately related to the state of hybridization of the carbon atoms. An analogous conclusion to the negligible relationship of isotopic effects to the molecular structure is also drawn from other approaches (Bigeleisen, 1975; Stern and Wolfsberg, 1966).

(6) For C, N, and O, the same type of systematic variation of the value of the isotopic number with variation of bond type is observed (Table 3.20). The number characterizing a $\sigma\pi$ (double) bond is larger than the number characterizing a single σ (ordinary) bond, but the increase is less than twofold. In other words, a compound containing a C=C bond shows less affinity to the ^{13}C isotope than a compound containing two C—C bonds. Also, the value of the isotopic number for a $\sigma\pi\pi$ bond is greater than that of a $\sigma\pi$ bond, but the increase is less than the increase in multiplicity.

(7) A systematic variation in the value of isotopic bond numbers of carbon is observed as a function of the partner element's position in the Mendeleyev table (Table 3.21). The values of the numbers increase from left to right and decrease from the top down. Consistent with this regularity, it should be expected that the value of the L_{C-S} bond numbers would be

TABLE 3.20

REGULARITIES OF VARIATION OF BOND NUMBERS OF DIFFERENT TYPE FOR N, C, O

Bond	Type of bond		
	σ	$\sigma\pi$	$\sigma\pi\pi$
Nitrogen—hydrogen	0.023	—	—
Nitrogen—carbon	0.040	—	0.065
Nitrogen—nitrogen	—	0.058	0.081
Nitrogen—oxygen	0.044	0.065	—
Carbon—hydrogen	0.0284	—	—
Carbon—carbon	0.0464	0.078	0.088
Carbon—nitrogen	0.050	0.079	0.090
Carbon—oxygen	0.055	0.096	—
Oxygen—hydrogen	0.033	—	—
Oxygen—carbon	0.065	0.105	—
Oxygen—nitrogen	—	—	—
Oxygen—oxygen	—	0.082	—

about 0.029 and of L_{C-Si} about 0.027. The L_{C-F} number has the maximum value. The minimal values should be expected for bonds of carbon with the heavy alkali metals. Correlations of isotopic effects with the position of elements in the periodic table were noted earlier (Varshavskiy and Vaysberg, 1957). However, it should be mentioned that the correlation actually would be valid for isotopic bond numbers rather than for isotopic effects (β factors of compounds). The clearest correlation was observed for the β factors of hydrogen (Varshavskiy and Vaysberg, 1957), which is understand-

TABLE 3.21

REGULARITY OF VARIATION OF BOND NUMBERS OF CARBON BY PERIOD AND GROUP OF THE MENDELEYEV TABLE[a]

H						
0.0284						
Li	Be	B	C	N	O	F
			0.0464	0.050	0.055	0.056
			(0.0785)	(0.079)	(0.096)	
			Si	P	S	Cl
					(0.050)	0.030
						Br
						0.027
						I
						0.018

able since, for hydrogen, the value of $\beta - 1$ is identical to the corresponding bond factor. It was thought, previously, that carbon in compounds with hydrogen was isotopically lighter than carbon in compounds with oxygen (Craig, 1953; Wilson *et al.*, 1960). This is valid to an extent, because the value of $L_{C\ H}$ is lower than the $L_{C\ O}$ value. However, a compound of carbon with oxygen such as CO has a lower value of the β factor than CH_4. This departs from the previously mentioned empirical rule.

(8) There is a correlation of the values of isotopic bond numbers with the electronegativity of partner atoms.

It is not difficult to see that the regularities listed are related in some way to the electron density near the nucleus. As is known from the LCAO-MO (linear combination of atomic orbitals, molecular orbital) theory, the electron density in the molecular orbit around the nucleus is less than in the corresponding atomic orbit (Pimental and Spratley, 1973). The more the chemical bond changes the electron density, the greater, obviously, the energy effect related to substitution of one isotope for another. Therefore, the increments pertaining to bonds formed by the given element with partner atoms capable of maximally elongating the electron cloud will make the greatest change in the value of the β factor. Note that the chemical bond itself is produced by overlap of the wave functions of the atoms (overlap integral) or by the increase in electron density between interacting atoms and, to a lesser extent, by compression of electron charges surrounding the nucleus (Shustarovich, 1973). Therefore, there is little correlation between isotopic numbers and values of bond energy.

A look at the isotopic bond number as a measure of the variation in electron density around the nucleus resulting from formation of a given chemical bond provides grounds both for explaining the observed regularities and for indicating some limitations of the additive method. In particular, phenomena related to delocalization of electrons by conjugation and superconjugation probably would lead to disruption of additivity. Possibly for this reason, application of this method to the description of isotopic effects in aromatic compounds, stressed cycles, and polyene structures requires the introduction of additional corrections.

X. OTHER ADDITIVE METHODS

A number of investigators have attempted to simplify the procedure of the classical method of calculating the β factor (β_{spectr}). The paradox of this method is that the force constants, the finding of which poses the major difficulties in solving the vibration problem, are the same for isotopic molecules. Therefore, it is logical to try to relate the values of isotopic effects directly to values of the force constants and elements, dependent on mass, thus avoiding calculations of vibrational frequencies. Bigeleisen, *et al.* (1971), Bigeleisen (1975), Bigeleisen and Ishida (1968), and Ishida *et al.* (1973) showed in several works that this is theoretically possible to do in the framework of approximate representations of quantum-statistical expression (3.30) in orthogonal polynomials. One of the main contributions of these works is that they discuss the relatively slight dependence of isotopic effects on molecular structure and the insignificant effect of remote frequencies of molecules (those removed by more than two bonds) on the isotopic effects for a given atom, i.e., the slight effect of the distant surroundings. The latter property was used by Stern and Wolfsberg (1966) to work out the procedure of the so-called cutoff calculations of isotopic effects. It was shown that, in investigating isotopic effects in systems including large molecules, it is possible to calculate isotopic effects with little error by using the interactions only in the immediate vicinity of the isotopically substituted position and ignoring the presence of the rest of the molecule.

It is understood that the properties mentioned are a manifestation of the same properties of localization, additivity, and transferability which make it possible to calculate β factors using the additive and universal increments in the method we have developed. The meaning of the increments (isotopic bond numbers), as discussed previously, is not related to the ideas of valency-force field theory. At the same time, the property of additivity is, to a certain extent, inherent in the interactions which are examined in the model of the valency-force field. Recently, Bigeleisen, Hom, and Ishida (1976) investigated the additivity of contributions to the total value of the thermodynamic isotopic factor resulting from the interaction of individual coordinates and found that additivity actually does occur. We note that the American authors use the concept of the derived ratio of partition functions of the constants $(s/s^*)f$, where f is the ratio of partition functions of the isotopic form constants of the molecule (s^* and s are the symmetry numbers of the latter) which corresponds to our thermodynamic factor. Usually the value $\ln(s/s^*)f$ is used, i.e., $\ln \beta$ for values of β close to 1 is approximately equal to $\beta - 1$.

The procedure worked out by Bigeleisen and his coauthors consists of the following: $(s/s^*)f$ is calculated on the basis of a sufficiently reliable system of force constants, i.e., β_{spectr} is calculated. Then the derived ratio of partition functions is estimated by extrapolation to a zero value of one of the force constants. The value obtained is subtracted from the total value of $(s/s^*)f$ and the contribution of the given force constant is found for all the interactions in the molecule. This procedure was called by its authors the "method of total effects."

Table 3.22 presents the results obtained for the methane molecule. Values are given for the force constants characterizing the interaction of the respective coordinates; for $\delta_{g_{ij}}$, equal to the difference between elements g^*_{ij} and g_{ij} of the **G** matrix (number of equivalent nonzero $\delta_{g_{ij}}$ elements in parentheses); for absolute values of $\ln(s/s^*)f$, i.e., values obtained by complete solution of the vibrational problem with a given **F** matrix for several temperatures; for the contribution of each kind of interaction; and finally, for the sum of contributions, in percentages. The closeness of the latter to 100% is a measure of the additivity of the contributions. Two versions are given, differing in the choice of **F** matrices on which the calculations are based. For the first, the force field model is that of Jones and McDowell, for the second, it is that of Hartshorn and Shiner (see the references in Bigeleisen *et al.*, 1976). Tables 3.23 and 3.24 present data obtained by Bigeleisen *et al.* (1976) for a number of compounds of carbon and oxygen at a temperature of 300 K. The authors consider the results obtained to be evidence of noteworthy additive contributions. The additivity is improved as the temperature rises, although it is fairly high at 300 K. The additivity property is manifested more strongly in the absence of larger nondiagonal terms in the **F** and **G** matrices. The presence of such terms in Jones's and McDowell's model of the force field of methane weakens the additivity observed in this case (see Table 3.22).

According to Bigeleisen, the best additive properties are observed for the hydrogen isotopes. Hydrogen occupies end positions in the molecule. In connection with this, δg_i $(i \neq j)$ is equal to zero for all types of coordinates except those immediately adjacent to the substituted atom. To a certain extent, this is characteristic of oxygen atoms that occupy an end position in the molecule). For carbon, the additivity is worse on the whole (cf. Tables 3.22 and 3.24). In addition to the nonadditivity which stems from the nondiagonal elements of the **G** matrix, substantial nonadditivity is introduced by extraplanar motions in the aromatic benzene molecule. According to Bigeleisen *et al.* (1976), "In the absence of large offdiagonal **F**-matrix elements, carbon isotope chemistry shows additivity at 300 K within 6%." It would be more precise to say that these authors investigated the additivity of the contributions of the force interactions.

The method developed by Bigeleisen *et al.* is an independent approach to the calculation of the properties of additivity. Therefore, the conclusion of the American investigators is very valuable: "The present examples supplement the previous theoretical justification of Galimov's method of calculating values of $\ln(s/s^*)f$ for a wide spectrum of carbon compounds by a summation of bond values" [Bigeleisen *et al.*, 1976, p. 3309].

An advantage of Bigeleisen's method is that autonomous calculation of the values of the additive increments is possible. However, it contains two essential limitations. First, the method is based on a model of the valency-force field model. Therefore, the values of the increments depend on the system of force constants adopted (see Tables 3.22 and 3.23). In this case the

TABLE 3.22

MAIN CONTRIBUTIONS TO THE VALUE $\ln(s/s^*)f$ FOR $^{13}CH_4/^{12}CH_4$ [a]

	Absolute value	*Total effects*	*Stretch of C—H bond*	*Bend of H—C—H angle*	*Stretch × stretch*	*Stretch × bend*	*Bend × bend*
F matrix of Jones and McDowel							
f_{ij}(mdyn Å)$^{-1}$			5.495	0.475	0.124	0.151	0.016
δg_{ij}(amu)$^{-1}$			0.0064(4)	0.0171(6)	−0.0021(6)	−0.0061(12)	−0.0171(3)
$T = 300$ K	1.1135–01	116.4%	82.7%	44.4%	−1.0%	−8.7%	−1.0%
760	3.0299–02	106.2%	84.1%	36.0%	−1.3%	−11.2%	−1.1%
3020	2.5476–03	100.5%	86.2%	30.4%	−1.9%	−13.2%	−1.0%
∞		100.0%	86.4%	29.9%	−1.9%	−13.4%	−1.0%
F matrix of Hartshorn and Shiner							
f_{ij}(mdyn Å)$^{-1}$			4.918	0.370	0.066	−0.0011	−0.058
δg_{ij}(amu)$^{-1}$			0.0064(4)	0.0171(6)	0.0021(6)	−0.0061(12)	−0.0171(3)
$T = 300$ C	1.1477–01	93.2%	60.3%	29.6%	−0.5%	0.1%	3.7%
760	3.1479–02	97.9%	68.8%	25.9%	−0.7%	0.1%	3.8%
3020	2.6413–03	99.7%	74.2%	22.8%	−1.0%	0.1%	3.6%
∞		100.0%	74.8%	22.5%	1.0%	0.1%	3.5%

[a] After Bigeleisen *et al.* (1976).

TABLE 3.23

CONTRIBUTIONS OF STRETCHING AND BENDING FORCES TO THE VALUE $10^3 \ln(s/s^*)f(^{13}C/^{12}C)$ AT 300 K[a]

Coordinates	CH_2O	CH_4(1)	CH_4(2)	C_2H_4	C_2H_6(3) IV	C_2H_6(3) V	 *Planar* (4)	 *Planar* (5)	 *Extraplanar* (4)
Stretching of C—H bond	30	92	69	32	49	49	14	16	—
Stretching of C—X bond	70	—	—	76	35	33	63	106	—
Bending of X—C—X angle	—	—	—	—	—	—	60	19	11
Bending of H—C—H angle	4	49	38	12	21	21	—	—	—
Bending of H—C—X angle	38	—	—	31	22	24	17	11	32
Total effects	141	130	108	143	126	123	157	141	36
Precise value	142	111	115	125	134	128	132	126	18

[a] The calculations were based on these matrices of force elements: Jones and McDowell (1); S. Hartshorn and V. Shiner (2); Schachtschneider and Snyder (3) V; Crawford–Miller (4); and Duinker–Mills (5). After Bigeleisen *et al.* (1976).

TABLE 3.24

CONTRIBUTION OF STRETCHING AND DEFORMING FORCES TO THE VALUE $10^3 \ln(s^*/s)f(^{18}O/^{16}O)$ AT 300 K[a]

Coordinates	H_2O	CO_2	CH_2O
Stretching of X—O bond	56.9	99.4	81.0
Bending of X—O—X angle	10.0	—	—
Bending of X—X—O angle	—	11.2	10.2
Total effects	66	111	91
Precise value	64	109	90

[a] After Bigeleisen *et al.* (1976).

uncertainty of the increments and, consequently, of the additive estimates based on them may prove to be substantially larger than the uncertainty of the β_{spectr} values calculated directly on the basis of various force field models. Second, the increments, which characterize the interactions with the coordinates in the molecule rather than the chemical bonds, are more closely related to its specific structure and therefore have worse transferability. Thus, to a certain extent, the most valuable property of the additive method—the universality of the system of increments—is lost.

XI. TEMPERATURE DEPENDENCE OF THERMODYNAMIC ISOTOPIC FACTORS

Like any other thermodynamic value, the thermodynamic isotopic factors are temperature dependent. As temperature drops, the values of the β factors and the values of the fractionation coefficients (α) increase. Table 3.25 gives the temperature dependences of the isotopic factors of simple carbon compounds.

In later chapters, in the examination of biological fractionation of isotopes, the numerical values of the β factors valid only for a temperature of 300 K are used. This of course is merely a convention simplifying the theoretical analysis. Strictly speaking, depending on the temperature conditions of biosynthesis, it is necessary to use the corresponding systems of β factors and isotopic bond numbers.

The values of L_{C-X} for $T = 300$ and 237.15 K are compared in the following tabulation.

T (K)	L_{C-H}	L_{C-C}	L_{C-N}	L_{C-O}	L_{C-F}	$L_{C=C}$	$L_{C\equiv C}$	$L_{C\equiv N}$	$L_{C=O}$
300	0.0284	0.046	0.050	0.055	0.056	0.078	0.088	0.091	0.095
273.15	0.033	0.055	0.059	0.064	0.065	0.095	0.100	0.103	0.108

TABLE 3.25

DEPENDENCE OF THE VALUES OF $\beta_\Sigma{}^{13}C$, α, AND Δ ON TEMPERATURE[a]

	β_Σ			CH_4/C_2H_6		C_2H_6/C_3H_8	
T (K)	CH_4	C_2H_6	C_3H_8	α	Δ	α	Δ
300	1.1136	1.1317	1.1385	0.9840	−16.0	0.9940	−6.0
400	1.0765	1.0861	1.0895	0.9912	−8.8	0.9969	−3.1
500	1.0554	1.0610	1.0630	0.9947	−5.3	0.9981	−1.9
600	1.0421	1.0455	1.0468	0.9967	−3.2	0.9988	−1.2
700	1.0331	1.0353	1.0361	0.9979	−2.1	0.9992	−0.8
800	1.0267	1.0282	1.0288	0.9985	−1.5	0.9994	−0.6

[a] See Galimov and Ivlev (1973–1974).

In many cases, however, an approximate calculation of the temperature dependence is sufficient. In the relatively narrow temperature range of 0° to 25°C at which living organisms typically exist, the dependence of L_{C-X} on T is nearly linear. Furthermore, an averaging of temperature dependences for the different bonds formed by the carbon tends to a certain extent to make these dependences invariant for various biochemical components.

It can be shown that, depending on the temperature, the coefficient is satisfactorily approximated by the expression

$$\Psi_t = 1 + \tfrac{2}{3}(27 - t^\circ) \times 10^{-2}, \tag{3.42}$$

where t° is the temperature of biosynthesis (in degrees Celsius).

The value of the β factor at a given temperature is determined as

$$\beta_t \simeq 1 + \Psi_t(\beta_{300\,K} - 1). \tag{3.43}$$

For molecules containing a small number of carbon bonds, for instance, for a one-carbon molecule, this formula may yield appreciable deviations from the true values. However, for polyatomic organic compounds, which are of interest in an examination of biological systems, the approximation is quite satisfactory. Even for the alanine molecule, which contains three carbon atoms, the value of the β_{273} factor obtained using the coefficient Ψ_t is 1.197, when the actual value is $\beta_{273} = 1.196$.

REFERENCES

Bachinski, D. J. (1969). Bond strength and sulfur isotope fractionation in coexisting sulfides. *Econ. Geol.* **69**, 56–65.

Bigeleisen, J. (1975). "Isotopes and Chemical Principles" (P.A. Roch, ed.). pp. 1–28. Amer. Chem. Soc., Washington, D.C.

Bigeleisen, J., and Ishida, T. (1968). Application of finite orthogonal polynomials to the thermal functions of harmonic oscillators. I. Reduced partition function of isotopic molecules. *J. Chem. Phys.*, **48**(3), 1311–1330.

Bigeleisen, J., and Ishida, T. (1973). Isotope chemistry and molecular structure: A simplified theory of end atom isotope effects. *J. Am. Chem. Soc.* **95**(9), 6155.

Bigeleisen, J., and Mayer, M. G. (1947). Calculation of equilibrium constánts for isotope exchange reactions. *J. Chem. Phys.* **15**, 261–257.

Bigeleisen, J., Ishida, T., and Spindel, W. (1971). Correlation of isotope effects with molecular forces. II. Triatomic molecules. *J. Chem. Phys.* **55**(10), 5021–5028.

Bigeleisen, J., Hom, R. C., and Ishida, T. (1976). Isotope chemistry and molecular structure: Carbon and oxygen isotope chemistry. *J. Chem. Phys.* **64**(8), 3303–3310.

Bottinga, Y. (1968). Calculation of fractionation factors for carbon and oxygen isotope exchange in the system calcite–carbon dioxide–water. *J. Phys. Chem.* **72**(2), 800–809.

Bottinga, Y. (1969). Calculated fractionation factors for carbon and hydrogen isotope exchange in the system calcite–carbon-dioxide–graphite–methane–hydrogen–water vapor. *Geochim. Cosmochim. Acta.* **33**(1), 49.

Craig, H. (1953). The geochemistry of the stable carbon isotopes. *Geochim. Cosmochim. Acta* **3**(2/3), 53–92.

Galimov, E. M. (1971). On the relationship of the fractionation coefficient of isotopes to the equilibrium constants of the isotope exchange reactions of carbon in hydrocarbon systems. *Zhurn. Fiz. Khimii* **45**(5), 1187–1191.

Galimov, E. M. (1972). Method of isotopic bond numbers. Application in biochemistry and geochemistry. *In* "Abstracts of IV All-Union Symposium on the Application of Stable Isotopes in Geochemistry." GEOKhI, Moscow.

Galimov, E. M. (1973a). Biogenic intermolecular and intramolecular carbon isotope effects: Method of "Isotopic numbers of bindings." Biochemical and geochemical applications. *In* "Intern. Meeting on Isotope Effects in Physical and Chemical Processes." Cluj, Romania.

Galimov, E. M. (1973b). "Carbon Isotopes in Oil and Gas Geology." Moscow, Nedra. English translation: NASA TT-682, Washington, D.C., 1975, 395 pp.

Galimov, E. M. (1974). Approximate method of finding the thermodynamic isotopic factors of carbon compounds. *Zhurn. Fiz. Khimii* **48**(2).

Galimov, E. M., and Ivlev, A. A. (1973–1974). Thermodynamic isotopic effects of organic compounds. I. Isotopic effects in normal alkanes. *Zhurn. fiz. khimii* **47**(11), 2787; II. Calculation of isotopic frequencies and the factors of carbon of normal alkanes. **47**(10), 2707; III. Hydrogen isotopic effect in *n*-alkanes. **47**(10), IV. Calculation of isotopic frequencies and β factors of carbon of benzene, toluene and cyclohexane. **47**(10), V. Isotopic effects of carbon in alicyclic and aromatic hydrocarbons. *Zhurn. Fiz. Khimii*, **48**(4).

Gribov, L. A. (1965). "Introduction to the Theory and Calculation of Vibrational Spectra of Polyatomic Molecules." Izd-vo Leningrad Univ., Leningrad.

Hartshorn, S. R., and Shiner, V. J. (1972). Calculation of H/D, $^{12}C/^{13}C$ and $^{12}C/^{14}C$ fractionation factors from force fields derived for a series of simple organic molecules. *J. Am. Chem. Soc.* **94**, 9002.

Ishida, T., Spindel, W., and Bigeleisen, J. (1969). Theoretical analysis of chemical isotope fractionation by orthogonal polynomial methods. *Adv. Chem. Am. Chem. Soc.* **89**, 1–65.

Ivlev, A. A., Koroleva, M. Ya., and Galimov, E. M. (1974a). Thermodynamic isotopic effects of organic compounds. VI. Acetic acid, acetaldehyde, methylamine, methyl alcohol. *Zhurn. Fiz. Khimii* **48**(1), 225.

Ivlev, A. A., Koroleva, M. Ya., Tuan Suankhoan, and Klina, M. (1974b). Thermodynamic isotopic effects of organic compounds. VII. Isotopic effects of nitrogen. *Zhurn. Fiz. Khimii* **48**(5), 1143–1146.

Johansen, H. (1964). Normal coordinate analysis of polyatomic molecules and statistical thermodynamics of isotope molecules, I. *Ztschr. Phys. Chem.* **227**(5/6), 306

Nakamoto, K. (1973). "Infrared Spectra of Inorganic and Coordinational Compounds." (Russian translation). Mir, Moscow.

Ohmoto, H, (1972). Systematics of sulfur and carbon isotopes in hydrothermal ore deposits. *Econ. Geol.* **67**, 551–578.

Pimental, G., and Spratley, P. (1973). "How Quantum Mechanics Explains the Chemical Bond" (Russian translation). Mir, Moscow.

Pustyl'nik, Ye. I. (1968). "Statistical Methods of Analysis and Processing of Observations." Nauka, Moscow.

Sakai, H. (1968). Isotopic properties of sulfur compounds on hydrothermal processes. *Geochem. J.* **2**, 29–49.

Shustarovich, Ye. M. (1973). "The Chemical Bond." Nauka, Moscow.

Singh, G., and Wolfsberg M. (1975). The calculation of isotopic partition function ratios by a perturbation theory technique. *J. Chem. Phys.* **62**, 4165–4180.

Stern, M. J., and Wolfsberg, M. (1966). Simplified procedure for the theoretical calculation of isotope effects involving large molecules. *J. Chem. Phys.* **45**(11), 4105–4124.

Sverdlov, L. I., Kovner, M. A., and Kondratov, Ye. P. (1970). "Vibration Spectra of Polyatomic Molecules." Nauka, Moscow.

Urey, H. G. (1935). Isotopic exchange equilibria. *J. Am. Chem. Soc.* **57**, 321–327.

Urey, H. G. (1947). The thermodynamic properties of isotopic substances. *J. Chem. Soc.*, part 1, 562–581.

Varshavskiy, L. N., and Vaysberg, S. E. (1957). Thermodynamic and kinetic particulars of isotope exchange reaction. *Uspekhi Khimii* **26**(12).

Vinogradov, A. P. (1954). Isotope geochemistry. *Izv. AN SSSR, Ser. Geol.* (3).

Vojta, G. (1960a). Approximate formulas for quantum-statistical calculation of chemical isotope exchange equilibrium and application to molecules with ^{18}O. *Kernenergie* **3**(9), 327

Vojta, G. (1960b). Fundamentals of statistical thermodynamics of isotopic systems with chemical reactions. *Kernenergie* **3**(9), 917.

Vojta, G. (1960c). On the statistical thermodynamics of ideal gases with isotopically substituted molecules. *Z. Phys. Chem.* **217**(5/6), 337.

Vol'kenshteyn, M. V., Yel'yashevich, M. A., and Stepanov, B. I. (1949). "Vibrations of Molecules," Vol. 1. GITTL, Moscow.

Wilson, E., Decius, J., and Cross, P. (1960). "Theory of Vibration Spectra of Molecules" (Russian translation). Izd-vo Inostr. Lit., Moscow.

CHAPTER 4

Experimental Evidence of Thermodynamically Ordered Distribution of Isotopes in Biological Systems

The procedure discussed in the preceding chapter can be used to calculate thermodynamic isotopic factors of biological compounds. When such calculations were first made (Galimov, 1972, 1973a, b, c), clear correspondence between the $\beta^{13}C$ values and the real isotopic composition was found in some cases. It must be stressed that the existence of such correspondence was not connected with the traditional idea of the conditions under which thermodynamic isotopic effects can be manifested. The values of the thermodynamic isotopic factors describe the distribution of isotopes in a state of isotope-exchange equilibrium. However, in the complex organic compounds typical of biological systems, isotope exchange (especially of carbon isotopes) is exceedingly slowly. Nevertheless, the facts indicate that in many cases, if not in most, there is a correlation between $\beta^{13}C$ and $\delta^{13}C$ values. Below we shall examine the pertinent experimental data, and in the next chapter we shall discuss the nature and possible mechanism of this phenomenon.

I. DISTRIBUTION OF ISOTOPES BETWEEN INDIVIDUAL AMINO ACIDS

Originally, the thermodynamic factors were calculated for those compounds for which data on their isotopic composition had been published. The most detailed date were the results of Abelson and Hoering (1961), obtained in a study of amino acids. These results are given in Table 2.5 of Chapter 2.

The calculation of $\beta_{\Sigma}{}^{13}C$ and $\beta_i{}^{13}C$ values of amino acids is given below (Table 4.1). The β factors were calculated for those amino acids whose

TABLE 4.1

CALCULATION OF THE THERMODYNAMIC ISOTOPIC FACTORS OF SOME AMINO ACIDS

Amino acid	*Structural formula*	β_i *factor*	*Set of bond numbers and numerical value of* β_i *factor*
Glutamic acid	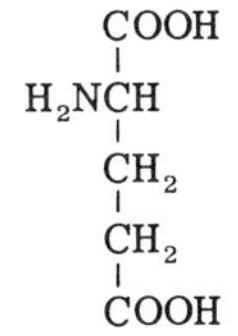	β_1	$1 + L_{C-C} + L_{C-O} + L_{C=O} + l_{C-C} + l_{C-N} = 1.2001$
		β_2	$1 + L_{C-H} + 2L_{C-C} + L_{C-N} + l_{C-O} + l_{C=O} + l_{C-C} = 1.1772$
		β_3	$1 + 2L_{C-H} + 2L_{C-C} + 2l_{C-C} + l_{C-N} = 1.1538$
		β_4	$1 + 2L_{C-H} + 2L_{C-C} + l_{C-C} + l_{C-O} + l_{C=O} = 1.1553$
		β_5	$1 + L_{C-C} + L_{C-O} + L_{C=O} + l_{C-C} = 1.1985$
		β_Σ	1.1769
Aspartic acid	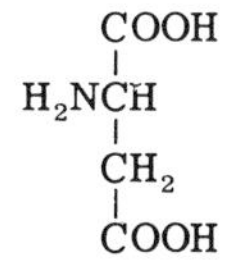	β_1	$1 + L_{C-C} + L_{C-O} + L_{C=O} + l_{C-C} + l_{C-N} = 1.2001$
		β_2	$1 + L_{C-H} + 2L_{C-C} + L_{C-N} + l_{C-O} + l_{C=O} + l_{C-C} = 1.1772$
		β_3	$1 + 2L_{C-H} + 2L_{C-C} + l_{C-C} + l_{C-O} + l_{C=O} + l_{C-N} = 1.1569$
		β_4	$1 + L_{C-C} + L_{C-O} + L_{C=O} + l_{C-C} = 1.1985$
		β_Σ	1.1832
Arginine	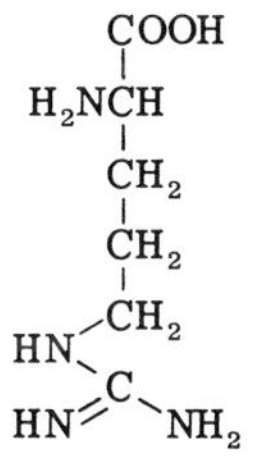	β_1	$1 + L_{C-C} + L_{C-O} + L_{C=O} + l_{C-C} + l_{C-N} = 1.2001$
		β_2	$1 + L_{C-H} + 2L_{C-C} + L_{C-N} + l_{C-O} + l_{C=O} + l_{C-C} = 1.1772$
		β_3	$1 + 2L_{C-H} + 2L_{C-C} + l_{C-N} + 2l_{C-C} = 1.1541$
		β_4	$1 + 2L_{C-H} + 2L_{C-C} + l_{C-N} + l_{C-C} = 1.1525$
		β_5	$1 + 2L_{C-H} + L_{C-C} + L_{C-N} + l_{C-C} = 1.1545$
		β_6	$1 + 2L_{C-N} + L_{C=N} + l_{C-C} = 1.1803$
		β_Σ	1.1698
Lysine	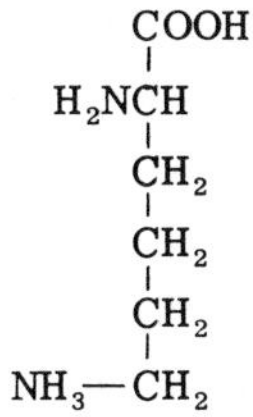	β_1	$1 + L_{C-C} + L_{C-O} + L_{C=O} + l_{C-C} + l_{C-N} = 1.2001$
		β_2	$1 + L_{C-H} + 2L_{C-C} + L_{C-N} + l_{C-O} + l_{C=O} + l_{C-C} = 1.1772$
		β_3	$1 + 2L_{C-H} + 2L_{C-C} + 2l_{C-C} + l_{C-N} = 1.1538$
		β_4	$1 + 2L_{C-H} + 2L_{C-C} + 2l_{C-C} = 1.1522$

(Table continues)

TABLE 4.1 (*continued*)

Amino acid	*Structural formula*	β_i *factor*	*Set of bond numbers and numerical value of* β_i *factor*
		β_5	$1 + 2L_{C-H} + 2L_{C-C} + l_{C-C} + l_{C-N}$ $= 1.1525$
		β_6	$1 + 2L_{C-H} + L_{C-C} + L_{C-N} + l_{C-C}$ $= 1.1545$
		β_Σ	1.1651
Alanine	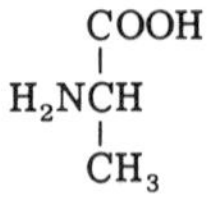	β_1	$1 + L_{C-C} + L_{C-O} + L_{C=O} + l_{C-C} + l_{C-N}$ $= 1.2001$
		β_2	$1 + L_{C-H} + 2L_{C-C} + L_{C-N} + l_{C=O} + l_{C-O}$ $= 1.1759$
		β_3	$1 + 3L_{C-H} + L_{C-C} + l_{C-C} + l_{C-N}$ $= 1.1345$
		β_Σ	1.1702
Leucine	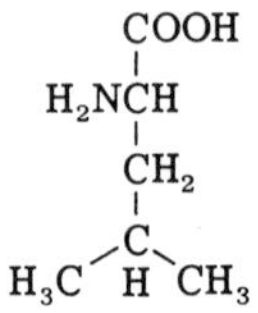	β_1	$1 + L_{C-C} + L_{C-O} + L_{C=O} + l_{C-C} + l_{C-N}$ $= 1.2001$
		β_2	$1 + L_{C-H} + 2L_{C-C} + L_{C-N} + l_{C=O} + l_{C-O} + l_{C-C}$ $= 1.1772$
		β_3	$1 + 2L_{C-H} + 2L_{C-C} + 3l_{C-C} + l_{C-N}$ $= 1.1538$
		β_4	$1 + L_{C-H} + 3L_{C-C} + l_{C-C}$ $= 1.1689$
		β_5, β_6	$1 + 3L_{C-H} + L_{C-C} + 2l_{C-C}$ $= 1.1342$
		β_Σ	1.1614
Isoleucine	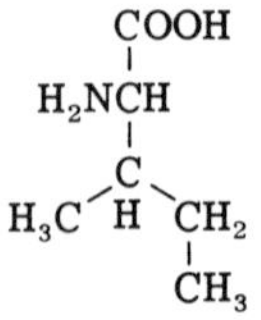	β_1	$1 + L_{C-C} + L_{C-O} + L_{C=O} + l_{C-C} + l_{C-N}$ $= 1.2001$
		β_2	$1 + L_{C-H} + 2L_{C-C} + L_{C-N} + l_{C-O} + l_{C=O} + 3l_{C-C}$ $= 1.1785$
		β_3	$1 + L_{C-H} + 3L_{C-C} + 2l_{C-C} + l_{C-N}$ $= 1.1718$
		β_4	$1 + 3L_{C-H} + L_{C-C} + 2l_{C-C}$ $= 1.1342$
		β_5	$1 + 2L_{C-H} + 2L_{C-C} + 2l_{C-C}$ $= 1.1509$
		β_6	$1 + 3L_{C-H} + L_{C-C} + l_{C-C}$ $= 1.1329$
		β_Σ	1.1614
Phenyl-alanine	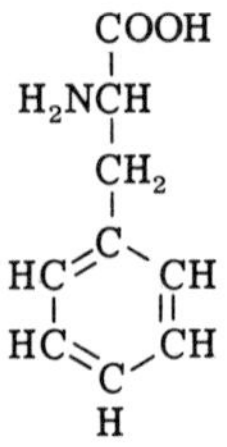	β_1	$1 + L_{C-C} + L_{C=O} + L_{C-O} + l_{C-N} + l_{C-C}$ $= 1.2001$
		β_2	$1 + L_{C-H} + 2L_{C-C} + L_{C-N} + l_{C=O} + l_{C-O} + l_{C-C}$ $= 1.1772$
		β_3	$1 + 2L_{C-H} + 2L_{C-C} + 2l_{C-C} + l_{C-N} + l_{C=C}$ $= 1.1554$
		β_4	$1 + 2L_{C-C} + L_{C=C} + 2l_{C-C} + l_{C=C} + l_O$ $= 1.1725$

Amino acid	*Structural formula*	β_i *factor*	*Set of bond numbers and numerical value of* β_i *factor*
		β_5, β_6	$1 + L_{C-H} + L_{C-C} + L_{C=C} + 2l_{C-C} + l_{C=C} + l_O = 1.1545$
		β_7, β_8	$1 + L_{C-H} + L_{C-C} + L_{C=C} + l_{C-C} + l_{C=C} + l_O = 1.1532$
		β_9	$1 + L_{C-H} + L_{C-C} + L_{C=C} + l_{C-C} + l_{C=C} + l_O = 1.1532$
		β_Σ	1.1637
Glycine	COOH | H_2NCH_2	β_1	$1 + L_{C-C} + L_{C=O} + L_{C-O} + l_{C-N} = 1.1988$
		β_2	$1 + 2L_{C-H} + L_{C-C} + L_{C-N} + l_{C=O} + l_{C-O} = 1.1579$
		β_Σ	1.1783
Serine	COOH | H_2NCH | $HOCH_2$	β_1	$1 + L_{C-C} + L_{C=O} + L_{C-O} + l_{C-N} + l_{C-C} = 1.2001$
		β_2	$1 + L_{C-H} + 2L_{C-C} + L_{C-N} + l_{C=O} + 2l_{C-O} = 1.1778$
		β_3	$1 + 2L_{C-H} + L_{C-C} + L_{C-O} + l_{C-C} + l_{C-N} = 1.1611$
		β_Σ	1.1797
Threonine	COOH | H_2NCH | HOCH | CH_3	β_1	$1 + L_{C-C} + L_{C=O} + L_{C-O} + l_{C-N} + l_{C-C} = 1.2001$
		β_2	$1 + L_{C-H} + 2L_{C-C} + L_{C-N} + l_{C=O} + 2l_{C-O} + l_{C-C} = 1.1791$
		β_3	$1 + L_{C-H} + 2L_{C-C} + L_{C-O} + l_{C-N} + l_{C-C} = 1.1791$
		β_4	$1 + 3L_{C-H} + L_{C-C} + l_{C-O} + l_{C-C} = 1.1363$
		β_Σ	1.1736
Tyrosine	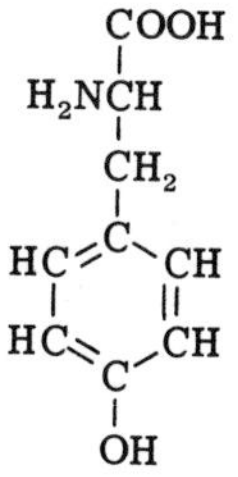	β_1	$1 + L_{C-C} + L_{C=O} + L_{C-O} + l_{C-N} + l_{C-C} = 1.2001$
		β_2	$1 + L_{C-H} + 2L_{C-C} + L_{C-N} + l_{C=O} + l_{C-O} + l_{C-C} = 1.1772$
		β_3	$1 + 2L_{C-H} + 2L_{C-C} + 2l_{C-C} + l_{C-N} + l_{C=C} = 1.1554$
		β_4	$1 + 2L_{C-C} + L_{C=C} + 2l_{C-C} + l_{C=C} + l_O = 1.1725$
		β_5, β_6	$1 + L_{C-H} + L_{C-C} + L_{C=C} + 2l_{C-C} + l_{C=C} + l_O = 1.1545$
		β_7, β_8	$1 + L_{C-H} + L_{C-C} + L_{C=C} + l_{C-C} + l_{C=C} + l_{C-O} + l_O = 1.1551$
		β_9	$1 + L_{C-C} + L_{C-O} + L_{C=C} + l_{C-C} + l_{C=C} + l_O = 1.1798$
		β_Σ	1.1671

TABLE 4.2

PARAMETERS CHARACTERIZING THE CORRELATION OF $\beta_\Sigma{}^{13}C$ AND $\delta^{13}C$ FOR THE AMINO ACIDS OF A NUMBER OF ORGANISMS

Organism	*Number of amino acids studied*	*Correlation coefficient* η	*Critical value of correlation coefficient*	κ
Chlorella pyrenoidosa	10	0.76	0.54	0.58
Euglena gracilis	11	0.82	0.54	0.60
Anacystis nidulans	6	0.59	0.69	—
Scenedesmus quadricauda	6	0.43	0.69	—
Gracilaria sp.	11	0.74	0.54	0.40
Chromatium sp.	4	0.45	0.81	—

isotopic compositions had been determined by Abelson and Hoering (1961). The bond numbers used were given in the preceding chapter (see Table 3.13).

In contrast to our first publications (Galimov, 1973b), in these calculations we used not only the main bond numbers L_{C-X} but also the bond numbers l_{C-X}, correcting for the distant surroundings. Entries of the β-factor values were derived with a precision to the fourth decimal place, although in individual cases the fourth, and sometimes the third place, may be meaningless (see the preceding chapter for a discussion on precision).

For each of the organisms studied by Abelson and Hoering (1961), the isotopic composition of the individual amino acids was compared with the β_Σ-factor values. Table 4.2 gives the results of a correlational analysis of the $\beta_\Sigma{}^{13}C$—$\delta^{13}C$ relationships obtained. For three of the six organisms a significant correlation is observed (the value of correlation coefficient n is higher than the critical value). In the three remaining cases, in which the number of amino acids investigated was small, the correlation is not significant.

Figures 4.1a–c graphically show the relationship between $\delta^{13}C$—$\beta_\Sigma{}^{13}C$ values for the amino acids of the three organisms. In the 10–11 amino acids investigated for each organism, the correlation coefficient has a value of 0.74 to 0.82, which indicates the nonrandom character of the relationship with a confidence probability of more than 0.99.

Analyzing the experimental data, Abelson and Hoering (1961) reported that aspartic acid, serine, and glycine are usually relatively depleted in the ^{12}C isotope in all organisms, whereas leucine, isoleucine, and the aromatic amino acids belong to the isotopically light compounds. This is consistent

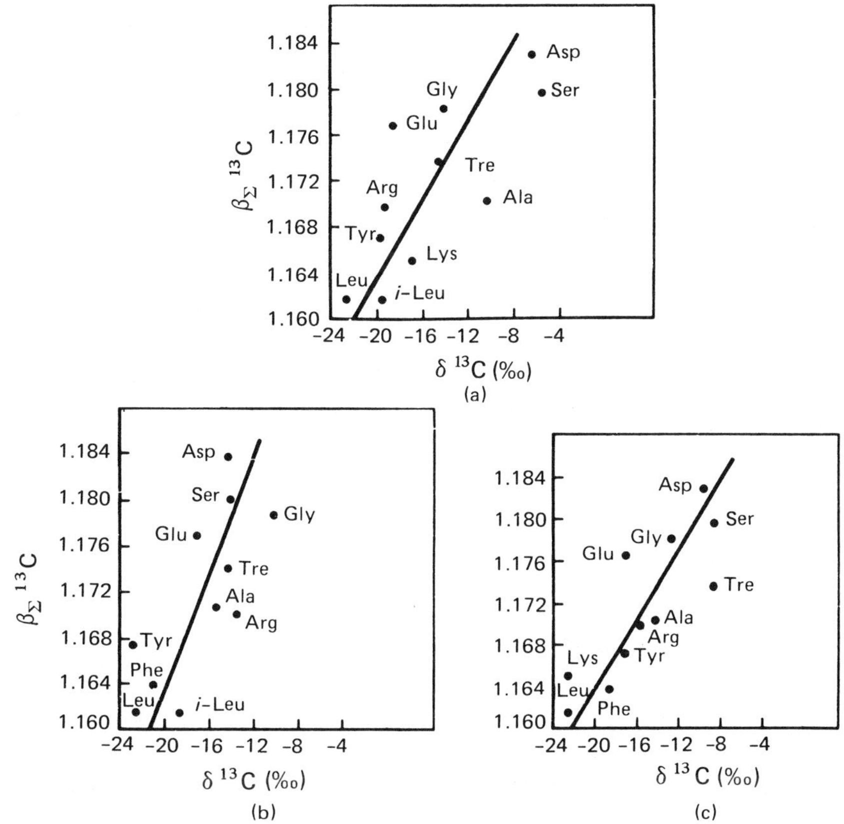

FIG. 4.1 Comparison of values of $\delta^{13}C$ and $\beta_\Sigma{}^{13}C$ of the amino acids for various organisms. The graphs show the value of the correlation coefficients η and give the regression equation. (a) *Chlorella pyrenoidosa*, $\delta^{13}C_X = -15.3 + 0.58(\beta_X - 1.1715) \times 10^3$, and $\eta = 0.76$; (b) *Gracilaria* sp., $\delta^{13}C_X = -16.8 + 0.40(\beta_X - 1.1714) \times 10^3$, and $\eta = 0.74$; (c) *Euglena gracilis*, $\delta^{13}C_X - 15.3 + 0.60(\beta_X - 1.1717) \times 10^3$, and $\eta = 0.82$.

with the high $\beta_\Sigma{}^{13}C$ values for the former group and the relatively low values for the latter.

The correlations, obviously, are linear because of the physical meaning of the $\delta^{13}C$ and $\beta^{13}C$ values. In an equilibrium system they are connected by the relationship

$$\delta^{13}C_1 - \delta^{13}C_2 = ((\beta_1/\beta_2) - 1) \times 10^3\,‰, \tag{4.1}$$

or

$$\delta^{13}C_1 - \delta^{13}C_2 \simeq (\beta_1 - \beta_2) \times 10^3\,‰. \tag{4.2}$$

Figures 4.1a,b,c show that the regression lines are characterized by

$$\delta^{13}C = \delta^{13}C_{av} + \kappa(\beta - \beta_{av}) \times 10^3, \tag{4.3}$$

where $\kappa < 1$. The coefficient κ indicates the fraction of the equilibrium value that the isotopic shifts represent, as determined directly from the ratio of their β factors. Such a reduction in isotopic effects is, as we shall see, typical of the distribution of isotopes in biological systems.

If the distribution of isotopes corresponds to the values of the β factors, this indicates that thermodynamic isotopic effects exist in the system, that is, an equilibrium distribution of isotopes exists in some form. However, use of the concept of isotopic equilibrium may produce the false idea that chemical equilibrium of the biological system is required, whereas in this case it is not. Therefore, we shall discuss the thermodynamically ordered distributions of isotopes in biological systems, not the equilibrium distribution.

II. INTRAMOLECULAR DISTRIBUTION OF ISOTOPES IN AMINO ACIDS

In their work dealing with isotopic analysis of amino acids, Abelson and Hoering (1961) measured the carbon isotopic composition of split-off carboxyl groups and whole amino acids. They were first to obtain data on the intramolecular distribution of isotopes in biological compounds. In the overwhelming majority of amino acids investigated, the carbon of the carboxyl group was appreciably depleted in the ^{12}C isotope compared with the carbon of the amino acids as a whole (see Table 2.5). On the average the ratio of the isotopes in the carboxyl carbon was close to that in the carbon of source carbon dioxide.

In thermodynamic isotopic analysis, it can easily be shown that the isotopic forms of amino acids substituted with respect to carbon of the carboxyl group have very high values of the β_i factors (see Table 4.1), close to the value of the thermodynamic isotopic factor of CO_2. Figure 4.2b is a histogram of the ratios of the β factors of the carboxyl carbon and decarboxylated residue of the amino acids analyzed by Abelson and Hoering (1961). Observed values of the isotopic shifts $\Delta^{13}C$ are given in Fig. 4.2a using the same scale. Both distributions have positive values for the respective isotopic shifts. The distribution of the observed values is characterized by a large dispersion, as would be expected. In addition, similiar to intermolecular isotopic effects, the observed isotopic shifts are less than equilib-

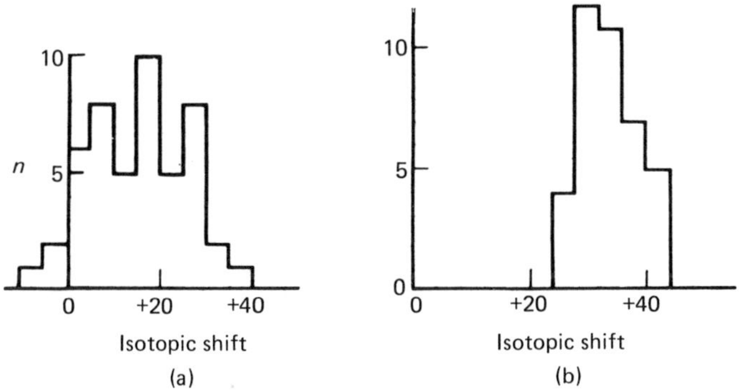

FIG. 4.2 Distribution of (a) experimentally observed and (b) theoretical equilibrium values of intramolecular isotopic shifts in biogenic amino acids. Data on the isotopic composition are given in Table 2.5, values of β factors in Table 4.1. (a) $\Delta^{13}C = \delta^{13}C_{COOH} - \delta^{13}C_R$; (b) $[(\beta^{13}C_{COOH}/\beta^{13}C_R) - 1] \times 10^3$.

rium values. The value of the coefficient κ, characterizing this change, averages 0.34.

Abelson and Hoering (1961) did not consider the enrichment of carboxyl carbon of amino acids in the ^{13}C isotope to be the result of any isotopic fractionation. They suggested that, in addition to the main pathway of assimilation of CO_2, via ribulose diphosphate and involving fractionation of isotopes, there is a secondary pathway (via the Krebs cycle) in which CO_2 is included first in the carboxyl group of aspartic and glutamic acids and then of other amino acids, without appreciable change in isotopic composition.

In other words, Abelson and Hoering (1961) do not suggest that the high ^{13}C content of the carboxyl carbon in amino acids is a consequence of enrichment. They suggest that this high content results because the carboxyl carbon is not depleted in ^{13}C, as is the rest of the biological carbon due to isotopic fractionation in the ribulose diphosphate stage.

If this consideration is valid, it is only partly so. It is, to some extent, inconsistent with the results obtained in that same work (Abelson and Hoering, 1961).

Investigating how the concentration of CO_2 in the external environment affects the isotopic composition of photosynthetic organisms, Abelson and Hoering established that, when the supply of CO_2 is limited, the cells of *Chlorella pyrenoidosa* fractionate isotopes much less strongly. Under favorable conditions (5% concentration of CO_2 in solution) the isotopic composition of the cells was characterized by a value of -25.8‰; when the supply of CO_2 was limited, the $\delta^{13}C$ value was -11.3‰.

TABLE 4.3

DISTRIBUTION OF CARBON ISOTOPES IN AMINO ACIDS GIVEN OFF BY *Chlorella pyrenoidosa* IN NORMAL AND LIMITED SUPPLY OF CO_2

	$\delta^{13}C$ (‰)	
Compound	*Normal*	*Limited*
Original CO_2	0.0	0.0
Total organic carbon	−25.8	−11.3
Glutamic acid		
Total	−18.7	−9.9
Carboxyl	−8.8	9.6
Residual	−21.1	−14.8
Aspartic acid		
Total	−6.6	−3.5
Carboxyl	−2.2	8.7
Residual	−15.6	−15.7
Glycine		
Total	−14.3	−2.1
Carboxyl	−5.3	−5.0
Residual	−23.3	0.8
Alanine		
Total	−10.3	−5.9
Carboxyl	−7.3	11.9
Residual	−11.8	−14.8

In Table 4.3 we have compared the results of these two experiments. The isotopic composition of both total carbon and carboxyl carbon of the amino acids characteristically is displaced toward more positive values. According to Abelson and Hoering this should not have occurred, as the isotopic composition of the fixed carbon dioxide did not change. The reduced concentration of CO_2 should have affected only carbon entering the organism via the stage of photosynthetic assimilation of CO_2. In addition, we see that it affected the isotopic composition of the carboxyl group of the amino acids. Consequently, the enrichment of the carbon of the carboxyl groups in the ^{13}C isotope cannot be considered merely the result of incorporation of exogenous CO_2. This phenomenon is explained by fractionation of isotopes in biological systems. The unusual character of this fractionation may seem difficult to explain. However, as we shall see, it is consistent with a thermodynamically ordered distribution of isotopes.

III. DISTRIBUTION OF CARBON ISOTOPES IN THE COMPONENTS OF THE LIPID FRACTION

Until 1972, when theoretical calculations of β factors of biomolecules using the additive method became possible, little experimental data on the isotopic composition of biomolecules had been accumulated. Other than the data for amino acids examined above, there had been no systematic investigations of the isotopic composition of biological compounds. Therefore, further study of the regularity that had been found required experimental data on the distribution of isotopes in biomolecules. It followed that the distribution of carbon isotopes should be studied in individual compounds and in narrow, chemically identifiable fractions, especially those belonging to the same class of biochemical compounds. This could determine whether the relationship established between the isotopic composition of biogenic compounds and the values of the thermodynamic isotopic factors is found at the level of compounds related in chemical structure, composition, and mechanism of biosynthesis.

The individual compounds and components of the lipid fraction of organisms would be a suitable subject for this purpose. To obtain generally applicable conclusions, it was important to investigate organisms in a broad taxonomic range, for instance, marine and freshwater organisms, autotrophs and heterotrophs, primitive and more highly organized forms, and algae and land plants.

In the work of Galimov and Shirinskiy (1975) we attempted to meet these requirements by choosing the following four organisms for study: marine plants (*Laminaria*), marine crayfish (krill), blue–green river algae, and meadow lupine.

The samples of marine algae (*Laminaria saccharina*) were collected in shelf zone of the White Sea, with a mean annual water temperature of 5°C. The krill—macrozooplankton represented by shrimp of the species *Euphausia superba*—were caught in the summer in the Antarctic waters of the Scotia Sea. The lupine plants (fast-growing yellow lupine, *Lupinus luteus*) were collected below Moscow. The blue-green algae, 96–98% represented by the species *Abavaena variabilis*, were taken from the Desna River, which has a water temperature of 28°C.

The lipids were extracted from the organisms by a chloroform–methanol mixture (2:1). The preparatory separation of the lipids into fractions was carried out by thin-layer chromatography on silica gel.

Identification of the fractions was accomplished using the individual R_f values and was confirmed by mass spectrometry. The details of the experimental procedure are discussed by Galimov and Shirinskiy (1975).

Tables 4.4–4.7 give the results of carbon isotope analysis of the components making up the lipid fractions of the organisms investigated. Calculated values of the β_Σ factors are also given. Sometimes the chromatographic fractions contained several compounds. In that case, the value of the thermodynamic factor was determined either from the composition of the fraction or from the predominant component (Galimov and Shirinskiy, 1975). It should be mentioned that, in view of the similarity of the structural formulas of the compounds making up the same chromatographic fraction, some uncertainty in their identification was slightly reflected in the value of the calculated β_Σ factor. The $\beta_\Sigma{}^{13}C$ values in Tables 4.4–4.7 differ somewhat from those given by Galimov and Shirinskiy (1975) inasmuch as here they were calculated using the l_x correction for the distant surroundings.

This investigation made it possible to establish the following:

(1) The results of the correlation analysis presented in Table 4.8 show high correlation coefficients between all measured values of $\delta^{13}C$ and calculated values of the β factors.

TABLE 4.4

ANALYSIS OF THE LIPID FRACTION OF BLUE-GREEN ALGAE

Order of position on chromatogram	*Component*	*Content* (%)	$\delta^{13}C$ (‰)	$\beta_\Sigma{}^{13}C$
	Total lipids			
12	Polar lipids	23.6	−31.6	—
11	Monoglycerides	6.8	−30.0	1.1562
10	Carotenoids (myxoxanthophyll)	4.8	−29.4	1.1607
9	Diglycerides	8.1	−30.1	1.1552
8	Chlorophyll	27.7	−28.4	1.1611
7	Unidentified	—	−30.0	—
6	Sterols (β-sitosterol)	2.7	−29.5	1.1590
5	Carotenoid (echinenone)	—	−29.5	1.1575
4	Fatty acids	4.9	−31.9	1.1540
3	Carotenoid (β-carotene)	2.2	−31.5	1.1567
2	Triglycerides	17.7	−30.5	1.1548
1	Waxes, hydrocarbons, esters, sterols	1.5	−32.3	1.1527
	Polar lipids			
5	Cephalin	—	−30.3	1.1553
4	Lecithin	—	−30.6	1.1538
3	Phosphatidylinositol	—	−29.1	1.1592
2	Sphingomyelin	—	−30.8	1.1530
1	Phosphatidylserine	—	−31.9	1.1571

TABLE 4.5

ANALYSIS OF THE LIPID FRACTION OF KRILL

Order of position on chromatogram	*Component*	*Content* (%)	$\delta^{13}C$ (‰)	$\beta_{\Sigma}{}^{13}C$
	Total lipids			
9	Polar lipids	2.6	−27.3	—
8	Monoglycerides	1.4	−27.0	1.1562
7	Carotenoid (astatine)	1.8	−25.7	1.1608
6	Diglycerides	1.7	−27.1	1.1552
5	Sterols (cholesterol)	8.0	−27.0	1.1590
4	Pigment (echinenone)	1.9	−26.9	1.1575
3	Fatty acids	64.5	−28.1	1.1540
2	Triglycerides	12.6	−27.3	1.1548
1	Waxes, hydrocarbons, esters, sterols	5.5	−29.8	1.1527
	Polar lipids			
3	Cardiolipin	—	−27.8	1.1561
2	Lecithin	—	−28.4	1.1538
1	Sphingomyelin	—	−28.6	1.1530

TABLE 4.6

ANALYSIS OF THE LIPID FRACTION OF LUPINE

Order of position on chromatogram	*Component*	*Content* (%)	$\delta^{13}C$ (‰)	$\beta_{\Sigma}{}^{13}C$
	Total lipids			
12	Polar lipids	36.1	−32.0	—
11	Monoglycerides	3.0	−31.4	1.1562
10	Pigment (lutein)	6.6	−31.1	1.1591
9	Diglycerides	3.6	−32.1	1.1552
8	Chlorophyll + xanthophyll	24.3	−29.7	—
7	Chlorophyll	10.2	−29.0	1.1611
6	Unidentified	0.7	−29.2	—
5	Unidentified	0.9	−31.4	—
4	Sterols (β-sitosterol)	7.1	−29.6	1.1590
3	Fatty acids	1.2	−32.7	1.1540
2	Triglycerides	0.3	−32.3	1.1548
1	Waxes, hydrocarbons, esters, sterols	6.0	−33.5	1.1527
	Polar lipids			
7	Monogalactosylglyceride	—	−32.0	1.1588
6	Cephalin	—	−32.7	1.1553
5	Lecithin	—	−32.8	1.1538
3	Phosphatidylserine	—	−33.1	1.1571

TABLE 4.7

ANALYSIS OF THE LIPID FRACTION OF LAMINARIA

Order of position on chromatogram	*Component*	*Content* (%)	$\delta^{13}C$ (‰)	$\beta_\Sigma{}^{13}C$
	Total lipids			
8	Polar lipids	45.6	−22.6	—
7	Monoglycerides	6.6	−25.0	1.1562
6	Carotenoid (fucoxanthine)	14.9	−23.7	1.1589
5	Chlorophyll	20.8	−21.6	1.1611
4	Sterols (fucosterol)	3.8	−24.5	1.1590
3	Fatty acids	3.8	−25.6	1.1540
2	Triglycerides	1.1	−24.8	1.1548
1	Waxes, hydrocarbons, esters, sterols	3.4	−27.8	1.1527

(2) The relationships depicted in Figs. 4.3a–b and 4.4a–b are significant at greater than the 99% confidence level.

(3) The values of the observed isotopic effects were reduced relative to the calculated value of the thermodynamic isotopic effects. The coefficient κ varies within 0.31–0.55.

Thus, thermodynamically ordered distribution of isotopes is typical of organisms in a broad taxonomic and ecological range. At least for all four

TABLE 4.8

CORRELATION OF $\beta_\Sigma{}^{13}C$ AND $\delta^{13}C$ VALUES OF THE COMPONENTS OF THE LIPID FRACTION OF SOME ORGANISMS

Organism	*Number of components of fraction studied*	*Correlation coefficient*	*Critical value of correlation coefficient*	$\bar{\kappa}$	*Regression equation*
Laminaria saccharina	7	0.90	0.65	0.55	$\delta^{13}C = -24.7 + 0.55(\beta - 1.1567) \times 10^3$
Lupinus luteus	12	0.82	0.82	0.52	$\delta^{13}C = -31.9 + 0.44(\beta - 1.1564) \times 10^3$
Euphausia superba	11	0.87	0.54	0.37	$\delta^{13}C = -27.6 + 0.37(\beta - 1.1557) \times 10^3$
Anabaena variabilis	15	0.73	0.47	0.31	$\delta^{13}C = -30.4 + 0.31(\beta - 1.1564) \times 10^3$

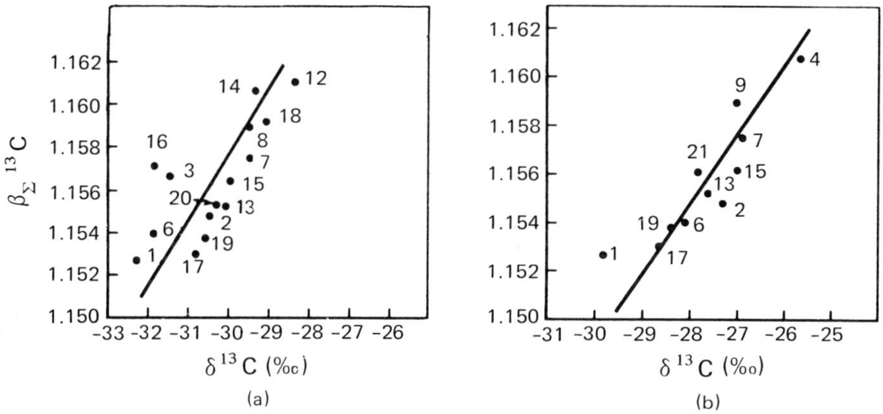

FIG. 4.3 $\beta_\Sigma{}^{13}C$–$\delta^{13}C$ correlation of components of the lipid fraction of (a) the blue-green alga (*Anabaena variabilis*) and (b) krill (*Euphausia superba*), with $\eta = 0.73$ and 0.87, respectively. 1, waxes, hydrocarbons, sterol alcohols; 2, triglycerides; 3, *n*-carotene; 4, astatine; 6, fatty acids; 7, carotenoid (echinenone); 8, sterol (β-sitosterol); 9, sterol (cholesterol); 12, chlorophyll; 13, diglycerides; 14, carotenoid (myxoxanthophyll); 15, monoglycerides; 16, phosphatidylserine; 17, sphingomyelin; 18, phosphatidylinositol; 19, lecithin; 20, cephalin; 21, cardiolipin.

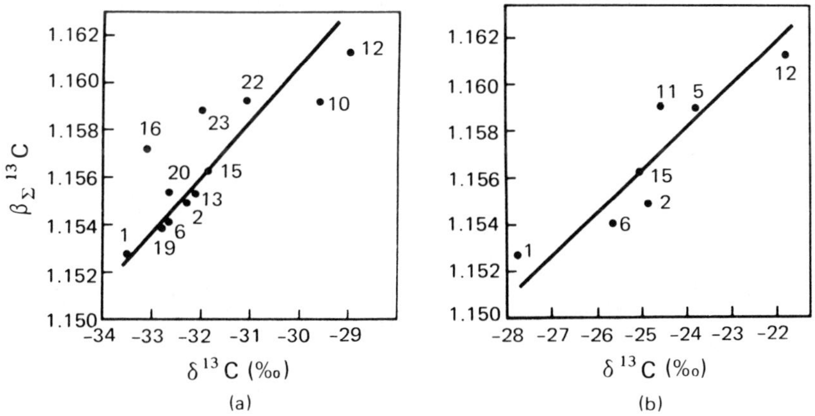

FIG. 4.4 $\beta_\Sigma{}^{13}C$–$\delta^{13}C$ correlation of the lipid fraction of (a) the lupine (*Lupinus luteus*) and (b) the laminarian (*Laminaria saccharina*) with $\eta = 0.82$ and 0.90, respectively. 1, waxes, hydrocarbons, sterol alcohols; 2, triglycerides; 5, carotenoid (fucoxanthine); 6, fatty acids; 10, sterol (β-sitosterol); 11, sterol (fucosterol); 12, chlorophyll; 13, diglycerides; 15, monoglycerides; 16, phosphatidylserine; 19, lecithin; 20, cephalin; 22, pigment (lutein); 23, monogalactosylglyceride.

investigated organisms, which, as previously mentioned, included representatives of both aquatic and terrestrial plants and animals, the distribution of carbon isotopes in the individual components of the lipid fraction and the relationship between the values of the β_Σ factors and of the $\delta^{13}C$ values are principally the same (see Fig. 4.3a–4.4b).

The isotopic compositions of the components of the lipid fractions of the different organisms investigated are characterized by somewhat different ranges of $\delta^{13}C$ values. This is a consequence of the differences in the isotopic composition of the carbon in the environment of the organisms studied. As has been mentioned, freshwater and terrestrial organisms are enriched in the ^{12}C isotope relative to marine organisms.

It should be noted that the data discussed pertain to a very narrow range of values of $\delta^{13}C$ and β_Σ factors, within which just the error of mass-spectrometer measurements (which is ± 0.2–$0.3\permil$), is capable of producing an appreciable share of the scatter of data relative to the regression line observed in the graphs (see Figs. 4.3a–b). A second problem is insufficiently precise identification of the component. Nevertheless, the relationship between β_Σ and $\delta^{13}C$ values is characterized by very high values of the correlation coefficients. This reconfirms the reliability of the additive method of calculating β_Σ factor values. Of course, the error of calculating β_Σ cannot exceed the scatter characterizing the relationship $\delta^{13}C$–$\beta_\Sigma{}^{13}C$. In the cases investigated, the spread, including all uncertainties, falls within ± 0.0008 on the $\beta_\Sigma{}^{13}C$ scale. Theoretical calculation of error also places it in the fourth decimal place of the β value. Because the components of the lipid fraction contain 20–30 carbon atoms, the error of determination of the β_Σ values would be approximately $\sqrt{n} \cong 5$ times less than the error of determination of the β_i factor. The β_i factor, according to the analysis given in the preceding chapter, is ± 0.003–0.005. Consequently, in agreement with experimental results, the error in calculating β_Σ factors involves the fourth decimal place. The error of calculating thermodynamic isotopic factors in this case obviously is comparable to the error of mass-spectrometric measurement of the isotopic composition.

On the other hand, because the scatter characterizing the relationship $\delta^{13}C$–$\beta_\Sigma{}^{13}C$ is caused, in part, by errors in calculations of $\delta^{13}C$ and $\beta_\Sigma{}^{13}C$, it is clear that the relationship between $\delta^{13}C$ and $\beta_\Sigma{}^{13}C$ is even more precise than would follow from the correlation analysis.

Above, we have examined the relationship $\delta^{13}C$–$\beta_\Sigma{}^{13}C$ using numerical values of the β factors at 27°C (300 K). Actually the organisms investigated developed in environments with different temperatures. If the true temperature of biosynthesis differs from 27°C, this affects the slope of the $\delta^{13}C$–$\beta^{13}C$ regression line, or the apparent value of the coefficient $\bar{\kappa}$. For organisms developing at temperatures lower than 27°C the coefficient κ is somewhat too high, and the respective regression line on the graph should be flatter; conversely, if biosynthesis takes place in an environment with a temperature above 27°C, the regression line will have a lower coefficient κ.

Using expressions (3.42) and (3.43) one can reduce the regression equation to one corresponding to the actual temperature. As an example, the relationship $\delta^{13}C$–$\beta^{13}C$ for the lipid components of *Laminaria* (see Fig. 4.4b) is characterized by the regression equation $\delta^{13}C = 0.55[\beta^{13}C_{(300\,K)} - 1.1567]10^3 - 24.7$. The *Laminaria* grew at $t = 5°C$. Reduction of the regression equation to that temperature yields $\delta^{13}C = 0.47[\beta^{13}C_{(278\,K)} - 1.1791] \times 10^3 - 24.7$. The apparent value of $\bar{\kappa}$ is lowered from 0.55 to 0.47.

IV. INTRAMOLECULAR DISTRIBUTION OF ISOTOPES IN ACETATE

Acetate is an important metabolite. The biosynthesis of the lipid components is accomplished chiefly via acetyl-CoA.

W. Meinschein *et al.* (1974) investigated the intramolecular distribution of carbon isotopes in acetate. They measured the isotopic composition of the carbon of the methyl and carboxyl groups of acetic acid isolated from apple cider. The acetic acid was the product of oxidation of ethanol by aerobic organisms. The alcohol was obtained as a result of catabolic breakdown of sugars in the course of glycolysis.

The isotopic composition of the carbon of the CH_3 group, specifically of methane given off upon pyrolysis of acetic acid, and the isotopic composition of acetic acid as a whole were measured. The isotopic composition of the carboxyl group was calculated from the difference in these $\delta^{13}C$ values. The results obtained are given in Table 4.9 (and in graphic form in Fig. 4.9). Also given are the values of the β factors of the isotopic forms substituted in the methyl and carboxyl groups.

TABLE 4.9

INTRAMOLECULAR DISTRIBUTION OF CARBON ISOTOPES IN BIOGENIC ACETIC ACID

	$\beta^{13}C$(‰)	$\beta^{13}C$
CH_3—	−34.9	1.1363
—COOH	−16.5	1.1972

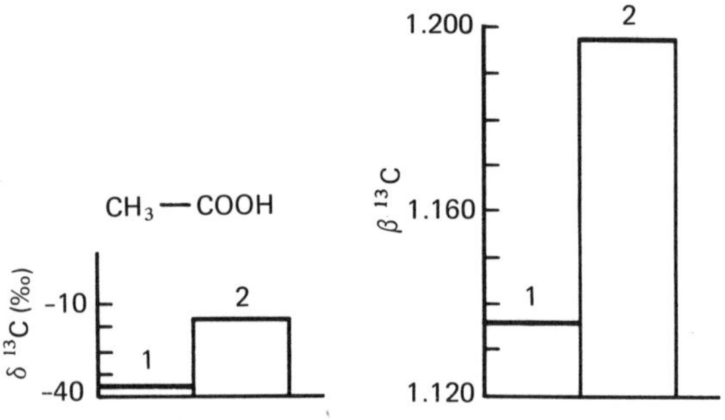

FIG. 4.5 Intramolecular isotopic shift in biogenic acetic acid: 1, methyl group; 2, carboxyl group.

The carboxyl group of acetic acid is enriched by 18.4‰ in the ^{13}C isotope relative to the methyl group. The value of the thermodynamic isotopic factor of the carboxyl group (isotopic form substituted in the carboxyl group) is larger than the value of the β_i factor of the methyl group. From the relationship

$$-16.5 = -34.9 + \kappa[(1.1972/1.1363) - 1] \times 10^3$$

the proportionality coefficient κ can be determined, which in this example turns out to be 0.34.

Thus within the acetic acid molecule the distribution of isotopes is similar to the distribution between the various compounds of the lipid fraction. The value of κ agrees with the average value of that quantity established for the intramolecular distribution of isotopes (between the carbons of the carboxyl and alkyl groups) in amino acids.

One of the interesting aspects of the investigation of the lipid fraction components, whose results have been discussed above, is the fact that their biosynthesis is accomplished chiefly from one original substrate—acetyl-CoA.

This clarifies several alternatives of intramolecular distribution of isotopes.

Figure 4.6a illustrates an example of the theoretically expected intramolecular distribution of isotopes in a constituent of the lipid fraction—the cholesterol molecule. The diameter of the circles corresponding to the carbon atoms is proportional to the calculated value of the β_i factor of the isotopic form substituted in that position, that is, to the probability of concentration of the ^{13}C in that position.

The difference between the intramolecular isotopic inhomogeneity which is thermodynamic and the possible intramolecular inhomogeneity caused

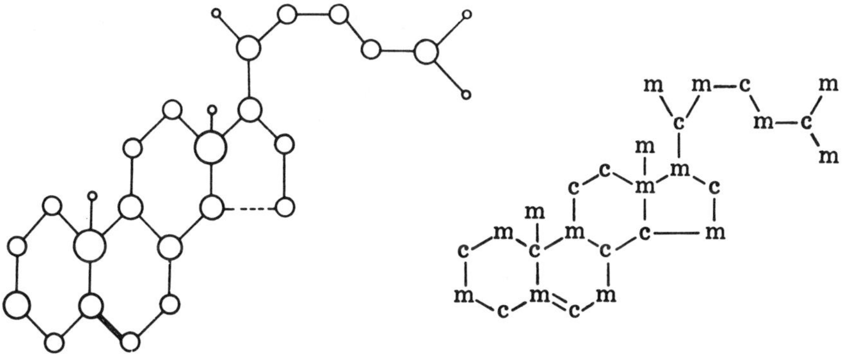

(a) (b)

FIG. 4.6 (a) Relative values of $\beta_i{}^{13}C$ factors of the cholesterol molecule and (b) distribution of carbon of methyl (m) and carboxyl (c) origin. The diameter of the circles is proportional to the value of $\beta_i{}^{13}C$ and in the case of thermodynamically ordered distribution of isotopes corresponds to relative enrichment of isotopes in ^{13}C of the carbon in a given structural position.

by a difference in isotopic composition of the predecessors of various atoms of carbon in the biomolecule should be stressed. In the first case, the isotopic composition of the carbon in the molecule is dependent on its chemical surroundings and is determined by the value of the β_i factor. The isotopic composition of the molecule, as a whole, is determined by the value of the β_Σ factor. In the second case, the intramolecular isotope inhomogeneity is purely inherited.

Thanks to the extensive investigation of biosynthetic pathways using ^{14}C as a tracer, the specific positions occupied by the carbon atoms derived from carboxyl and methyl groups of acetate have been established for many

TABLE 4.10

COMPARISON OF THE CARBON ISOTOPE COMPOSITION OF SOME INVESTIGATED COMPONENTS OF BLUE-GREEN ALGAE.

Compound	β_Σ *Factor*	$\delta^{13}C$ (‰)	m/c
Palmitic acid	1.1540	−31.9	1.00
β-Carotene	1.1567	−31.5	1.50
β-Sitosterol	1.1590	−29.5	1.25
Myxoxanthophyll	1.1607	−29.4	1.50

compounds. An example of such a distribution in the cholesterol molecule is given in Fig. 4.6b. Whereas the isotopic composition of the carbon atoms is determined by their origin, the isotopic composition of the compound as a whole should be determined by the m/c ratio, the ratio of the number of isotopically light carbon atoms inherited from the methyl group to the number of isotopically heavy atoms inherited from the carboxyl group of acetate.

In Table 4.10 we give the m/c of compounds whose carbon atoms come only from acetate. In the fatty acids, atoms of m and c origin alternate (Mahler and Cordes, 1968; Sprinson and Coulon, 1954). An analogous alternating distribution of m- and c-type atoms occurs in squalene and in the carotenoids originating from it. The carbon of the CH_3 branches in carotenoids is of methyl-acetate origin (Grob and Butler, 1956; Staple, 1963). The distribution of atoms of m and c type in β-sitosterol is analogous to the distribution in cholesterol (Clayton, 1965; Staple, 1963). Not all the lipids investigated have acetate as the only low-molecular predecessor. In triglycerides and phosphoglycerides, the glycerine residue is derived from α-glycerophosphate, formed as an intermediate product of glycolysis (Buell and Reiser, 1959). In sphingolipids, the first and second atoms of sphingosine are formed from the second and third carbon atoms of serine (Sprinson and Coulon, 1951). In chlorophyll, eight carbon atoms of the porphyrin nucleus come from the α carbon of glycine, the rest from acetate (Shemin and Wittenberg, 1951).

In chlorophyll, methyl atoms predominate appreciably (m/c = 1.62) if the α carbon atoms of glycine are identified with the methyl atoms of the acetate. However, chlorophyll is one of the isotopically heaviest components of the lipid fraction.

In order to avoid additional assumptions and stipulations, in Table 4.10 only those compounds for which the m/c value was determined by the ratio of methyl and carboxyl atoms of acetate were taken. These values are compared with the isotopic composition of the respective compounds in the specific organism investigated. Although the fatty acids should have been heavier than the carotenoids, they are isotopically the lightest compounds. Comparison of two carotenoids isolated from the blue–green alga, β-carotene, and myxoxanthophyll, is informative. Both components have similar carbon skeletons in which atoms derived from the methyl and carboxyl groups of acetate alternate, m − c − m − c.... The difference is that in myxoxanthophyll, the hydroxyl groups are attached to the 12th–15th, 17th, and 18th carbon atoms, and the 39th atom forms the carbonyl group. The presence of the C—O bonds, characterized by high values of the isotopic numbers ($L_{C-O} = 0.0550$, $L_{C=O} = 0.0958$) should lead to an increase in the β_i factors of the corresponding carbon atoms, and hence in the β_Σ factor as a whole. Calculation yields a value of 1.1607 for the β_Σ factor of myxoxanthophyll, and 1.1567 for the β_Σ factor of β-carotene (see Table 4.10). The results of investigation of the isotopic composition of these components shows that myxoxanthophyll ($\delta^{13}C = -29.4‰$) actually is isotopically heavier than β-carotene ($\delta^{13}C = -31.5‰$). At the same time, the m/c ratio in both compounds is the same, therefore, from this point of view they should have the same isotopic composition.

From what has been said, it follows that the isotopic composition of biomolecules in the lipid fraction and the intramolecular distribution of isotopes in them depend less on the isotopic composition of the precursors and the manner of their incorporation in a given compound than on the fractionation processes occurring during biosynthesis, which lead to distribution of isotopes in accordance with the thermodynamic isotopic factors of the product.

V. INTRAMOLECULAR DISTRIBUTION OF ISOTOPES IN AROMATIC MONOMERS OF LIGNIN

It has been shown that the carboxyl group in the biomolecules investigated, as a rule, is enriched in the heavy carbon isotope. This corresponds to the high value of the β_i factor of carboxyl carbon, which is generally higher than for total organic carbon averaged.

The aromatic monomers of lignin contain the methoxyl group which like the carboxyl contains oxygen, but it is characterized by a substantially lower value of the β_i factor, less than the average for the carbon skeleton of the biomolecule.

It is of interest to measure the intramolecular isotopic shift for methoxyl carbon. Such investigations were undertaken by Galimov *et al.* (1976) and Vinogradov *et al.* (1976). In Galimov *et al.* (1976), three biogenetically and structurally related aromatic aldehydes—*p*-hydroxybenzaldehyde (I), 4-hydroxybenzaldehyde (II) (vanillin), and 4-hydroxy-3,5-dimethoxybenzalde hyde (III) (syringaldehyde)—were analyzed.

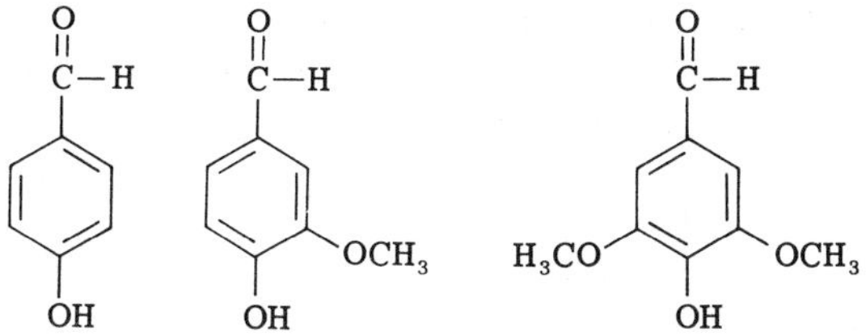

I II III

The aldehydes are products of alkali-nitrobenzene oxidation of phenylpropane structures of type Ia, IIa, IIIa, which are monomeric units of lignin (Manskaya and Kodina, 1975) and some other natural compounds (flavonoids, hydroxycinnamic acid, etc.):

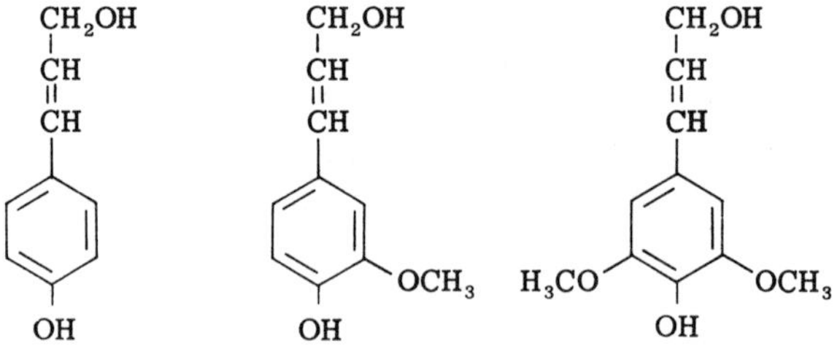

Ia IIa IIIa

To isolate the aromatic aldehydes, degreased biomass of stems of the wild Central Asian reed *Phragmites communis* and winter wheat of the "Mironovskaya 808" variety were used. Oxidation of the plant biomass and isolation of the aromatic aldehydes were carried out by the Hibbert method. The mixture of aldehydes was separated by dry distillation at low pressure (10–12 mm) with subsequent repeated recrystallization of the preparates and use of preparatory thin-layer chromatography. The details of the analytical and preparatory procedures are discussed in Galimov *et al.* (1976). The preparates obtained for the individual aldehydes contained not more than 5% of contaminating substances.

The classical method of splitting simple ether bonds using hydriodic acid was employed to separate the carbon of the methyl group. At the boiling temperature of hydriodic acid, the carbon of the methoxyl group is quantitatively given off in the form of methyl iodide, the carbon of which is then oxidized to CO_2 and analyzed in a mass spectrometer.

Table 4.11 gives data on the isotopic composition of both the total carbon of *p*-hydroxybenzaldehyde, vanillin, and syringaldehyde from the reed and of the carbon of the methoxyl group of vanillin from both plants. The carbon of the methoxyl group separated from the vanillin molecule was isotopically lighter in both plants than the total carbon of the aromatic ring. Thus, here too a correspondence of the isotopic distribution is manifested with respect to the thermodynamic isotopic factors (Fig. 4.7a).

In addition to this, two other aromatic compounds isolated from plants were investigated: transanethole and isoeugenol (Vinogradov *et al.*, 1976). The isotopic shift between the carbon from the methoxyl group and from the rest of the molecule had the same sign and approximately the same value as vanillin (Fig. 4.7b and c).

TABLE 4.11

CARBON ISOTOPES IN AROMATIC MONOMERS OF LIGNIN

Sample	$\delta^{13}C$ (‰) Reed	Wheat
p-Oxybenzaldehyde	−29.5	—
4-Oxy-3-methoxybenzaldehyde (vanillin)	−29.9	−30.2
4-Oxy-3,5-dimethoxybenzaldehyde (syringaldehyde)	−30.6	−31.3
OCH_3 group of 4-oxy-3-methoxybenzaldehyde (vanillin)	−32.7	−32.4

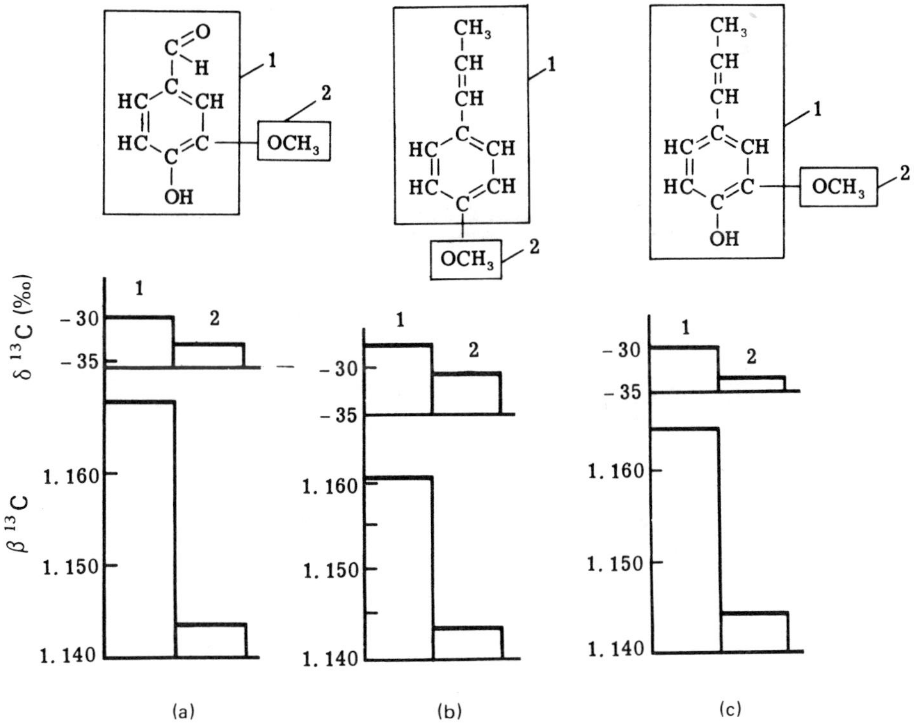

FIG. 4.7 Intramolecular isotopic shift and its correlation with the ratio of $\beta^{13}C$ values in (a) biogenic vanillin, (b) biogenic anethole, and (c) biogenic eugenol.

The coefficient of reduction of the isotopic effect κ was approximately the same in all cases and equal to 0.12 for the vanillin from the reed, 0.19 for the transanethole, and 0.17 for the isoeugenol.

VI. ISOTOPIC COMPOSITION OF THE CARBON OF PHYTOL AND THE PORPHYRIN SERIES OF CHLOROPHYLL

The chlorophyll molecule is an excellent object for intramolecular isotopic analysis. It contains carbon atoms with appreciably different β-factor values. Therefore perceptible isotopic shifts between carbons in different structural positions can be expected. Chlorophyll is an important biological compound. It also has considerable geochemical importance. Its phytol is the predecessor of isoprenoid hydrocarbons, and the tetrapyrrole nucleus is the predecessor of fossil porphyrins.

In our laboratory the isotopic composition of carbon in two fragments of the chlorophyll molecule, the phytol chain and the porphyrin nucleus, were investigated (Bogacheva and Galimov, 1979; Galimov *et al.*, 1977).

The initial chlorophyll $a + b$ for the experiment was isolated from nettle. The phytol was isolated by saponifying the chlorophyll with alkali in methanol and extracting the reaction mixture with ether. After distillation of the ether in a vacuum, the phytol was purified using distillation at a pressure of 10 Torr and temperatures of 202.5 to 204.0°C. The distillation procedure was used twice. The porphyrin residue was extracted from the alkaline methanol solution after its pH was reduced to 4.5. The suspension that formed was extracted with ether. The residue after evaporation of the ether was the porphyrin nucleus of the chlorophyll. The purity of the preparates was monitored using ultraviolet spectroscopy. It should be mentioned that the

TABLE 4.12

ISOTOPIC COMPOSITION OF CARBON OF CHLOROPHYLL AND OF FRAGMENTS OF THE CHLOROPHYLL MOLECULE

		$\delta^{13}C$ (‰)	
	$\beta^{13}C$	*Experiment I*	*Experiment II*
Chlorophyll $a + b$	1.1617	−32.50	−31.45
Phytol	1.1525	−33.00	−33.88
Porphyrin nucleus	1.1669	−31.40	−28.77

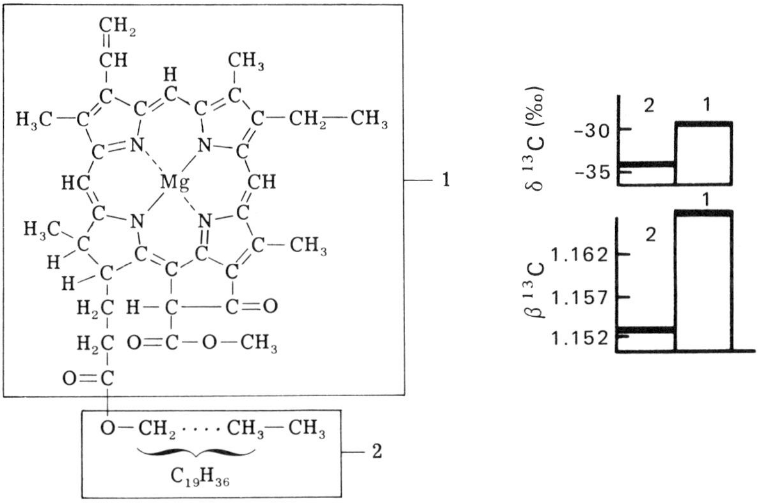

FIG. 4.8 Intramolecular isotopic shift in chlorophyll: 1, porphyrin nucleus; 2, phytol (experiment II).

phytol purification procedure by distillation is an important condition for obtaining reproducible results of isotopic analysis. Table 4.12 gives the results of two independent experiments.

The carbon of the phytol chain proved to be isotopically lighter than the porphyrin nucleus, which corresponds to the relationship of $\beta^{13}C$ values characterizing these fragments (Figs. 4.8).

VII. ISOTOPIC COMPOSITION OF FRAGMENTS OF THE HEMIN MOLECULE

Study of the isotopic composition of fragments of the hemin molecule was essentially a continuation of the investigation of the intramolecular distribution of isotopes in chlorophyll (Table 4.13). Hemin was used as a convenient initial compound for studying the isotopic distribution within the porphyrin nucleus (Bogacheva and Galimov, 1979).

The fragmentation of the porphyrin nucleus was obtained by splitting mesoporphyrin IX. It was first prepared from hemin isolated from fresh defibrinated beef blood. The hemin was reduced to mesohemin by hydrogenation in 90% formic acid using a palladium oxide catalyst. The mesohemin was converted by treatment with disodium phosphate in an ammonium hydroxide solution into disodium mesoporphyrin which, in turn,

TABLE 4.13

ISOTOPIC COMPOSITION OF THE CARBON OF HEMIN AND FRAGMENTS OF ITS PORPHYRIN NUCLEUS

Investigated compound (fragment)	*Analyzed reaction product*	$\beta^{13}C$ *of fragment*	$\delta^{13}C$ (‰)	
			Experiment I	*Experiment II*
Hemin	Hemin	1.1654	—	−25.74
Mesoporphyrin IX	Mesoporphyrin IX		−26.58	−26.18
Pyrrole rings *A* and *B*	Methylethyl maleinimide	1.1626	−25.99	−25.54
Pyrrole rings *C* and *D*	Hematinic acid	1.1691	−25.08	−24.37
CH group of methine bridges	CO_2	1.1592	—	−32.52

was converted into the dimethyl ester of mesoporphyrin by saturating the chloroform–methanol solution of the disodium salt with dry hydrogen chloride gas. After purifying on Al_2O_3 columns and recrystallizing it, the dimethyl ester of mesoporhyrin was converted by hydrolysis to free mesoporphyrin IX. The mesoporphyrin inherited all the carbon atoms of the hemin. However, it proved to be more enriched in the ^{12}C isotope than the hemin. This was related either to a kinetic isotope effect in the process of hydrogenation of the hemin and methylation of the mesoporphyrin disodium salt, or to residual impurities of carbon-containing reagents used in the procedure described for reduction of the hemin.

Decomposition of the mesoporphyrin was accomplished using chromic acid in a reaction vessel adapted for catching the carbon dioxide produced. The reaction was carried out at room temperature. As a result of the reaction, three products were formed: carbon dioxide—the product of oxidation of carbon of the methine bridges; methylethyl maleimide—the product of the splitting off of the *A* and *B* rings (Fig. 4.9); and hematinic acid—the product of the splitting off of the *C* and *D* rings. The carbon dioxide was trapped by barium hydroxide. The methylethyl maleimide and hematinic acid were isolated from the reaction mixture by extraction with ethyl acetate and dissolved in boiling water. This water solution was cooled to 0°C and neutralized to pH 7.0 by adding $NaHCO_3$. The methylethyl maleimide was extracted from this neutral solution by chloroform. The residual solution was acidified to a pH of 5 and hematinic acid was ex-

TABLE 4.14

ISOTOPIC COMPOSITION OF THE CARBON OF CO_2 DERIVED FROM OXIDATION OF METHINE BRIDGES OF MESOPORPHYRIN FOR DIFFERENT YIELDS OF THE PRODUCT

Yield of product (%)	$\delta^{13}C$ (‰)
50	−35.22
60	−36.21
90	−32.52
97	−32.92

tracted from it with ethyl acetate. The preparates were purified by recrystallization.

We have discussed the details of the chemical procedure because it is rather complicated. At some stages a quantitative output is not guaranteed, therefore there is a danger of introducing distortions into the original isotopic distribution.

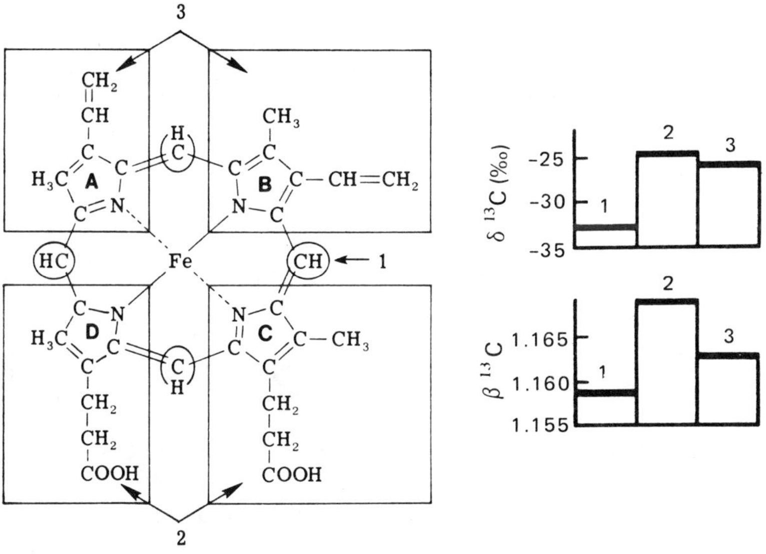

FIG. 4.9 Intramolecular isotopic shifts in the hemin molecule; 1, methine bridges; 2, fragments transformed into hematinic acid; 3, fragments transformed into methylethyl maleimide (experiment II).

For nonquantitative output, the carbon isotope composition of the methine bridges, for instance, may be appreciably changed by kinetic isotopic effects, as is seen from Table 4.14, where the results of analysis for various outputs of the product to be analyzed are given.

From Fig. 4.9, it is seen that the isotopic compositions of the investigated fragments are related to each other as the values of their β factors are. Although this is fairly obvious, let us note that the isotopic composition of a fragment of a biomolecule should be compared with the value of its β factor in the original compound. If no isotopic effects accompany a reaction, the products of the fragmentation reaction inherit the isotopic composition of the corresponding fragment, but the values of their β factors may be substantially altered.

VIII. INTRAMOLECULAR ISOTOPIC EFFECTS IN OTHER BIOGENIC COMPOUNDS

Above, we discussed the results of the experiments which we made to check the hypothesis of thermodynamically ordered distribution of isotopes in biosystems. In the literature there are data which also can be interpreted to support this view. Some of them have been discussed. Several other examples merit mention.

Rinaldi, Meinschein, and Hayes (1974)† investigated the intramolecular distribution of isotopes in the molecule of biogenic acetoin. Their data are compared with the calculated values of the β factors in Fig. 4.10. Agreement is observed between the measured isotopic effects and the thermodynamic factors.

In Abelson and Hoering (1961) arginine isolated from chlorella was enzymatically split into ornithine and urea:

$$\underset{\text{arginine}}{NH_2NHCNH(CH_2)CHNH_2COOH} + H_2O \longrightarrow \underset{\text{urea}}{(NH_2)_2CO} + \underset{\text{ornithine}}{NH_2(CH_2)CHNH_2COOH}$$

In this reaction the urea formed at the expense of the guanidine group and the ornithine at the expense of the rest of the molecule. Therefore the isotopic composition of the reaction products should reflect the intramolecular distribution of isotopes in the arginine if the arginine is completely

† I thank Prof. J. Hayes for making it possible for me to become acquainted with this work (1974) before its publication.

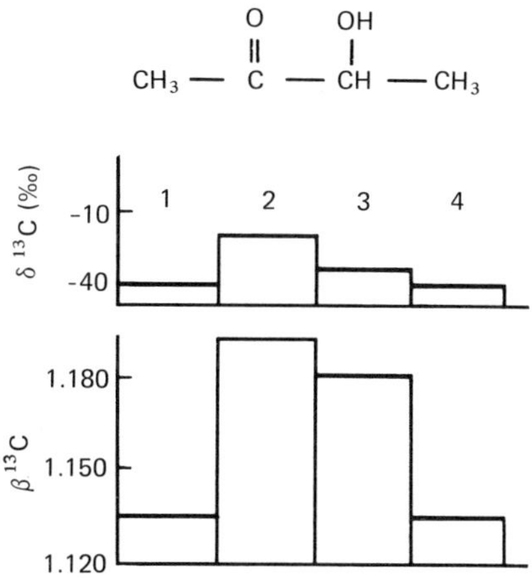

FIG. 4.10 Intramolecular isotopic shift in the acetoin molecule.

converted. Unfortunately, in Abelson and Hoering (1961) there is no discussion of the conditions under which the reaction took place. Nor is the isotopic composition of the original arginine indicated. If there was little conversion of the arginine, then the isotopic composition of the urea should correlate with the value of the β factor of urea, and if conversion is complete, with the $\beta_i{}^{13}C$ value of guanidine carbon. In this experiment the urea carbon had $\delta^{13}C = +7.1‰$ and the ornithine carbon had $\delta^{13}C = -15.2‰$. The value of the β factor of urea is 1.1958, of guanidine carbon 1.1802, and of ornithine, 1.1675. Despite the mentioned uncertainty, it can be seen that the observed enrichment of urea in the ^{13}C isotope relative to ornithine agrees with the ratio of the corresponding β factors.

Di Marco *et al.* (1977) investigated the isotopic composition of the carbon of malic and tartaric acids isolated from grape juice. The chemical composition of tartaric acid differs from that of malic only by the presence of an additional OH group. Correspondingly, its β factor is somewhat higher. It proved to be enriched in the ^{13}C isotope by 0.5‰ (Table 4.15).

From the experimental data, it can be concluded that a connection between the isotopic composition of the components and the values of the thermodynamic isotope factors is generally typical of many of the biosystems examined. It is characteristic of compounds of different structures and of components belonging to different biochemical fractions; it is observed in organisms of different ecological and taxonomic appurtenance, and most importantly—it is manifested both on the intermolecular and on the intramolecular level.

TABLE 4.15

CORRELATION OF $\delta^{13}C$ AND $\beta^{13}C$ VALUES OF MALIC AND TARTARIC ACIDS

Acid	*Structural formula*	$\delta^{13}C$(‰)	$\beta^{13}C$
Malic	COOH—CH_2—CH(OH)—COOH	−23.5	1.1844
Tartaric	COOH—CH(OH)—CH(OH)—COOH	−23.0	1.1918

It must be emphasized that cases may be encountered where the relationship between the observed isotopic effects and values of the β factors is absent. Even in the fairly good correlations, for instance, between the lipid components, one can always pick out a pair of components whose isotopic compositions do not agree with the β values. In other words, thermodynamically ordered distribution of isotopes is a statistical regularity most clearly manifested when a large number of cases or components is considered. The high correlation coefficients indicate the significance of the relationship, i.e., the existence in biological systems of a mechanism producing the corresponding distribution of isotopes. But in individual cases the relationship $\delta^{13}C$–β is disrupted. This means that either the regularity requires certain conditions, or there is a superposition of different isotopic effects, kinetic, diffusional, etc.

It must also be mentioned that the correlation appears to a different extent in different biosystems. For the components of the lipid fraction of the several organisms investigated, the $\delta^{13}C$–β relationship is very close. For amino acids it is less marked.

An essential property of the observed distribution of isotopes is that the isotopic effects are nearly always less than equilibrium values, i.e., values determined directly by the ratio of β factors. The isotopic effect is reduced by half or more, i.e., the coefficient κ is less than 0.5 in most cases.

In the following chapter, we attempt to consider the possible mechanism of fractionation of isotopes in biosystems caused by thermodynamically ordered distribution of isotopes and explain the observed properties of this distribution.

REFERENCES

Abelson, P. H., and Hoering, T. C. (1961). Carbon isotope fractionation in formation of amino acids by photosynthetic organisms. *Proc. Nat. Acad. Sci. USA* **47**(5), 632.

Bogacheva, M. P., and Galimov, E. M. (1979). Intramolecular distribution of carbon isotopes in chlorophyll and hemin. *Geokhimiya* (7), 1166–1172.

Buell, G. C., and Reiser, R. (1959). Glyceride–glycerol precursors in the intestinal mucosa. *J. Biol. Chem.* **234**(2), 217–219.

Clayton, R. B. (1965). Biosynthesis of sterols, steroids, and terpenoids. *Q. Rev.* **19**, 168–230.

Di Marco, G., Grego, S., Tricoli, D., and Turi, B. (1977). Carbon isotope ratios $^{13}C/^{12}C$ in fractions of field-grown grape. *Physiol. Plant* **41**(2), 139–141.

Galimov, E. M. (1972). Method of isotopic bond numbers. Application in biochemistry and geochemistry. *In* "Abstracts of Papers at the IV All-Union Symposium on the Use of Stable Isotopes in Geochemistry." Moscow.

Galimov, E. M. (1973a). Biogenic intermolecular and intramolecular carbon isotope effects: Method of isotopic number of bindings. Biochemical and geochemical applications. *In* "Intern. Meeting on Isotope Effects in Physical and Chemical Processes." Cluj, Romania.

Galimov, E. M. (1973b). "Carbon Isotopes in Oil and Gas Geology." Nedra, Moscow. English translation: NASA TT-682, Washington, D.C., 1975, 395 pp.

Galimov, E. M. (1973c). Organic geochemistry of carbon isotopes. *In* "Advances in Organic Geochemistry," pp. 439–452. Technip, Paris.

Galimov, E. M., and Shirinskiy, U. G. (1975). Ordered distribution of carbon isotopes in individual compounds and components of the lipid fraction of organisms. *Geokhimiya* (4), 157–180.

Galimov, E. M., Kodina, L. A., and Generalova, V. N. (1976). Experimental investigation of intra- and intermolecular isotopic effects in aromatic compounds of biological Nature. *Geokhimiya* (1), 9–13.

Galimov, E. M. *et al.* (1977). Investigation of the distribution of carbon isotopes in biogenic compounds. *In* "Proc. VIII Internat. Cong. on Organic Geochemistry." Nauka, Moscow.

Grob, E. C., and Butler, R. (1956). On the biosynthesis of the β-carotene in *Mucor hiemalis* Wehmer. *Helv. Chim. Acta* **39**, 439–452.

Mahler, H., and Cordes, E. (1968). "Basic Biological Chemistry." Harper and Row, New York.

Manskaya, S. M., and Kodina, L. A. (1975). "Geochemistry of Lignin." Nauka, Moscow.

Meinschein, W. G., Rinaldi, G. G. L., Hayes, J. M., and Schoeller, D. A. (1974). Intramolecular isotopic order in biologically produced acetic acid. *Biomed. Mass Spectrometry* **1**(3), 172–174.

Rinaldi, G. G., Meinschein, W. G., and Hayes, J. M. (1974). Intramolecular carbon isotopic distribution in biologically produced acetoin. *Biomed. Mass Spectrometry* **1**(6), 415–417.

Shemin, D., and Wittenberg, J. (1951). The mechanism of porphyrin formation. *J. Biol. Chem.* **192**, 315.

Sprinson, D. B., and Coulon, A. (1954). The precursors of sphingosine in brain tissue. *J. Biol. Chem.* **207**(2), 585–592.

Staple, E. (1963). Biosynthesis of steroids. *In* "Biogenesis of Natural Compounds" (P. Bernfeld, ed.), ch. 4. Pergamon, New York.

Strickland, K. P. (1963). Biogenesis of lipids. *In* "Biogenesis of Natural Compounds" (P. Bernfeld, ed.), ch. 3. Pergamon, New York.

Vinogradov, A. P., Galimov, E. M., Kodina, L. A., and Generalova, V. N. (1976). Intramolecular distribution of carbon isotopes as a criterion of the biological origin of organic compounds. *Geokhimiya* (7), 1068–1074.

CHAPTER 5

Theoretical Model of Biological Fractionation of Isotopes

I. MECHANISM OF PRODUCTION OF THE THERMODYNAMIC ISOTOPIC EFFECT IN ENZYMATIC CATALYSIS

At first the idea of thermodynamically ordered distribution of isotopes in biological systems seems equivalent to the idea of complete chemical equilibrium of biomolecules and thus not applicable, in principle, to living systems. Such doubts have sometimes been raised in discussing the thermodynamic concept.

It seems that some of the difficulties in accepting the idea of thermodynamically ordered distribution of isotopes stems, first, from underestimating the importance of the enzymatic nature of biochemical reactions and, second, from overemphasizing the idea that living systems are not in equilibrium. That idea concerns the organization of biological systems rather than the content of elementary chemical events; the latter may be in perfect equilibrium precisely because the process, as a whole, is highly organized.

To explain the nature of thermodynamically ordered distribution of isotopes in biomolecules, it is necessary to consider the consequences of the essential feature of the chemistry of living systems, the fact that in living systems (in contrast to nonliving systems) all chemical reactions occur with the participation of enzymes. A specific interpretation, corresponding to the observed isotope fractionation, has to be given.

An enzymatic reaction is usually written as

$$\mathrm{E + S} \rightleftharpoons \mathrm{ES} \longrightarrow \mathrm{E + P}, \tag{5.1}$$

where E is the free enzyme, S the substrate, P the product, and ES the enzyme–substrate complex.

The beginning of the reaction is the formation of the enzyme–substrate complex. This stage includes orientation of the substrate, inductional conformation of the enzyme, adaptation of the active center, and other actions preceding the actual chemical reorganization of the substrate. We do not link any substantial isotopic effect to this stage.

The second process occurs in the active center of the enzyme-substrate complex. This stage results in the chemical reorganization of the substrate. Usually there is no need to give the details of this stage, as the presence or absence of any intermediate states does not alter the final chemical process, which is conversion of substrate S into product P. However, to understand the mechanism of biological fractionation of isotopes, it is essential to discuss the details of this stage.

We shall show that the isotope distribution observed in biosystems can be explained and quantitatively described if the idea of microscopic reversibility of substrate conversion in the enzymatic complex is introduced. As a whole, however, the enzymatic reaction may be irreversible.

As is known, the formation of product P is preceded by conversion of the enzyme–substrate complex into the enzyme–product complex EP. We assume that this conversion is reversible. In fact, if the lifetime of the enzyme–substrate complex is long relative to the reorganization of the bonds, then it is reasonable to expect that after conversion of substrate to product the latter may convert back and that this process may be repeated many times during the lifetime of the substrate–enzyme complex.

The appeal of that interpretation is that the idea of reversibility refers to the process of conversion of one molecule (or when several molecules are reacting together, it refers to the elementary event). In an ordinary chemical reaction the conversion of one molecule is always irreversible. This is because the energy released in reorganization of the bonds (formation of the product) is dissipated in the environment. It is believed that the transitional complex either decomposes to the original state or is irreversibly transformed into product $A \rightleftarrows [A]\# \rightarrow B$. When we say that a reaction as a whole is reversible ($A \rightleftarrows B$), we mean that at another time, in another place of the system, another molecule of type B may be converted into A ($B \rightleftarrows [B]\# \rightarrow A$); the reversibility of the reaction is macroscopic. Microscopic reversibility of the conversion of a molecule of the substrate in an enzyme complex presumes that the energy released in the formation of the product is not dissipated into the surrounding environment but may be immediately realized again with transformation of the product back into the initial state; the enzyme behaves like a spring. A reaction of that type should be written in the form $A \rightleftarrows [A \rightleftarrows B]\# \rightarrow B$.

The corresponding scheme of an enzymatic reaction can be depicted in a better-developed form, in the accepted notation, as

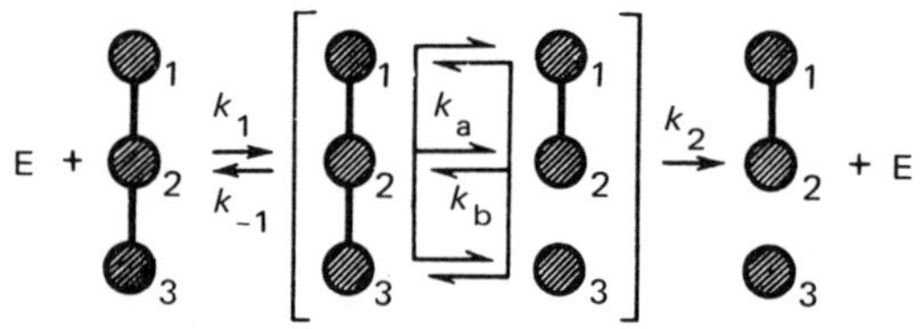

$$E + S_0 \rightleftharpoons E\left(S \rightleftharpoons P\right) \longrightarrow P + E. \qquad (5.2)$$

The process of conversion of the transformed compound from state S to state P, and vice versa, is interrupted at any moment by decomposition of the complex into either free enzyme and substrate (from the ES state) or free enzyme and product (from the EP state). We consider the last stage to be chiefly irreversible.

Owing to the steric specificity of an enzyme, carbon in position (1) of the product will always originate from carbon in position (1) of the substrate, carbon of $P_{(2)}$, from carbon of $S_{(2)}$, etc. In a single enzyme–substrate complex the carbon in the *i*th position in a molecule of the substrate can be either in the form of the ^{13}C isotope or the ^{12}C isotope. In the first case

$$S_i(^{13}C) \rightleftharpoons P_i(^{13}C), \qquad (5.3)$$

and in the second

$$S_i(^{12}C) \rightleftharpoons P_i(^{12}C). \qquad (5.4)$$

Let β_i be the factor of the isotopic form replaced by the *i*th atom of carbon, which is larger in the P state than in the S state ($\beta_{P(i)} > \beta_{S(i)}$). Applied to a single molecule, this means that the probability of the occurrence of the isotopic form $P_i(^{13}C)$ relative to $P_i(^{12}C)$ is greater than that of the isotopic form $S_i(^{13}C)$ relative to $S_i(^{12}C)$. The ratio of these relative probabilities by definition is the ratio of β-factor values (β_{P_i}/β_{S_i}). From this it follows that the moment of dissociation of the enzyme–substrate complex will find EP in the $P_i(^{13}C)$ state relatively more often than ES in the $S_i(^{13}C)$ state. For this probability to be realized, the number of conversions $S \rightleftharpoons P$ during the lifetime of the enzyme–substrate complex should be rather large. In fulfilling this condition the product will be enriched in the ^{13}C isotope. If the β_{P_i}/β_{S_i} ratio is less than 1, the distribution of S_i and P_i isotopic compositions is the opposite. On the whole it is determined by the relationship

$$\delta^{13}C_{P_i} - \delta^{13}C_{S_i} = [(\beta_{P_i}/\beta_{S_i}) - 1] \times 10^3\,‰. \qquad (5.5)$$

Summing up processes (5.3) and (5.4) occurring in a set of single enzyme–substrate complexes for all such complexes present in the cells, i.e., changing to macroscopic consideration, we obtain a formal relationship fully analogous to the usual equation of isotope exchange:

$$S_i(^{13}C) + P_i(^{12}C) \rightleftharpoons S_i(^{12}C) + P_i(^{13}C). \tag{5.6}$$

Therefore, if it is assumed that the reversible reaction $S \rightleftarrows P$ takes place in the active center of the enzyme, there is fractionation of isotopes between the corresponding structural positions S_i and P_i which corresponds to the thermodynamic isotopic effect. In this case the latter is not the result of isotopic exchange in the literal sense. Within a single enzyme–substrate complex, S_i and P_i are two forms of existence of the same carbon (or other element). Because they exist alternately in time, there cannot be an exchange of substance between them (including isotopic exchange). The process in a single enzyme–substrate complex is selection of isotopic forms by energy. Here, as in isotopic exchange, the process proceeds toward a ratio of isotopic forms that corresponds to the minimum free energy of the system. Expression (5.5) determines the amount of the thermodynamic isotopic effect in the system $S_i \rightleftarrows P_i$. Taking into account what has been said and remembering the statistical sense of the thermodynamic isotopic effect, we can understand that effect to be the sum total of similar enzyme–substrate complexes in which a process of $S_i \rightleftarrows P_i$ type is accomplished.

Thus, we have the reaction

$$E + S \underset{k_{-1}}{\overset{k_1}{\rightleftharpoons}} \boxed{ES \underset{k_b}{\overset{k_a}{\rightleftharpoons}} EP} \xrightarrow{k_2} P + E, \tag{5.7}$$

where k_a and k_b are constants of the rate of conversion of the enzyme–substrate complex from the S to the P state and vice versa.

Let us focus our attention on the process occurring in the enzyme–substrate complex and examine the case where the concentration of free enzyme E is low and the rates k_a and k_b are greater than k_1, k_{-1}, and k_2. Then

$$[ES] + [EP] = 1 \tag{5.8}$$

and

$$k_a[ES] = k_b[EP]. \tag{5.9}$$

Therefore,

$$k_a[ES] = k_b - k_b[ES] \tag{5.10}$$

and

$$[ES] = k_b/(k_a + k_b). \tag{5.11}$$

Although the substrate (S) liberated in the decomposition of the enzyme–substrate complex ES [left-hand side of Eq. (5.7)] is chemically fully analogous to the original substrate S_0, they are not isotopically identical. As a result of isotopic fractionation in the system $S \rightleftarrows P$, the liberated substrate differs from the original in isotopic composition. Therefore it will be more precise to write Eq. (5.7) in the form

$$\begin{array}{ccccc} S_0 + E & \rightleftharpoons & ES_0 & & \\ & & \downarrow & & \\ S + E & \longleftarrow & E(S \rightleftarrows P) & \longrightarrow & E + P. \end{array} \qquad (5.12)$$

Because the carbon of both the product and the liberated substrate originate from the carbon of the original substrate, there should always be the isotopic balance

$$\delta^{13}C_{S_{0_i}} = \delta^{13}C_{S_i}[ES] + \delta^{13}C_{P_i}(1 - [ES]). \qquad (5.13)$$

The isotopic composition of the substrate in complex with the enzyme (i.e., the value of $\delta^{13}C_{S_i}$), cannot be measured. To obtain the relationship between the measured values, one should exclude $\delta^{13}C_{S_i}$ from expression (5.5), using (5.13):

$$\delta^{13}C_{P_i} - \frac{1}{[ES]}[\delta^{13}C_{S_{0_i}} - (1 - [ES])\delta^{13}C_{P_i}] = \left(\frac{\beta_{P_i}}{\beta_{S_i}} - 1\right) \times 10^3 ‰. \qquad (5.14)$$

From this we obtain

$$\delta^{13}C_{P_i} = \delta^{13}C_{S_{0_i}} + [ES]\left(\frac{\beta_{P_i}}{\beta_{S_i}} - 1\right) \times 10^3 ‰. \qquad (5.15)$$

The value [ES] before the bracket in Eq. (5.14), which contains the ratio of the thermodynamic isotopic factors, can be considered a reduced multiplier analogous to the coefficient κ introduced previously. Let us designate it as κ. The value of κ, in correspondence to (5.11), is determined by the ratio of the rate constants and is always less than unity.

Thus

$$\delta^{13}C_{P_i} = \delta^{13}C_{S_{0_i}} + \kappa\left(\frac{\beta_{P_i}}{\beta_{S_i}} - 1\right) \times 10^3 ‰. \qquad (5.16)$$

According to the definition of the coefficient of isotopic fractionation, for the product–substrate system that parameter is

$$\alpha_i = (\delta^{13}C_{S_{0_i}} - \delta^{13}C_{P_i}) \times 10^{-3} + 1. \qquad (5.17)$$

Therefore,

$$\alpha_{\kappa_i} = -\kappa[(\beta_{P_i}/\beta_{S_i}) - 1] + 1. \tag{5.18}$$

The subscript κ to α will be used later to designate the fractionation coefficient determined by the thermodynamic isotopic effect in the enzyme–substrate complex, in contrast to other types of isotopic effects.

If the isotopic compositions are not expressed in $\delta^{13}C$ values but directly via the ratio of abundance of the heavy and light isotope (R), then using the relationship

$$R_{P_{(i)}}/R_{So_{(i)}} = -(\delta^{13}C_{P_{(i)}} - \delta^{13}C_{So_{(i)}}) \times 10^{-3} + 1 \tag{5.19}$$

(see Chapter 1), we obtain

$$R_{P_{(i)}} = R_{So_{(i)}}(1/\alpha_{\kappa_{(i)}}). \tag{5.20}$$

Equation (5.16) obtained above is very significant. It conveys the three most important features inherent in the observed distribution of isotopes in biological systems.

(1) It links the isotopic composition of the product $\delta^{13}C_{P_i}$ to the isotopic composition of its immediate predecessor $\delta^{13}C_{So_{(i)}}$.

(2) It conveys the relationship of isotopic compositions to the values of the thermodynamic isotopic factors (i.e., it explains, in principle, the observed thermodynamic ordering of isotopes in biological systems).

(3) It shows that the observed isotopic effects should always be less than the equilibrium effects determined directly by the ratio of β factors (because κ is always less than unity). We introduced the coefficient κ empirically, as the value showing how much less the observed isotopic effect is than the equilibrium effect. Expressions (5.11) and (5.15) demonstrate its meaning, as values depending on the rate constant of the enzymatic process.

It should be emphasized that the presumed isotopic fractionation in the elementary act of enzymatic catalysis can lead to a thermodynamic isotopic effect between the S_1 positions of the substrate and the P_1 positions of the product, between the S_2 positions of the substrate and the P_2 positions of the product, etc. but cannot cause thermodynamically ordered distribution of isotopes between the P_1, P_2, etc., positions of the product. As we shall show later, the intramolecular thermodynamic ordering of isotopes observed in biomolecules is the result of a higher-order process rather than an isolated enzymatic reaction. It is caused by cyclic transformations inherent in metabolic processes. But first we shall obtain a more complete formula of isotopic fractionation in an enzymatic reaction.

II. FRACTIONATION OF ISOTOPES IN AN ENZYMATIC REACTION

Above, we considered a simplified version of the process in order to show how the thermodynamic isotopic effect can be realized in an enzymatic reaction. In principle, that effect results when microscopic reversibility of the conversion of the substrate in an enzyme–substrate complex takes place. Now we shall investigate fractionation of isotopes in an enzymatic reaction in a more general aspect.

We shall return to writing the enzymatic reaction in the form

$$\mathrm{E} + \mathrm{S}_0 \underset{k_{-1}}{\overset{k_1}{\rightleftharpoons}} \boxed{\mathrm{ES} \underset{k_b}{\overset{k_a}{\rightleftharpoons}} \mathrm{EP}} \xrightarrow{k_2} \mathrm{P} + \mathrm{E}. \tag{5.21}$$

The following kinetic equations are valid in the process in progress:

$$[\mathrm{ES}](k_a - k_{-1}) = ([\mathrm{E}_T] - [\mathrm{ES}] - [\mathrm{EP}]) \times [\mathrm{S}_0]k_1 + [\mathrm{EP}]k_b, \tag{5.22}$$

$$[\mathrm{ES}]k_a = [\mathrm{EP}](k_2 + k_b), \tag{5.23}$$

$$v = [\mathrm{EP}]k_2, \tag{5.24}$$

where v is the rate of the process (rate of formation of the product), and E_T the total concentration of the enzyme:

$$[\mathrm{E}_T] = [\mathrm{E}] + [\mathrm{ES}] + [\mathrm{EP}].$$

Substituting in (5.22) the value of [ES] from (5.23) and of [EP] from (5.24), we obtain

$$\frac{v(k_2 + k_b)(k_a + k_{-1})}{k_2 k_a} = \mathrm{E}_T \mathrm{S}_0 k_1 - \frac{v(k_2 + k_b)}{k_a k_2} \mathrm{S}_0 k_1 - \frac{v \mathrm{S}_0 k_1}{k_2} - \frac{v k_b}{k_2}. \tag{5.25}$$

The kinetics of the enzymatic process, as is known, are described by the Michaelis–Menten equation (Lehninger, 1974)

$$v = V_{\mathrm{max}}[\mathrm{S}_0]/(K_{\mathrm{m}} + [\mathrm{S}_0]), \tag{5.26}$$

where V_{max} is the maximum rate of the reaction and K_{m} the Michaelis constant.

We can transform (5.25) to a form corresponding to the Michaelis–Menten equation:

$$v = \frac{[\mathrm{E}_T][\mathrm{S}_0]k_1 k_2 k_a}{(k_2 + k_b)(k_a + k_{-1}) - k_b k_a + [(k_2 + k_b)k_1 + k_1 k_a]\mathrm{S}_0}. \tag{5.27}$$

Therefore,

$$V_{\max} = \frac{k_2 k_a[\mathrm{E}_T]}{k_2 + k_b + k_a}, \tag{5.28}$$

$$K_m = \frac{k_2 k_{-1}[1 + (k_b/k_2) + (k_a/k_{-1})]}{k_1(k_2 + k_b + k_a)}. \tag{5.29}$$

In general the isotopic effect of a reaction described by the equation

$$-d[\mathrm{S}_0]/dt = V_{\max}[\mathrm{S}_0]/(K_m + [\mathrm{S}_0]) \tag{5.30}$$

is difficult to investigate. Therefore, we shall examine two extreme cases corresponding to the Michaelis–Menten equation, reactions with a low concentration of the substrate ($K_m \gg \mathrm{S}_0$) and with a high concentration of the substrate ($K_m \ll \mathrm{S}_0$).

With a low concentration of the substrate the Michaelis–Menten equation takes the form (Lehninger, 1974)

$$v = (V_{\max}/K_m)[\mathrm{S}_0]. \tag{5.31}$$

The reaction is of first order. Relationship (5.27) for this case is transformed to the form

$$-\frac{d[\mathrm{S}_0]}{dt} = \frac{k_a k_1[\mathrm{E}_T]}{k_{-1}(1 + k_b/k_2 + k_a/k_{-1})}[\mathrm{S}_0]. \tag{5.32}$$

The term in the right-hand side of the expression before $[\mathrm{S}_0]$ corresponds to the rate constant of the reaction. Let us designate it k_e, then

$$-d[\mathrm{S}_0]/dt = k_e[\mathrm{S}_0]. \tag{5.33}$$

For the isotopically substituted form, the kinetic equation is

$$-d[\mathrm{S}_0^*]/dt = k_e^*[\mathrm{S}_0^*] \tag{5.34}$$

and

$$\frac{d[\mathrm{S}_0]}{[\mathrm{S}_0]} = \frac{k_e}{k_e^*}\frac{d[\mathrm{S}_0^*]}{[\mathrm{S}_e^*]}, \tag{5.35}$$

where k_e/k_e^* as usual characterizes the isotopic effect of the reaction

$$\alpha_e = k_e/k_e^*. \tag{5.36}$$

Thus the isotopic effect of the process in question is determined by the relationship

$$\alpha_e = \frac{k_a k_1 k_{-1}^*(1 + k_b^*/k_2^* + k_a^*/k_{-1}^*)}{k_a^* k_1^* k_{-1}(1 + k_b/k_2 + k_a/k_{-1})}. \tag{5.37}$$

The isotopic effects of the individual stages are expressed by relationships of (5.36) type, namely,

$$k_1^* = k_1/\alpha_1; \qquad k_{-1}^* = k_{-1}/\alpha_{-1}; \qquad k_2^* = k_2/\alpha_2;$$
$$k_a^* = k_a/\alpha_a; \qquad k_b^* = k_b/\alpha_b.$$

Introducing these relationships into expression (5.37) we obtain

$$\alpha_e = \frac{\alpha_a\alpha_1/\alpha_{-1} + k_b\alpha_a\alpha_2\alpha_1/k_2\alpha_b\alpha_{-1} + k_a\alpha_1/k_{-1}}{1 + k_b/k_2 + k_a/k_{-1}} \tag{5.38}$$

The ratio of kinetic isotopic effects of the normal and reverse reaction α_a/α_b is equal to the thermodynamic isotopic effect, which in this case is determined by the ratio of β-factor values of S and P, i.e.,

$$\alpha_a/\alpha_b = \beta_S/\beta_P. \tag{5.39}$$

The values α_2 and α_1 are the kinetic isotopic effects of dissociation of the enzyme–substrate complex. The probability of physical breakdown of the complex apparently depends little on the isotopic composition of the substrate. Therefore $\alpha_2 = 1$ and $\alpha_{-1} = 1$.

As a result, expression (5.38) can be represented in the following form:

$$\alpha_e = \alpha_1\left[1 + \frac{k_b/k_2}{1 + k_b/k_2 + k_a/k_{-1}}\left(\frac{\beta_S}{\beta_P} - 1\right) + \frac{1}{1 + k_b/k_2 + k_a/k_{-1}}(\alpha_a - 1)\right]. \tag{5.40}$$

The multiplier in front of the bracketed term describing the thermodynamic isotopic effect we as before, designate as κ:

$$\kappa = \frac{(k_b/k_2)}{(1 + k_b/k_2 + k_a/k_{-1})}. \tag{5.41}$$

The value α_a is the kinetic isotopic effect of transformation of the substrate into the product in the enzyme–substrate complex. The multiplier in front of the parentheses $(\alpha_a - 1)$ in (5.40) we designate λ as follows:

$$\lambda = \frac{1}{1 + k_b/k_2 + k_a/k_{-1}}. \tag{5.42}$$

With these notations expression (5.40) takes the form

$$\alpha_e = \alpha_1\left[1 + \kappa\left(\frac{\beta_S}{\beta_P} - 1\right) + \lambda(\alpha_a - 1)\right]. \tag{5.43}$$

We invert the β_S/β_P ratio. This changes the sign before κ. As both addends (not counting the units) in the brackets are much less than unity because of their physical meaning, (5.43) can be transformed as

$$\alpha_e = \alpha_1 \left[1 - \kappa \left(\frac{\beta_P}{\beta_S} - 1 \right) \right] [1 + \lambda(\alpha_a - 1)]. \tag{5.44}$$

The first bracketed term is the reduced thermodynamic isotopic effect of the substrate–product system in enzyme–substrate complex α_κ [see expression (5.18)]. The second bracketed term is the reduced kinetic isotopic effect of the substrate transformation in the enzyme–substrate complex. We shall designate it α_λ.

Thus we arrive at the relationship

$$\alpha_e = \alpha_1 \alpha_\kappa \alpha_\lambda. \tag{5.45}$$

From this relationship it follows that the isotopic effect of an enzymatic reaction is determined generally by the product of the values of the isotopic effect of the adsorption of the substrate on the enzyme (α_1), the reduced thermodynamic (α_κ) and reduced kinetic isotopic (α_λ) effects.

The isotopic effect of adsorption of the substrate on the enzyme apparently is insignificant and can be neglected. That effect is of possible importance only for hydrogen, owing to both the formation of hydrogen bonds and steric differences in the normal and isotopically replaced form, which are larger for hydrogen than for other elements. However, in view of the conformational mobility of the enzyme's active center such effect, even for hydrogen, can hardly be appreciable. The relative contribution of the thermodynamic and kinetic isotopic effects depends on the values of the corresponding reduced multipliers, κ and λ. The values of these in turn depend on the ratio of rate constants of the individual stages of the process.

As is seen from expressions (5.41) and (5.42), the relationships between k_b and k_2 and between k_{-1} and k_2 are important for the values of κ and λ. Therefore, the following extreme cases are possible:

(1) if $k_b \ll k_2$ and $k_a \gg k_{-1}$, then $\lambda \simeq 0$ and $\kappa \simeq 0$;
(2) if $k_b \ll k_2$ and $k_a \ll k_{-1}$, then $\lambda \simeq 1$ and $\kappa \simeq 0$;
(3) if $k_b \gg k_2$ and $k_a \ll k_{-1}$, then $\lambda \simeq 0$ and $\kappa \simeq 1$;
(4) if $k_b \gg k_2$ and $k_a \gg k_{-1}$, then $\lambda \simeq 0$ and

$$\kappa \simeq (k_b/k_2)/(1 + k_b/k_2 + k_a/k_{-1}). \tag{5.46}$$

The condition $k_b \ll k_2$ when $k_a \gg k_{-1}$ means virtual irreversibility of the next stages, because no isotopic effects (except α_1) can take place.

The second case, when $k_b \ll k_2$ and $k_a \ll k_{-1}$, corresponds to the maximum manifestation of the kinetic isotopic effect. In this case, the process is reversible in the stage of formation of the enzyme–substrate complex, but the S $\rightleftarrows$ P conversion is practically irreversible, as the rate of decomposition of the EP complex is greater than the rate of the reverse conversion. In other words, the lifetime of the enzyme–substrate complex is too short for repeated S $\rightleftarrows$ P conversion. Under these conditions the thermodynamic isotopic effect is suppressed ($\kappa = 0$). At the same time the kinetic isotopic effect is minimally reduced.

In the third and fourth cases, when $k_b \gg k_2$, the kinetic isotopic effect is suppressed and the thermodynamic isotopic effect appears. The conjunction of conditions $k_b \gg k_2$ and $k_a \gg k_{-1}$ corresponds to the situation in which repeated S $\rightleftarrows$ P conversion is certain during the lifetime of the enzyme–substrate complex. In this case, the thermodynamic isotopic effect is reduced by the value of the coefficient κ, determined by expression (5.46).

Thus, depending on the nature of the enzymatic process, either a kinetic or a thermodynamic isotopic effect can take place separately or in combination.

Now let us consider another extreme enzyme reaction—the case with a high concentration of the substrate. The Michaelis constant is numerically equal to the concentration of the substrate at which the rate is half the maximum. When the condition $K_m \ll S_0$ is fulfilled, the rate of the reaction reaches a maximum. In view of saturation of the enzyme it ceases to depend on concentration of the substrate. The reaction becomes first-order, and

$$v = V_{\max}, \tag{5.47}$$

and therefore, in correspondence with (5.28):

$$-\frac{d[S_0]}{dt} = \frac{k_2 k_a [E_T]}{k_2 + k_b + k_a}. \tag{5.48}$$

The magnitude of isotopic effect of the reaction in this case is determined by the relationship

$$\alpha_e = \frac{\alpha_2 \alpha_a [(k_2/\alpha_2) + (k_b/\alpha_b) + (k_a/\alpha_a)]}{k_2 + k_b + k_a}. \tag{5.49}$$

Neglecting the isotopic effect in the dissociation of the enzyme–substrate complex ($\alpha_2 = 1$), as in the preceding case, we obtain

$$\alpha_e = \left[1 - \frac{k_b/k_2}{1 + k_b/k_2 + k_a/k_2}\left(\frac{\beta_P}{\beta_S} - 1\right)\right]\left[1 + \frac{1}{1 + k_b/k_2 + k_a/k_2}(\alpha_a - 1)\right]. \tag{5.50}$$

Expression (5.50) on the whole is fully analogous to expression (5.40) except for the absence of α_1 and the fact that the constant k_{-1} in the expressions for κ and λ is replaced by the constant k_2. However, above we have assumed $\alpha_1 = 1$, and replacing k_{-1} by k_2 does not significantly change the expressions for κ and λ as the constants that characterize the rate of dissociation of the enzyme–substrate complex. The relationship between κ and λ and the general structure of their dependence on the ratio of the rate constants remain the same as in the preceding case.

Because expressions (5.40) and (5.50), corresponding to two extreme cases of application of the Michaelis–Menten equation, are similar, it can be concluded that an expression of the form

$$\alpha_e = \left[1 - \kappa\left(\frac{\beta_P}{\beta_S} - 1\right)\right][1 + \lambda(\alpha_a - 1)] \tag{5.51}$$

is sufficiently general to describe the fractionation of isotopes in the enzymatic process.

If the kinetic isotopic effect is small, and it occurs under the condition $k_b \gg k_2$, i.e., when the rate of the reversible $S \rightleftarrows P$ conversion is higher than the rate of decomposition of the complex, expression (5.51) is reduced to (5.18) and to (5.16). These were obtained under the assumption that reversibility of the catalytic stage existed.

The thermodynamic component in expression (5.51) can be calculated and, converting to δ values, can be represented in the form

$$\delta^{13}C_P = \delta^{13}C_{S_0} + \kappa\left(\frac{\beta_P}{\beta_S} - 1\right) \times 10^3 + \Delta_{kin}, \tag{5.52}$$

where Δ_{kin} is the kinetic component of the isotopic effect.

In natural biological materials we often find thermodynamically ordered distribution of isotopes, which suggests that the thermodynamic component of enzymatic fractionation of isotopes is important.

Experimental data indicate that in most cases the value of the empirical coefficient κ is close to 0.5. From the theoretical expression for the reduced multiplier κ in (5.41) and (5.46) it follows that its value closed to 0.5 corresponds to the case $k_b/k_2 \gg 1$ and $k_b/k_2 \approx k_a/k_{-1}$. Thus the observed values of the coefficient κ indicate not only reversibility of the catalytic stage of the enzyme reaction but also in many cases its symmetry with respect to dissociation of the complex, both toward the formation of the product and toward liberation of the substrate [the latter is determined by the ratio $(k_b/k_2)/(k_a/k_{-1})$].

It is expedient to introduce special notations characterizing these properties of the catalytic stage:

$$\frac{k_b}{k_2} = g, \qquad \frac{k_b}{k_2}\bigg/\frac{k_a}{k_{-1}} = \zeta. \tag{5.53}$$

The relationships obtained above make it possible to relate the value of κ to the fundamental parameters (V_{max} and K_m) characterizing the kinetics of the enzymatic process on the one hand and to the parameters characterizing the catalytic stage (g and ζ) on the other.

Comparison of (5.41) with (5.28) and (5.29) makes it possible to obtain the following expression for the value of κ ($E_T = 1$ and $K_m \gg S_0$):

$$\kappa = (k_b/k_2)(k_{-1}/k_a)(1/k_1). \tag{5.54}$$

Keeping (5.53) in mind, we obtain

$$\kappa = (\zeta/k_1)(V_{max}/K_m). \tag{5.55}$$

The values of the reduced multipliers of the thermodynamic and kinetic isotopic effects are related to each other, as is seen from (5.40) and (5.50), by the simple relationship

$$\kappa = g\lambda. \tag{5.56}$$

Substituting (5.55) and (5.56) in (5.51), we obtain

$$\alpha_e = \left[1 - \frac{\zeta}{k_1}\frac{V_{max}}{K_m}\left(\frac{\beta_P}{\beta_S} - 1\right)\right]\left[1 + \frac{\zeta}{gk_1}\frac{V_{max}}{K_m}(\alpha_a - 1)\right]. \tag{5.57}$$

Expression (5.57) determines the magnitude of the isotopic effect as a function of the parameters characterizing the enzymatic process. It follows that, in principle, one can approach an analysis of the catalytic stage of an enzymatic reaction on the basis of a combination of kinetic and isotopic investigations.

III. THE ISOTOPIC TRANSFER EFFECT

In addition to chemical transformation, transfer of matter occurs in organisms. The passage of matter through inter- and intracellular membranes can be accompanied by separation of isotopes, because the mobilities of different isotopic forms of a compound differ.

In this case, isotopic fractionation is usually proportional to the ratio of the diffusion coefficients of the isotopic forms. The expression for those contains the square root of the masses of diffusing molecules. Therefore the corresponding coefficient of isotopic fractionation is expressed by the relationship

$$\alpha = \sqrt{M/M^*}.$$

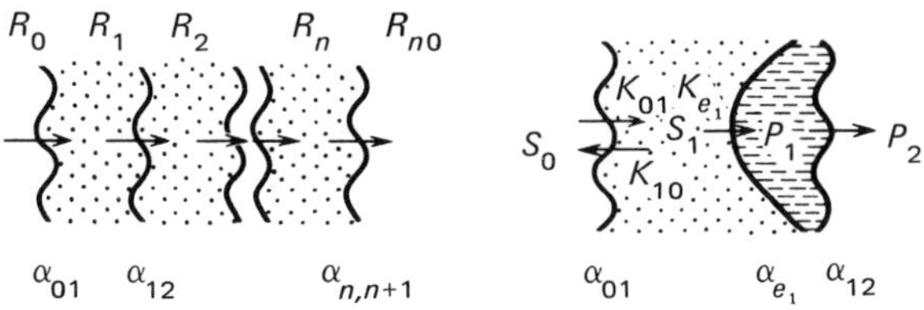

FIG. 5.1 An illustration of isotopic transfer effects: (a) passage of a flow of matter through successive fractionating membranes; (b) passage through a cell in which its chemical conversion is accomplished.

In the actual process the isotopic effect depends on the relationship between the rates of normal and reverse diffusion, on the presence and value of activational barriers in the process of transfer, on the presence of competing processes of transfer of matter in flux and in complex with other compounds, and on the presence of intermolecular relations and other factors. In combination they complicate numerical calculation of the values of the isotopic effect related to the transfer process.

Isotopic fractionation in an established flow through a volume bounded by membranes and containing internal partitions (Fig. 5.1) is determined by the relaticnships

$$R_1 = R_0 \frac{\alpha_{12}}{\alpha_{01}}; \qquad R_2 = R_0 \frac{\alpha_{23}}{\alpha_{01}}; \qquad R_n = R_0 \frac{\alpha_{n,n+1}}{\alpha_{01}}; \qquad R_{n0} = R_0 \frac{1}{\alpha_{01}}; \quad (5.58)$$

where R_0 is the isotopic composition of a substance in a medium before flow through the volume bounded by the membrane begins, R_n the isotopic composition of the substance in unit cells bounded by membranes, R_{n0} the isotopic composition of a substance liberated into the external environment, and $\alpha_{n,n+1}$ the coefficient of fractionation on the $(n + 1)$th membrane.

As in (5.58), the case of an established flow, the fractionation of isotopes between the arriving and departing substance is determined by the value of the coefficient of single separation of isotopes regardless of the number of internal membranes.

The isotopic composition of the substance in a given cell compared with the preceding one is determined by the ratio of fractionation coefficients on entering $(\alpha_{n-1,n})$ and leaving $(\alpha_{n,n+1})$ the cell:

$$R_n = R_{n-1}(\alpha_{n-1,n}/\alpha_{n,n+1}).$$

If the coefficients of fractionation of isotopes on entrance and exit are the same, as can be assumed for many cases, the isotopic composition of the substance in a given cell would not differ from the previous composition. If this condition can be extended to a combination of cells, then the isotopic distribution would not depend on transfer effects in the corresponding multicellular structure.

However, in the general case, the separation of isotopes in the process of arrival and removal of matter may not be the same for the following reasons: a difference in mechanism of normal and reverse passage through the wall (for instance, arrival is accomplished by diffusion and release in discrete batches or via a highly permeable interface); a difference in form of transfer (for instance, arrival in the form of complexes, removal in the free state); a change in chemical form of a compound as a result of chemical reaction that changes the diffusional properties of the compound and the relative difference in masses of its isotopic forms.

If $\alpha_{n-1,n}$ is not equal to $\alpha_{n,n+1}$, a certain resultant isotopic transfer effect occurs:

$$\alpha_{\mathrm{TR}} = \alpha_{n-1,n}/\alpha_{n,n+1}.$$

The case where the substrate comes to the site of the reaction through a membrane, undergoes transformation, and the product leaves through a membrane is of practical interest (see Fig. 5.1). It is not difficult to see that in this case R_{P_2}, the isotopic composition of the product given off, will be related to the isotopic composition of the substrate R_{S_0} before its arrival in the cell by the relationship

$$R_{\mathrm{P}_2} = R_{\mathrm{S}_0} \frac{(k_{10} + k_{\mathrm{e}})}{\alpha_{01}\alpha_{\mathrm{e}_1}(k_{10}/\alpha_{10} + k_{\mathrm{e}}/\alpha_{\mathrm{e}})}. \tag{5.59}$$

From this expression isotopic fractionation between the product liberated and the substance consumed can be seen to depend on the ratio of rate constants characterizing the circulation of the consumed substance through the membrane (k_{10}), and its chemical transformation (k_{e_1}). Two limiting cases are possible.

1. $k_{\mathrm{e}_1} \gg k_{10}$. Application of this condition to expression (5.59) gives the relationship

$$R_{\mathrm{P}_2} = R_{\mathrm{S}_0}(1/\alpha_{01}). \tag{5.60}$$

Thus, in the case of high rates of transformation, the isotopic effect is determined only by the fractionation of isotopes on the membrane. The isotopic effect of chemical transformation is eliminated.

2. $k_{\mathrm{e}_1} \ll k_{10}$. In this case we obtain

$$R_{\mathrm{P}_2} = \alpha_{10} R_{\mathrm{S}_0}/\alpha_{01}\alpha_{\mathrm{e}},$$

or

$$R_{\mathrm{P}_2} = R_{\mathrm{S}_0}/\alpha_{\mathrm{TR}}\alpha_{\mathrm{e}}. \tag{5.61}$$

From the relationships obtained it follows that, during sufficiently intensive exchange of a substance through a membrane (compared with the rate of its chemical transformation), the isotopic effect depends on the product of the fractionation coefficients characterizing isotopic fractionation both in the process of transfer and in the chemical reaction proper:

$$\alpha_{\text{sum}} = \alpha_{\text{e}}\alpha_{\text{TR}} = \alpha_{\kappa}\alpha_{\lambda}\alpha_{\text{TR}}. \tag{5.62}$$

Converting to the designation of the isotopic composition in δ values and taking

$$-\Delta_{\text{TR}} = (\alpha_{\text{TR}} - 1) \times 10^3,$$

we obtain a general formula in the form

$$\delta^{13}\text{C}_{\text{P}_2} = \delta^{13}\text{C}_{\text{S}_0} + \kappa\left(\frac{\beta_{\text{P}}}{\beta_{\text{S}}} - 1\right) \times 10^3 + \Delta_{\text{kin}} + \Delta_{\text{TR}}. \tag{5.63}$$

Thus if a biochemical reaction is associated with transfer of matter through the cell wall or membrane, the formula for biological fractionation would contain an additional term taking into account a possible isotopic effect in the transfer process.

IV. REMARKS ON THE PROCEDURE OF KINETIC ANALYSIS OF ENZYMATIC REACTIONS

In recent years Cleland (1974), Shimerlik *et al.* (1975, 1977), O'Leary and Limburg (1977), and Northrop (1975) have developed a procedure for estimating the relative reaction rates of individual stages of the enzymatic process on the basis of isotopic effects.

The procedure consists essentially of the following. A model is described by a certain ratio of the rate constants at which the individual steps of the process take place. It is assumed that an isotopic effect occurs at one of the steps. It is accepted that this effect is a kinetic one. The observed isotopic effect is reduced to some extent, depending on the relationship between the rate-limiting and effect-producing stages. The extent of reduction depends on the ratio of the respective rate constants. Ultimately, by varying the conditions of the experiment and analyzing the results, one can come to a conclusion concerning the limiting stage of the process investigated.

The method of Cleland and others is the inverse of the problem of determining the isotopic effect of chemical reactions, but this conceptually appealing method does not permit determination of the value for the characteristic isotopic effects of the reaction step investigated.

As an example, let us examine a study of that type. O'Leary and Limburg (1977) investigated the NADF-dependent isocitrate dehydrogenase catalyzed oxidational decarboxylation of isocitric acid to α-ketoglutaric acid in the presence of a divalent metal ion, according to the following scheme:

$$\text{isocitrate} + \text{NADF}^+ + \text{E} \underset{k_2}{\overset{k_1}{\rightleftharpoons}} \text{E} - \text{isocitrate} - \text{NADF}^+$$

$$\underset{k_4}{\overset{k_3}{\rightleftharpoons}} \text{E} - \text{oxalosuccinate} - \text{NADF}$$

$$\underset{k_6}{\overset{k_5}{\rightleftharpoons}} \text{CO}_2 + \text{E} + \alpha\text{-ketoglutarate} - \text{NADF}\dagger$$

$$\underset{k_8}{\overset{k_7}{\rightleftharpoons}} \text{CO}_2 + \text{NADF} + \text{E} + \alpha\text{-ketoglutarate}. \qquad (5.64)$$

The system was coupled with glutamate dehydrogenase in the presence of NH_3 to guarantee irreversibility of the reaction and regeneration of the NADF.

The relationship of the isotopic effect observed in enzymatic reaction (5.64) is given by the equations below (O'Leary and Limburg, 1977):

$$\frac{k^{12}}{k^{13}}(\text{obs}) = \frac{k_5^{12}/k_5^{13} + R_1}{1 + R_1}, \qquad (5.65)$$

$$R_1 = \frac{k_5}{k_4}\left[1 + \frac{k_3}{k_2}\right]. \qquad (5.66)$$

The authors of the work cited assumed that $k_5^{12}/k_5^{13} = 1.03$–1.06 and that only this stage shows a substantial isotopic effect. They also assumed that there was no interaction of enzyme–nucleotide–α-ketoglutarate with CO_2, i.e., $k_6 = 0$ under the conditions of the experiment. The factor R_1 is equal to the ratio of the rate of decarboxylation to the rate of the reverse reaction for the enzyme–oxalosuccinate–NADF complex.

First, we note that under the condition $k_6 = 0$ (adopted in O'Leary and Limburg, 1977) a formal analogy arises between the scheme of the enzymatic reaction of oxidative decarboxylation of isocitric acid and the model of enzymatic reaction (5.21) investigated by us. The subscripts of the respective constants are $k_1(k_1)$, $k_2(k_{-1})$, $k_3(k_a)$, $k_4(k_b)$, and $k_5(k_2)$. [The constants with the subscripts adopted in Eq. (5.21) are indicated in parentheses.]

We convert the relationships derived in the paper by O'Leary and Limburg (1977) to a form suitable for comparison with the relationship derived for process (5.21):

$$\frac{k^{12}}{k^{13}}(\text{obs}) = \frac{k_5^{12}/k_5^{13} + k_5/k_4(1 + k_3/k_2)}{1 + k_5/k_4(1 + k_3/k_2)}$$

$$= \frac{(k_5^{12}/k_5^{13})(k_4/k_5) + 1 + k_3/k_2}{k_4/k_5 + 1 + k_3/k_2}. \tag{5.67}$$

Substituting the subscripts of the constants, we obtain

$$\alpha_{\text{obs}} = \frac{1 + k_b/k_2\alpha_2 + k_a/k_{-1}}{1 + k_b/k_2 + k_a/k_{-1}}.$$

This expression is analogous in form to relationship (5.38). It can be further transformed to

$$\alpha_{\text{obs}} = 1 + \frac{k_b/k_2}{1 + k_b/k_2 + k_a/k_{-1}}(\alpha_2 - 1). \tag{5.68}$$

Comparison of (5.68) with (5.40) reveals that the important thermodynamic isotopic effect term is missing in (5.68). Also, the coefficient α_2 in (5.68) replaces α_a in (5.40). This is explained by assigning the kinetic isotopic effect to different stages in Eqs. (5.21) and (5.64).

Formulas (5.65) and (5.68) are incomplete due to the original postulation of the purely kinetic character of the isotopic effect of an enzymatic reaction. Formulas (5.40) and (5.57) a sounder basis for use in the isotopic method for studying the kinetics of enzymatic reactions.

V. FRACTIONATION OF ISOTOPES IN THE CASE OF VARIABLE VOLUMES OF SUBSTRATE AND PRODUCT

In deriving the formulas in the preceding section, the isotopic composition of the initial substrate was considered invariant. In reality, the isotopic composition of the substrate will vary in time. As the product accumulates, the substrate, owing to the enzymatic isotopic effect, will become more and more depleted in the isotope in which the product is enriched. This circumstance should be taken into account by introducing into the appropriate formulas a time factor or some coordinate defining the development of the process in time, for instance, the degree of completion of the reaction.

Generally, a change in isotopic composition is described by differential equations. The latter have been investigated for various types of processes in general isotopic chemistry. If the process takes place so that accumulation of the product is accompanied by a gradual disappearance of the substrate, then the relationship of the isotopic compositions of the substrate R_S and product R_P is determined by the relationship (Brodskiy, 1957)

$$\frac{1}{\alpha_e} = \frac{\ln[1 - (R_P/R_S)f]}{\ln(1 - f)}, \tag{5.69}$$

where f is the coordinate of the process which, in the given case, is characterized by a fraction of the transformed substrate.

Expression (5.69) can be transformed to

$$R_P/R_{S_0} = (1/f)[1 - (1 - f)^{1/\alpha_e}]. \tag{5.70}$$

The term $(1/f)[1 - (1 - f)^{1/\alpha_e}]$ behaves as follows. In the beginning of the process, when f is close to zero, it approaches the value $1/\alpha_e$.

Thus initially the relationships of the isotopic compositions of the product and substrate are determined by the value of the fractionation coefficient

$$R_{P(t=0)}/R_{S_0} = 1/\alpha_e. \tag{5.71}$$

When the reaction is nearly completed, i.e. $f \to 1$, the value of $(1/f)[1 - (1 - f)^{1/\alpha_e}]$ is close to unity. Consequently,

$$R_{P(t=0)} = R_{S_0}. \tag{5.72}$$

When the substrate has been substantially converted into the product, the isotopic composition of the product approximates the isotopic composition of the original substrate. In this case there is no resultant isotopic fractionation.

At intermediate moments the fractionation of isotopes is less than that described by expression (5.71).

In terms of δ values, expression (5.70) takes on the following notation:

$$\delta^{13}C_P = \delta^{13}C_{S_0} + \{(1/f)[1 - (1 - f)^{1/\alpha_e}] - 1\} \times 10^3\,‰. \tag{5.73}$$

Because α approximates 1 (except for the isotopic effects for hydrogen) equation (5.51) can be presented in the form

$$\frac{1}{\alpha_e} = \left[\kappa\left(\frac{\beta_P}{\beta_S} - 1\right) + 1\right]\frac{1}{\alpha_\lambda},$$

and then (5.73) can be rewritten as

$$\delta^{13}C_P = \delta^{13}C_{S_0} + \mu(f)\kappa\left(\frac{\beta_P}{\beta_S} - 1\right) + \mu(f)\Delta_{kin}, \tag{5.74}$$

where $\mu(f)$ is some coefficient $0 < \mu < 1$, depending on $f:\mu = 0$ when $f = 1$, and $\mu = 1$ when $f = 0$. From this it follows that a non-steady-state character of the process lowers the value of the reduced multipliers of the enzymatic fractionation of isotopes.

Such a case may be encountered in biochemical reactions in a living organism if the process occurs with intense consumption of the substrate (when its decrease is not compensated by its arrival in the reaction zone.

As an approximation to the process in an organism, consider a model in which the decrease in the substrate is compensated for by continuous synthesis of substrate, so that the volume of the latter remains constant.

This case corresponds to the model known in isotopic chemistry of the fractionation of isotopes as the process of "distillation with total reflux," for which the following relationship is known (Brodskiy, 1957):

$$\ln\left(\frac{\alpha - 1}{\alpha - N/N_0}\right) = \frac{V}{\alpha V_0}, \tag{5.75}$$

where α is the coefficient of fractionation of isotopes (the ratio of the percentage of the heavy isotope in the original volume to the percentage of the heavy isotope in the distillate), N_0 the content of the heavy isotope in the original volume (or in matter arriving from outside), N the present content of the heavy isotope in the original volume, V_0 the original volume of the substance kept at a constant level, and V the distilled volume (volume of evaporated, reacted, etc., substance).

Expression (5.75) is conveniently transformed to the following form by substituting the symbols used by their logical analogs with respect to an enzymatic reaction:

$$R_P/R_{S_0} = (\alpha_e - 1)(1 - \exp(-[P]/[S_0])) + 1. \tag{5.76}$$

In δ values it takes the following form, if limited to the thermodynamic component of the enzymatic isotopic effect:

$$\delta^{13}C_P = \delta^{13}C_{S_0} + \kappa \exp(-[P]/[S])(\beta_P/\beta_S - 1) \times 10^3\,‰. \tag{5.77}$$

Analysis of this expression shows that, depending on the ratio of concentrations of product and substrate, the apparent value of the reduced multiplier, as in the previous case, varies from 1 to 0. If $[P] \ll [S_0]$, the fractionation of isotopes is determined by the value of the fractionation coefficient; if the amount of product formed exceeds the amount of substrate, the product's isotopic composition approximates the original composition of the substrate. Thus, the isotopic fractionation may again be described by an equation of the type (5.74).

The "distillation with constant level" model, which accounts for constancy of the volume of substrate during unlimited increase in volume of the product, can be important in some special cases, for instance, if intense biosynthesis of some components occurs in substantial volume.

VI. MODEL OF FRACTIONATION OF ISOTOPES IN A METABOLICALLY RELATED SYSTEM OF ENZYMATIC REACTIONS

So far we have discussed the fractionation of isotopes in an isolated enzymatic system, but as Lehinger (1974, p. 321) points out, "Nearly all metabolic reactions are interrelated inasmuch as the product of one enzymatic reaction serves as the substrate for another reaction which is the next stage of a given process. The existence of such succession is caused by the specific features of enzymes."

In a living organism the processes of anabolism are counterbalanced by the processes of catabolism. Therefore the concentrations of intermediate metabolities can be considered constant within certain limits. Many metabolic pathways are interrelated so that a tracer introduced into some compound can be found after a specified time in the immediate predecessor of that compound. A generalized model of such a process is depicted in Fig. 5.2, which illustrates an idealized metabolic system which, in addition to the direct pathway of the successive transformations of intermediate metabolites (accompanied by isotopic effects $\alpha_{1,2}$, $\alpha_{2,3}$, etc.), includes reverse pathways of transformation from products to predecessors.

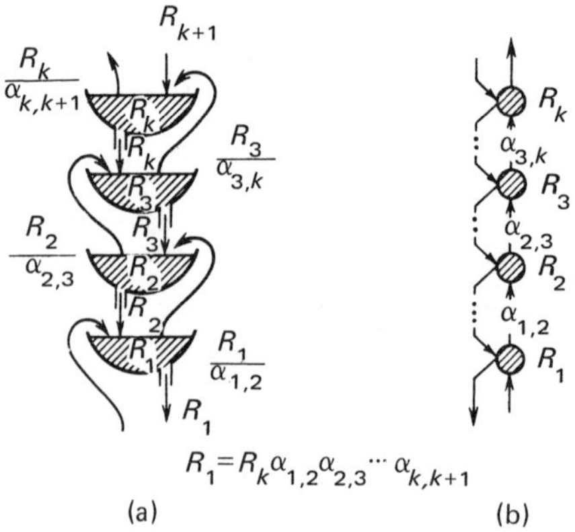

FIG. 5.2 The analogy between fractionation of isotopes in (a) a fractionating column and (b) an ideal metabolic system.

Comparing this model with known mechanisms of isotopic fractionation, it is obvious that there is an analogy between the process of metabolism and the process occurring in a fractionating column. In the steady state the volume of substance on every "plate" of the fractionating column remains constant, and a typical feature of the operation of a fractionating column is "reflux" (Brodskiy, 1957).

As is known, the ratio of isotopic compositions of substances between any two "plates" of the column in a steady-state regime is determined by the product of the isotopic fractionation coefficients of all the fractionating stages between the "plates" in question. As applied to biological systems, this means that the difference in isotopic composition of any two metabolities belonging to a given metabolic pathway will be determined by the product of the isotopic fractionation coefficients of all intermediate enzymatic reactions on the pathway of their biosynthesis:

$$R_k/R_1 = \alpha_{1,2}^{-1}\alpha_{2,3}^{-1}, \cdots, \alpha_{k-1,k}^{-1} = \prod_{i=1}^{k-1} \alpha_{i,i+1}^{-1}. \tag{5.78}$$

The coefficient of enzymatic fractionation is determined at each stage by the product of the thermodynamic and kinetic components of the isotopic effect:

$$\alpha_{e(k-1,k)} = \alpha_{\kappa(k-1,k)}\alpha_{\lambda(k-1,k)}. \tag{5.79}$$

In case the enzymatic process is preceded by a process of transfer of matter, for instance through a cell membrane, the coefficient of enzymatic fractionation α_e would be multiplied by the value of the isotopic transfer effect. Therefore,

$$R_k/R_1 = \prod_{i=1}^{k-1} \alpha_{\kappa(i,i+1)}^{-1} \prod_{i=1}^{k-1} \alpha_{\lambda(i,i+1)}^{-1} \prod_{i=1}^{k-1} \alpha_{\mathrm{TR}(i,i+1)}^{-1}. \tag{5.80}$$

The thermodynamic component of the isotopic effect between the carbon of product P_k and the carbon of product P_{k-1} immediately preceding it is expressed, as was shown above, by the relationship

$$\alpha_{\kappa(k-1,k)} = -\kappa_{k-1,k}(\beta_k/\beta_{k-1} - 1) + 1 \tag{5.81}$$

or

$$\alpha_{\kappa(k-1,k)}^{-1} = -\kappa_{k-1,k}(\beta_{k-1}/\beta_k - 1) + 1.$$

The latter relationship can be given in the following form:

$$\alpha_{\kappa(k-1,k)}^{-1} = \frac{\beta_{k-1}}{\beta_k}\left[1 + (1 + \kappa_{k-1,k})\left(\frac{\beta_k}{\beta_{k-1}} - 1\right)\right]. \tag{5.82}$$

Correspondingly

$$\prod_{i=1}^{k-1} \alpha_{\kappa(i,i+1)}^{-1} = \prod_{i=1}^{k-1} \frac{\beta_i}{\beta_{i+1}} \prod_{i=1}^{k-1} \left[1 + (1 + \kappa_{i,i+1})\left(\frac{\beta_{i+1}}{\beta_i} - 1 \right) \right]. \tag{5.83}$$

The right-hand side of expression (5.83) is represented by two products. The first is simply the ratio of β factors of the first and kth carbons:

$$\prod_{i=1}^{k-1} \frac{\beta_i}{\beta_{i+1}} = \frac{\beta_1}{\beta_2}\frac{\beta_2}{\beta_3} \cdots \frac{\beta_{k-2}}{\beta_{k-1}}\frac{\beta_{k-1}}{\beta_k} = \frac{\beta_1}{\beta_k}. \tag{5.84}$$

In the second product the bracket is close to 1 for every value of i (this condition is fulfilled for isotopic effects in carbon and heavier atoms). Starting from this, the product can be expressed via the sum

$$\prod_{i=1}^{k-1} \left[1 + (1 + \kappa_{i,i+1})\left(\frac{\beta_{i+1}}{\beta_i} - 1 \right) \right] = 1 + \sum_{i=1}^{k-1} (1 + \kappa_{i,i+1})\left(\frac{\beta_{i+1}}{\beta_i} - 1 \right). \tag{5.85}$$

Consequently,

$$\prod_{i=1}^{k-1} \alpha_{\kappa(i,i+1)}^{-1} = \frac{\beta_1}{\beta_k} \left[1 + \sum_{i=1}^{k-1} (1 + \kappa_{i,i+1})\left(\frac{\beta_{i+1}}{\beta_i} - 1 \right) \right]. \tag{5.86}$$

We represent the value $\kappa_{i,i+1}$, which, in principle, can be different for each elementary act of biosynthesis, in the form

$$\kappa_{i,i+1} = \bar{\kappa} + \Delta\kappa_{i,i+1}, \tag{5.87}$$

where $\bar{\kappa}$ is some average value of the coefficient and $\Delta\kappa_{i,i+1}$ the deviation from it.

We rewrite (5.86) taking into account (5.87):

$$\prod_{i=1}^{k-1} \alpha_{\kappa(i,i+1)}^{-1} = \frac{\beta_1}{\beta_k} \left[1 + (1 + \bar{\kappa}) \sum_{i=1}^{k-1} \left(\frac{\beta_{i+1}}{\beta_i} - 1 \right) \right] + \frac{\beta_1}{\beta_k} \sum_{i=1}^{k-1} \Delta\kappa_{i,i+1}\left(\frac{\beta_{i+1}}{\beta_i} - 1 \right). \tag{5.88}$$

Using the fact that the sign $\sum_{i=1}^{k-1}$ in the bracket sums values constituting a small percentage of unity, we represent the corresponding sum via the product

$$\sum_{i=1}^{k-1} \frac{\beta_{i+1}}{\beta_i} - 1 = \prod_{i=1}^{k-1} \frac{\beta_{i+1}}{\beta_i} - 1. \tag{5.89}$$

The product in the right-hand part reduces to the β_k/β_1 ratio:

$$\prod_{i=1}^{k-1} \frac{\beta_{i+1}}{\beta_i} - 1 = \frac{\beta_k}{\beta_1} - 1. \tag{5.90}$$

Substituting (5.89) in form (5.90) in expression (5.88), we obtain

$$\prod_{i=1}^{k-1} \alpha_{\kappa(i,i+1)}^{-1} = \frac{\beta_1}{\beta_k}\left[1 + (1+\bar{\kappa})\left(\frac{\beta_k}{\beta_1} - 1\right)\right] + \frac{\beta_1}{\beta_k}\sum_{i=1}^{k-1} \Delta\kappa_{i,i+1}\left(\frac{\beta_{i+1}}{\beta_i} - 1\right). \quad (5.91)$$

The first term on the right-hand side of this expression has the structure of the value α_κ as in relationship (5.82). Hence

$$\prod_{i=1}^{k-1} \alpha_{\kappa(i,i+1)}^{-1} = -\bar{\kappa}\left(\frac{\beta_1}{\beta_k} - 1\right) + 1 + \frac{\beta_1}{\beta_k}\sum_{i=1}^{k-1} \Delta\kappa_{i,i+1}\left(\frac{\beta_{i+1}}{\beta_1} - 1\right). \quad (5.92)$$

The last term of expression (5.92) plus 1 we again represent in the form of a product:

$$\prod_{i=1}^{k-1} \alpha_{\kappa(i,i+1)}^{-1} = -\bar{\kappa}\left(\frac{\beta_1}{\beta_k} - 1\right) + \frac{\beta_1}{\beta_k}\prod_{i=1}^{k-1}\left[\Delta\kappa_{i,i+1}\left(\frac{\beta_{i+1}}{\beta_i} - 1\right) + 1\right]. \quad (5.93)$$

We return to the original expression (5.80):

$$\frac{R_k}{R_1} = \left\{-\bar{\kappa}\left(\frac{\beta_1}{\beta_k} - 1\right) + \frac{\beta_1}{\beta_k}\prod_{i=1}^{k-1}\left[\Delta\kappa_{i,i+1}\left(\frac{\beta_{i+1}}{\beta_i} - 1\right) + 1\right]\right\}$$
$$\times \prod_{i=1}^{k-1} \alpha_{\lambda(i,i+1)}^{-1} \prod_{i=1}^{k-1} \alpha_{\mathrm{TR}(i,i+1)}^{-1}. \quad (5.94)$$

Using the rule of transformation of a product into a sum, we develop expression (5.94) in the following form:

$$\frac{R_k}{R_1} = -\bar{\kappa}\left(\frac{\beta_1}{\beta_k} - 1\right) - \left\{\sum_{i=1}^{k-1}[(\alpha_{\lambda(i,i+1)}^{-1} - 1) + (\alpha_{\mathrm{TR}(i,i+1)}^{-1} - 1)]\right.$$
$$\left. + \sum_{i=1}^{k-1}(\alpha_{\lambda(i,i+1)}^{-1} - 1)\sum_{i=1}^{k-1}(\alpha_{\mathrm{TR}(i,i+1)}^{-1} - 1)\right\}\bar{\kappa}\left(\frac{\beta_1}{\beta_k} - 1\right)$$
$$+ \frac{\beta_1}{\beta_k}\prod_{i=1}^{k-1}\left[\Delta\kappa_{i,i+1}\left(\frac{\beta_{i+1}}{\beta_i} - 1\right) + 1\right]\alpha_{\lambda(i,i+1)}^{-1}\alpha_{\mathrm{TR}(i,i+1)}^{-1}. \quad (5.95)$$

The second term on the right-hand side is a small second-order term, compared with the first term. Therefore it can be neglected or taken into account by means of a corresponding change in the value $\bar{\kappa}$ in the first term. The last term we again represent in the form of a sum:

$$\frac{R_k}{R_1} = -\bar{\kappa}\left(\frac{\beta_1}{\beta_k} - 1\right) + 1$$
$$+ \frac{\beta_1}{\beta_k}\sum_{i=1}^{k-1}\left\{\Delta\kappa_{i,i+1}\left(\frac{\beta_{i+1}}{\beta_i} - 1\right) + 1\right.$$
$$\left. + (\alpha_{\lambda(i,i+1)}^{-1} - 1) + (\alpha_{\mathrm{TR}(i,i+1)}^{-1} - 1)\right\}. \quad (5.96)$$

Considering that

$$R_k/R_1 = (\delta^{13}C_k - \delta^{13}C_1) \times 10^{-3} + 1$$

and converting to $\delta^{13}C$ values, we obtain

$$\delta^{13}C_1 - \delta^{13}C_k = \bar{\kappa}\left(\frac{\beta_1}{\beta_k} - 1\right) \times 10^3 - \frac{\beta_1}{\beta_k}\sum_{i=1}^{k-1}\left\{\Delta\kappa_{i,i+1}\left(\frac{\beta_{i+1}}{\beta_i} - 1\right) + [\alpha^{-1}_{\lambda(i,i+1)} - 1] + [\alpha^{-1}_{\mathrm{TR}(i,i+1)} - 1]\right\} \times 10^3. \quad (5.97)$$

Let us designate the second term in expression (5.97) as Δ. Depending on the relationship of the values forming it, Δ may have either a positive or a negative sign. Ultimately we have the relationship

$$\delta^{13}C_1 - \delta^{13}C_k = \bar{\kappa}\left(\frac{\beta_1}{\beta_k} - 1\right) \times 10^3 \pm |\Delta|. \quad (5.98)$$

Thus the difference in isotopic composition between any two atoms of carbon can be expressed as the sum of two terms. The first is the value of the reduced thermodynamic isotopic effect, expressed by the ratio of thermodynamic isotopic factors of the components being compared and the average value $\bar{\kappa}$. The second characterizes the deviation of the observed isotopic effect from the purely thermodynamic effect and includes the algebraic sum of the deviations caused by the difference between the actual values $\kappa_{i,i+1}$ from the average value $\bar{\kappa}$. It also includes the algebraic sum of kinetic isotopic effects and transfer effects on the pathways separating the biosynthesis of the first and *k*th carbon.

The relationships obtained, as we shall see later, create a basis for understanding the origin of intramolecular isotopic effects in biomolecules and the correlations between $\delta^{13}C$ and β_Σ values observed in biological systems.

It should be mentioned that, in deriving the theoretical relationships in this section, no conditions were used which would be specifically characteristic of carbon isotopes alone. Therefore, although for considerations already referred to, it was convenient to conduct the discussion with respect to a specific element, and still more convenient to choose carbon as the example, there is no reason why the validity of the relationships obtained should be limited just to isotopes of that element. Obviously, where the symbols $\delta^{13}C$ and $\beta^{13}C$ were used, the corresponding symbols $\delta^{18}O$ and $\beta^{18}O$, $\delta^{15}N$ and $\beta^{15}N$, and $\delta^{34}S$ and $\beta^{34}S$ could have been used. In some cases, specific channels of fractionation are possible. Hydrogen and oxygen in some cases probably can exchange isotopes with water in addition to enzymatic reactions. For hydrogen, isotopic effects related to the tunnel effect may play a part. More than for other elements, the relative differences in mass of protium and deuterium might cause isotopic effects produced by physiological inequality of the isotopic forms of molecules containing these isotopes. Some of the transformations discussed above were made under the assumption that the value of α was close to unity. However, in the case of larger isotopic effects, which are especially typical of isotopic effects in hydrogen, this assumption becomes unsatisfactory. Still another limitation which must be kept in mind is that, in deriving a number of theoretical relationships, one of the isotopes was considered to be rare compared with the other.

REFERENCES

Brodskiy, A. I. (1957). "Isotope Chemistry." Izd-vo AN SSSR, Moscow.

Cleland, W. W. (1975). Partition analysis and the concept of net rate constants as tools in enzyme kinetics. *Biochemistry* **14**(14), 3220.

Lehninger, A. (1974). "Biochemistry." Mir, Moscow [Russian translation].

Northrop, D. B. (1975). Steady-state analysis of kinetic isotope effects in enzymatic reactions. *Biochemistry* **14**(12), 2644.

O'Leary, H., and Limburg, J. A. (1977). Isotope effect studies of the role of metal ions in isocitrate dehydrogenase. *Biochemistry* **16**(6), 1129.

Schimerlik, M., Rife, J. E., and Cleland, W. W. (1975). Equilibrium perturbation by isotope substitution. *Biochemistry* **14**(24), 5347.

Schimerlik, M., Grimshaw, C. E., and Cleland, W. W. (1977). Determination of the rate-limiting steps for malic enzyme by the use of isotope effects and other kinetic studies. *Biochemistry* **16**(4), 571.

CHAPTER 6

Explanation of the Regularities of the Biological Distribution of Isotopes in the Framework of the Proposed Theory

I. $\delta^{13}C$–$\beta_{\Sigma}{}^{13}C$ CORRELATION

The main experimental observation, which again raises the question of the origin of the biological isotopic effects, is the correlation between the $\delta^{13}C$ and $\beta^{13}C$ values which has been found for many biological systems.

The correspondence of the actual isotopic distribution to the thermodynamic isotopic factors of biomolecules suggests that thermodynamic isotopic effects of some kind are realized in biological systems. Previously, we postulated that the mechanism of production of thermodynamic isotopic effects is related to the mechanism of catalytic action of enzymes. The universality of thermodynamic ordering of isotopes, which appears in organisms belonging to different species, environments, and degree of organization, is observed in various biochemical fractions both between and within molecules and is consistent with the universal role played by enzymes in the chemistry of life.

Using a model, we showed that, with certain assumptions and conditions, isotopic fractionation in enzymatic reactions can cause separation of isotopes between the product and the substrate. This separation is proportional to the values of the thermodynamic isotopic factors which either characterize the carbon positions in the product and in the substrate (β_i factors) or characterize the molecules of the product and substrate as a whole (β_{Σ} factors).

But the existence of correlations of $\delta^{13}C$–$\beta^{13}C$ type suggests something more than a thermodynamic isotopic effect between the product and its immediate predecessor. The correlation implies a relationship between the isotopic compositions and thermodynamic isotopic factors of any compounds belonging to the system in question, not just of the participants in one enzymatic reaction. The significance of expression (5.97), obtained in the preceding section, is that it shows both the necessary conditions and how the relationship between isotopic compositions of any arbitrarily selected compounds of a given system is accomplished. Using variations of expression (5.97) or its abbreviated form (5.98), one can arrive at a regression-type equation characterizing the correlational relationship $\delta^{13}C$–$\beta^{13}C$.

Let us rewrite (5.98) in the following form:

$$\delta^{13}C_k = \delta^{13}C_1 + \bar{\kappa}/\beta_k(\beta_k - \beta_1) \times 10^3 \pm |\Delta|. \qquad (6.1)$$

Because $\bar{\kappa}/\beta \simeq \bar{\kappa}$ and $\delta^{13}C_1$ can mean the isotopic composition of any carbon including $\delta^{13}C_{av}$, relationship (6.1) takes the form

$$\delta^{13}C_k = \delta^{13}C_{av} + \bar{\kappa}(\beta_k - \beta_{av}) \times 10^3 \pm |\Delta|. \qquad (6.2)$$

Turning to analysis of relationship (5.97) we see that the complex term Δ in (5.98), and correspondingly in (6.2), becomes zero in the absence of both kinetic isotopic effects and isotopic transfer effects (when the condition $\kappa_{1,2} = \kappa_{2,3} = \cdots = \kappa_{k-1,k}$). Then the equation

$$\delta^{13}C_k = \delta^{13}C_{av} + \bar{\kappa}(\beta_k - \beta_{av}) \times 10^3 \qquad (6.3)$$

results. This is the regression equation in the form we obtained by directly comparing the experimentally measured isotopic compositions of biological compounds with the values of their β factors.

This reveals the meaning of the stochastic component of the observed correlational relationships of $\delta^{13}C$–$\beta^{13}C$. The relationship between $\delta^{13}C$ and $\beta^{13}C$ corresponds to the ideal state in which the system is completely steady-state, the pathways of biosynthesis are reversible, transfer effects and kinetic isotopic effects are uncommon, and the values of coefficient κ (the reduced thermodynamic isotopic effect at different stages) are the same.

Remarkably, these conditions are satisfied to a substantial extent for at least some biological systems. We recall that, in the investigation of the lipid components, the strength of the relationship between the $\delta^{13}C$ and β_{Σ} values was characterized by correlation coefficients of 0.87 to 0.94.

Expression (6.2) also explains the random component of the experimental relationship of $\delta^{13}C$–$\beta^{13}C$. In addition to the error of measurement of $\delta^{13}C$ and an uncertainty inherent in calculation of the β value, the reduction of the $\delta^{13}C$–$\beta^{13}C$ correlation coefficient can be explained by factors whose effect is described by the Δ term in expression (6.2). Variations in the values $\Delta\kappa_{k,k-1}$ are a cause for deviation of the $\delta^{13}C$ values from the regression equation. Another reason is the manifestation of a kinetic component of the enzymatic isotopic effect which differs from zero and the appearance of isotopic transfer effects. As a result of disruption of the steady-state condition, the model expressed by the original relationship (5.78) becomes only partially valid. In this case the partial models described by expressions (5.73) and (5.77), which in extreme cases correspond to the absence of isotopic effects at the respective stages, become important. The value Δ in expression (6.2) is indicative of the imprecision characterizing the correlation between the values of $\delta^{13}C$ and $\beta^{13}C$. Consequently, the imprecision corresponding to the experimental relationship of the $\delta^{13}C$ and $\beta^{13}C$ values for a given system defines the combined contribution of all the previously mentioned factors that change the isotopic distribution from that of a strictly thermodynamic one.

The fact that in many cases one can find a good correlation between the values of $\delta^{13}C$ and $\beta^{13}C$ indicates that kinetic and transport isotopic effects are of minor importance in these biological systems. The predominant type of isotopic effect in these cases is the thermodynamic isotopic effect. With correlation coefficients on the order of 0.9 typical of the investigated lipid systems, isotopic fractionation is at least 90% thermodynamic in nature.

In some other systems the correlation is poorer. For instance, amino acids liberated by way of hydrolysis of total proteins show a $\delta^{13}C$–$\beta^{13}C$ relationship with much lower correlation coefficients than the lipids (see Chapter 4). The reason for this generally may be related to variations in the κ value. From analysis of relationship (5.97) it follows that the value of the dispersion $|\Delta|$ increases with an increase in the value of the deviation $\Delta\kappa_k$ and in the number of increments $\Delta\kappa_k$ summed under the sign $\sum$. Evidently, when other conditions are equal, different enzymes can have different κ values. It is obvious that the number of summed up increments $\Delta\kappa_k$ is determined by how far apart the compounds in question are on the metabolic pathways. Therefore, it is understandable that the clearest $\delta^{13}C$–$\beta^{13}C$ relationships would be observed for systems of compounds whose biosynthesis is similar and based on common substrates, e.g., the components of the lipid fraction. In contrast to the lipids, which nearly always form by uniform condensation of the two-carbon fragment of acetyl-CoA, amino acids of proteins are formed from various predecessors that are produced by different biosynthetic pathways. In such cases a weaker correlation of $\delta^{13}C$ and $\beta^{13}C$ values would be expected.

The validity of Eq. (5.97) is limited by the conditions from which it was obtained, namely the the steady-state nature of the process and the reversibility of metabolic pathways. Violation of these conditions upsets the validity of the relationships between the $\delta^{13}C$ and $\beta^{13}C$ values conveyed by this equation. Consequently if the metabolic process becomes non–steady-state, because of an abrupt change in external conditions or some other reason, this would lead to a lessening of the $\delta^{13}C$–$\beta^{13}C$ correlation and, under some conditions, to total loss of thermodynamically ordered distribution of isotopes.

Expression (5.97) apparently generalizes the conditions of biological fractionation of isotopes. It reveals the nature of the $\delta^{13}C$–$\beta^{13}C$ correlation and links the $\delta^{13}C$ and $\beta^{13}C$ values to the isotopic effects and the state of the metabolic system.

We note that a relatively poor correlation or an absence of a correlation does not necessarily mean that isotopic fractionation in biosynthesis does not include thermodynamic isotopic effects. A thermodynamically ordered distribution may not be manifested for reasons related to the state of the metabolic system which are independent of the nature of the isotopic effects inherent in enzymatic reactions per se. In other words, there are two levels of isotopic fractionation in biological systems. The first is related to the enzymatic process. The nature of the isotopic effect is determined on this first level. The second is related to the state of the metabolic system as a whole. On this level the regularities relating the isotopic composition of different components of the biological system are formed; in particular, the extent of thermodynamically ordered distribution of isotopes is determined.

II. INTRAMOLECULAR THERMODYNAMICALLY ORDERED DISTRIBUTION OF ISOTOPES

The production of intramolecular ordering of isotopes is related to the second level of isotopic fractionation—the metabolic.

Isotopic fractionation in a single enzymatic reaction can produce thermodynamic separation of isotopes between the carbon in a given position in the product and the carbon in a corresponding position in the substrate, but it does not determine the distribution of isotopes between the carbon positions within a compound, i.e., it does not determine the intramolecular thermodynamic distribution of isotopes. How does intramolecular thermodynamic ordering of isotopes arise in biological systems?

In a general way the answer is contained in formula (5.97), which determines the possibility and extent of the appearance of a thermodynamic isotopic effect between any two carbon atoms belonging to the biological system in question. When the conditions that initiate it are fulfilled, it is not limited by whether the atoms having the isotopic compositions being compared are constituents of a single molecule or of different compounds. Thus if a biochemical pathway exists which links two atoms of the same molecule, then a thermodynamic isotopic effect can appear between them, as between any first and kth atoms in the chain of transformations as shown by (5.97).

In many cases the metabolic processes in an organism would satisfy this condition. Let us consider one of the most important processes from both the biochemical point of view and the point of view of isotopic fractionation—the process of primary inclusion of CO_2 in the metabolism of plants. As is known, it occurs via two pathways: by dark fixation and by photosynthetic assimilation of CO_2. Dark fixation results from carboxylation reactions: carboxylation of pyrotartaric acid (Wood–Workman reaction), α-ketoglutaric acid (Ochoa reaction), γ-aminobutyric acid, and others (Rubin and Gavrilenko, 1977). In this case absorbed CO_2 is used in the synthesis of glucose. Regeneration of the acceptors is accomplished in the process of respiration. In photosynthesis the initial act also is dark fixation of CO_2 using suitable acceptors such as ribulose diphosphate or phosphoenolpyruvate. In the light, regeneration of the acceptors is related to a reduction process. In this case glucose and other organic compounds are formed from the CO_2 carbon. The reducing stage is accomplished by reactions of the Calvin cycle—pentosophosphatic pathway in C_3-type plants. The Hatch–Slack cycle—the dicarboxylic acids pathway—precedes the Calvin cycle in C_4-type plants.

Figure 6.1 is a flowchart of the reactions of the Calvin cycle. The main scheme of the process consists of the following. The first acceptor of CO_2 is ribulose 1,5-diphosphate (Rb 1,5-DP). As a result of its carboxylation and a dismutation reaction, two molecules of 3-phosphoglyceric acid (PGA) are formed. One of these contains carbon originating from CO_2. In the next enzymatic reaction the PGA is reduced to glyceraldehyde-3-phosphate (GAP). It is here that the ATP and NADF, formed in the light cycle of photosynthesis, are used (Lehninger, 1974). In turn, the GAP is isomerized to dihydroxyacetone phosphate (DHAP), following which both trioses are condensed into the fructose 1,6-diphosphate molecule. The subsequent release of phosphate yields fructose 6-phosphate, which is changed into glucose 6-phosphate. Thus the initial units are formed for the synthesis of sucrose; starch and metabolites of other important biosynthetic pathways are formed by the splitting of glucose. At the same time, in interconnected reactions of phosphorylated sugars with a different number of carbon atoms (e.g., triose, tetrose, pentose, hexose, and sedoheptulose), the ribulose 1,5-diphosphate is regenerated, signifying the closing of the cycle. Carboxylation of the regenerated ribulose 1,5-diphosphate again yields phosphoglyceric acid. In this case the carbon of exogenous CO_2 is absorbed in the carboxyl group of PGA, and the carbon of CO_2, assimilated in a previous

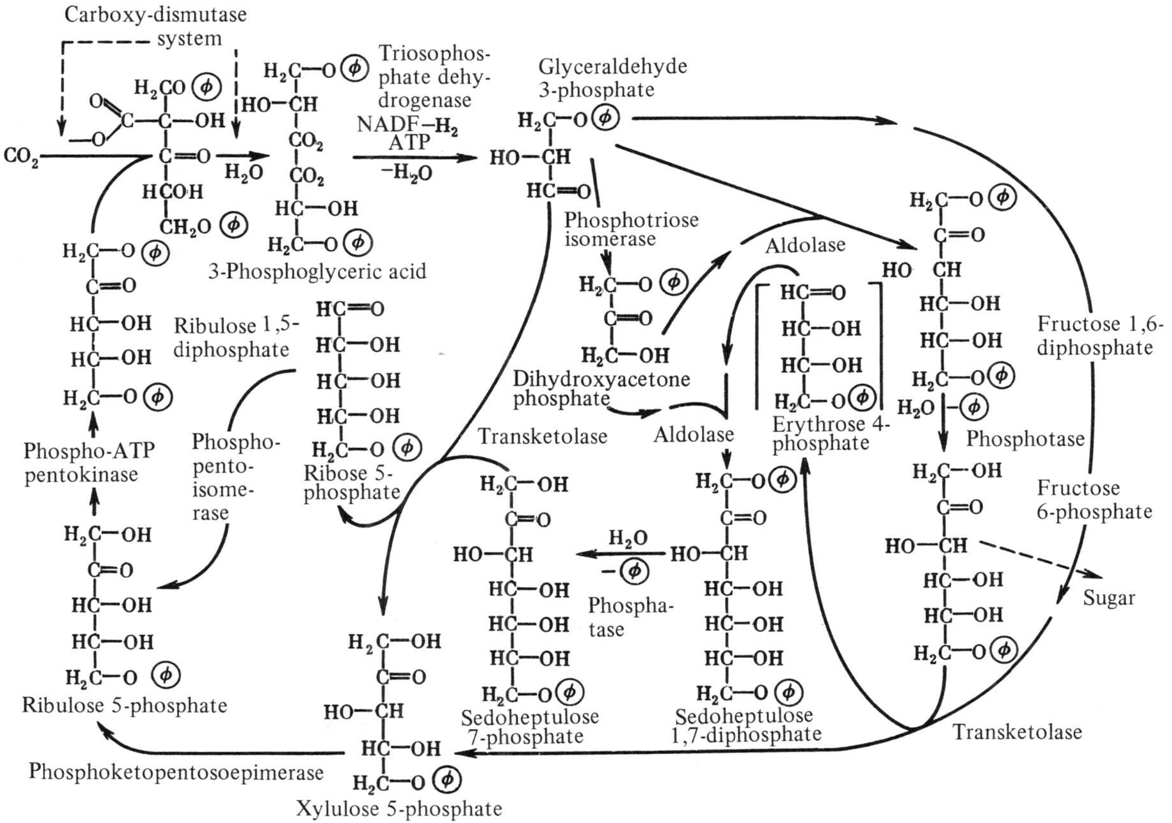

FIG. 6.1 Reactions of the Calvin cycle. From Rubin and Gavrilenko (1977).

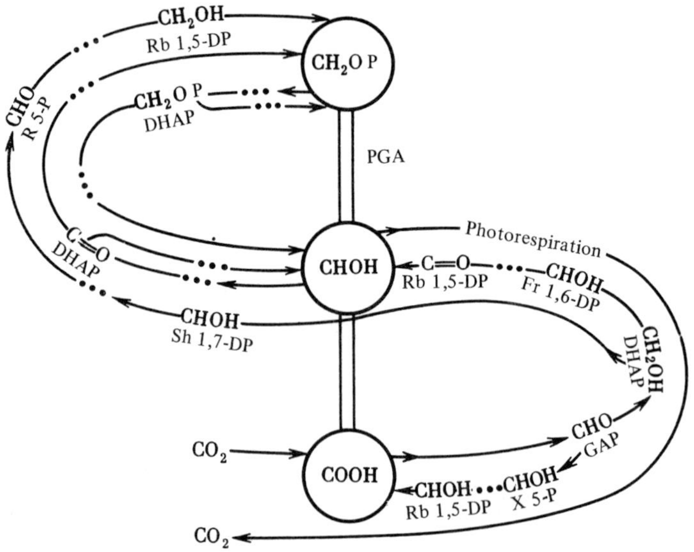

FIG. 6.2 Pathways of carbon transformation in the pentose phosphate cycle, genetically relating the carbon atoms of phosphoglyceric acid. Rb 1,5-DP (ribulose 1,5-diphosphate), X 5-P (xylose 5-phosphate), GAP (glyceraldehyde 3-phosphate), DHAP (dihydroxyacetone phosphate), Fr 1,6-DP (fructose 1,6-diphospate), R 5-P (ribose 5-phosphate), Sh 1,7-DP (sedoheptulose 1,7-diphosphate), and PGA (phosphoglyceric acid).

cycle, finds itself in the CHOH or CH_2 attached to the phosphate group of PGA.

The brief description of this well-known process is used to point out the fact that in the intramolecular picture a closed cycle does not exist, rather we have a *spiral*. If it is unwound, a successive chain of transformations linking carbon atoms in different positions within a given molecule is formed. This is illustrated in Fig. 6.2, the Calvin cycle. An analysis of this illustration shows that the carbon atoms of phosphoglyceric acid are genetically related to one another by the reactions of the pentose pathway. It also shows that the metabolic pathways connecting these atoms are reversible. Thus the conditions under which formula (5.97) (which describes the thermodynamic isotopic effect between two arbitrarily chosen carbon atoms of the system) is valid are fulfilled. In that example this pertains to the carbon atoms of one molecule. Consequently, in phosphoglyceric acid there theoretically can be established intramolecular thermodynamically ordered distribution of isotopes.

This conclusion does not pertain only to PGA. If the carbon atom of CO_2 is traced from the moment it enters the cycle, one can see that its transformation successively passes through all the carbon positions of each intermediate metabolite of the cycle.

Intramolecular thermodynamic ordering of isotopes in the product requires establishment of intramolecular ordering in the substrate. This is easy to demonstrate in a generalized form. As a result of isotopic fractionation in an enzymatic reaction, the following relationships are established between the isotopic compositions of the corresponding carbon atoms of the product and substrate:

$$\delta^{13}C_{P_i} = \delta^{13}C_{So_{(i)}} + \kappa_k\left(\frac{\beta_{P_i}}{\beta_{S_i}} - 1\right) \times 10^3 + \Delta_i, \tag{6.4}$$

$$\delta^{13}C_{P_{i+1}} = \delta^{13}C_{So_{(i+1)}} + \kappa_k\left(\frac{\beta_{P_{i+1}}}{\beta_{S_{i+1}}} - 1\right) \times 10^3 + \Delta_{i+1}.$$

Within a molecule of the product the isotopic effect is determined by the difference between the values of $\delta^{13}C$ in the ith and $(i + 1)$th positions:

$$\delta^{13}C_{P_i} - \delta^{13}C_{P_{i+1}} = \delta^{13}C_{So_{(i)}} - \delta^{13}C_{So_{(i+1)}} + \kappa_k\left(\frac{\beta_{P_i}}{\beta_{S_i}} - \frac{\beta_{P_{i+1}}}{\beta_{S_{i+1}}}\right) \times 10^3 + (\Delta_i - \Delta_{i+1}). \tag{6.5}$$

If the isotopic distribution is thermodynamically ordered within the substrate molecule, then

$$\delta^{13}C_{So_{(i)}} = \delta^{13}C_{So_{(i+1)}} + \kappa_{k-1}\left(\frac{\beta_{S_i}}{\beta_{S_{i+1}}} - 1\right) \times 10^3. \tag{6.6}$$

Substituting (6.6) in (6.5), we obtain

$$\delta^{13}C_{P_i} = \delta^{13}C_{P_{i+1}} + \kappa_{k-1}\left(\frac{\beta_{S_i}}{\beta_{S_{i+1}}} - 1\right) \times 10^3 + \frac{\kappa_k \beta_{P_{i+1}}}{\beta_{S_i}}\left(\frac{\beta_{P_i}}{\beta_{P_{i+1}}} - \frac{\beta_{S_i}}{\beta_{S_{i+1}}}\right) \times 10^3 + (\Delta_i + \Delta_{i+1}) \tag{6.7}$$

or

$$\delta^{13}C_{P_i} = \delta^{13}C_{P_{i+1}} + \bar{\kappa}\left(\frac{\beta_{P_i}}{\beta_{P_{i+1}}} - 1\right) \times 10^3 + \Delta\kappa_{k,k-1}\left(\frac{\beta_{S_i}}{\beta_{S_{i+1}}} - 1\right) \times 10^3 + (\Delta_i - \Delta_{i+1}). \tag{6.8}$$

The last expression is similar to formula (5.97) but it is of more limited scope. From Eq. (6.8), it follows that, under the conditions of thermodynamically ordered distribution of isotopes within a molecule of the substrate, intramolecular ordering in the product is possible. Realization of this possibility, as in the case described by expression (5.97), depends on the values of $\Delta\kappa_k$ and Δ_{kin}. If these are small, i.e., if the value of κ is similar for the corresponding enzymatic reactions and kinetic isotopic effects are of minor importance, then the intramolecular distribution of isotopes in the product may be thermodynamically ordered, provided that it is so in the substrate.

We would like to stress that intramolecular thermodynamic ordering of isotopes is a property of molecules which are synthesized not merely in the course of an enzymatic reaction but also in the course of the metabolic process accomplished in the system of enzymes, i.e., in organisms.

III. CASES OF THE ABSENCE OF THERMODYNAMICALLY ORDERED DISTRIBUTION OF ISOTOPES

The theoretical model previously discussed makes it possible not only to explain thermodynamically ordered distribution of isotopes but also to indicate possible reasons for its absence.

The necessary prerequisite for the thermodynamically ordered distribution of isotopes in biological systems is the thermodynamic isotopic effect in the enzymatic reaction. Its appearance is contingent upon microscopic reversibility of the process of chemical reorganization of the substrates in the enzyme–substrate complex. Generally, enzymatic reactions may be characterized both by thermodynamic and kinetic isotopic effects. Which type of effect prevails depends on the ratio of the rates of the individual steps of the enzymatic reaction, especially on the ratio of rates of disruption and formation of the complex and transformation within the complex. Evidently these ratios, and, consequently, the quantitative and qualitative biomolecular distribution of isotopes, vary depending on the conditions of biosynthesis.

Also, as has been shown, a thermodynamic isotopic effect in an enzymatic reaction is a necessary but not exclusive condition for the establishment of thermodynamically ordered distribution of isotopes both in an arbitrarily chosen system of biomolecules and within biomolecules. An essential requirement of the model examined above is the need at each stage for a reversible pathway of transformation from product to predecessor. As previously stated, cyclicity of metabolic pathways is typical of biological systems; it is a conjunction of the processes of anabolism and catabolism. This justifies the use of the model previously discussed as a basis for the description of the general regularities of distribution of isotopes in biological systems. However, when limited parts of the pathways of biosynthesis

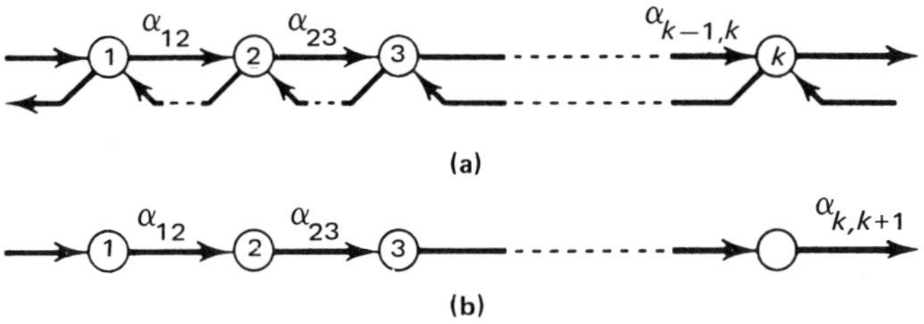

FIG. 6.3 Model of the sequence of reactions: (a) including reverse pathways of transformation at each stage; (b) reverse pathways absent.

are examined and when some biosystems are investigated *in vitro*, the condition that the individual stages be closed may not be fulfilled. Figure 6.3 shows the two extreme cases—part (a) corresponds to the model examined above, and (b) to the absence of reverse pathways. Whereas in the first case the isotopic compositions of the first and kth components are related by the already-mentioned relationship

$$R_k = R_1 \frac{1}{\alpha_{12}} \frac{1}{\alpha_{23}} \cdots \frac{1}{\alpha_{k-1,k}},$$

in the second case this relationship has the form

$$R_k = R_1 \frac{1}{\alpha_{12}} \alpha_{k,k+1},$$

i.e., the isotopic effects accompanying the formation of intermediate metabolites drop out. Consequently, the chain of fractionation coefficients that relate isotopic compositions to their β factors is broken. Therefore, in the second case a correlation between the isotopic compositions and β values may be absent. Thus, despite the thermodynamic character of isotopic effects of individual enzymatic reactions, thermodynamically ordered distribution of isotopes in the system as a whole may not be realized.

The next important condition is the steady-state condition. Formula (5.78), which describes the separation of isotopes in a fractionating column, is valid only in an established regime of operation of the column. Therefore it can be applied only to steady-state metabolic systems. If a substantial change occurs in the rates of individual processes or ratios of concentrations of the metabolites in the course of biosynthesis, this disrupts the distribution of isotopes conveyed by formula (5.98) and thus disrupts the thermodynamically ordered distribution of isotopes. Therefore, in extreme settings, organisms may show a distribution of isotopes that is not typical of them under conditions of normal development. Such facts have actually been observed.

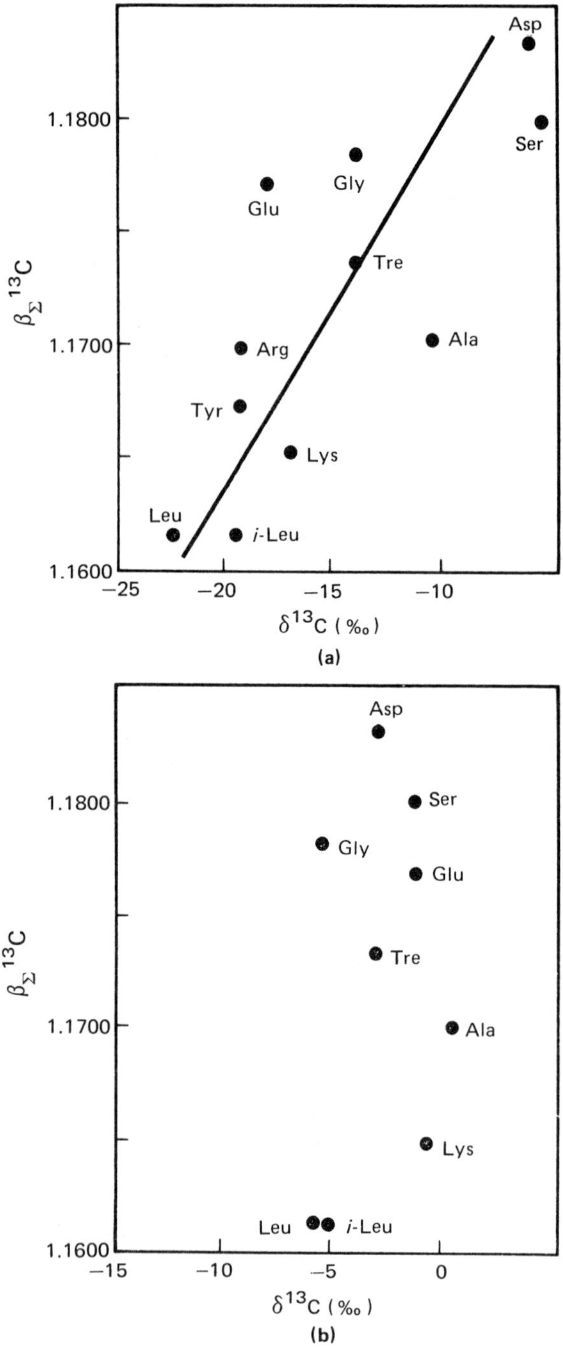

FIG. 6.4 Relationship of $\beta_\Sigma{}^{13}C$ and $\delta^{13}C$ values of amino acids (a) isolated from photosynthetic cultures of *Chlorella pyrenoidosa* and (b) from the same culture grown heterotrophically on glucose.

In heterotrophic assimilation, fragments of organic compounds synthesized outside a given organism are directly included in its biological systems. Obviously this should lead to disruption of thermodynamically ordered distribution of isotopes between and within the biomolecules of the heterotrophs. Available data (as follows) confirm this. For example, amino acids from *Chlorella pyrenoidosa* and *Escherichia coli*, grown heterotrophically on glucose, do not show the presence of a $\delta^{13}C$–$\beta_{\Sigma}{}^{13}C$ relationship, whereas that relationship does occur for photosynthetic organisms, including photosynthetic *Chlorella pyrenoidosa*, the isotopic composition of the amino acids of which was studied by Abelson and Hoering (1961) (Fig. 6.4).

Investigation of the causes of departure from a thermodynamically ordered isotopic distribution may, in certain cases, be a means of studying the details of biosynthesis, e.g., investigating the pathological development of an organism.

On the other hand, in the investigation of a large number of cases the tendency toward thermodynamically ordered distribution of isotopes should become more and more systematic. For the reasons mentioned above, a thermodynamically ordered distribution of isotopes may often be absent within molecules of a given kind in a given organism. However, there is much greater likelihood of its appearance in systems of different biomolecules of a given organism or within molecules of a given kind from a multitude of organisms. That is why the thermodynamically ordered distribution of isotopes is especially important for geochemistry, which averages the material composition of many individuals from different populations.

REFERENCES

Lehninger, A. (1974). "Biochemistry" (Russian translation). Mir, Moscow.

Rubin, B. A., and Gavrilenko, V. F. (1977). "Biochemistry and Physiology of Photosynthesis." Izd-vo Mosk. Univ., Moscow.

CHAPTER 7

Interpretation of the Results of Some Biochemical Experiments

I. ENZYMATIC DECARBOXYLATION OF PYRUVATE

In connection with the causes of biological fractionation of isotopes, it is of interest to consider the experimental data obtained by DeNiro and Epstein (1977) on the fractionation of isotopes in the biosynthesis of lipids. These data, as we shall show, can be given a different interpretation than that given by the authors of the work.

As early as 1960, Abelson and Hoering (1961) had established that lipids synthesized by microorganisms growing on glucose were impoverished in the ^{13}C isotope relative to the original glucose carbon. In this case the biosynthesis of lipids is accomplished by glycolitic decomposition of glucose to pyruvate as a result of reactions on the Embden–Meyerhof pathway, followed by decarboxylation and oxidation of the pyruvate to acetate, and condensation of the latter into the molecular structures of lipids.

DeNiro and Epstein attempted to establish at which stage during the transformation of glucose carbon into lipid carbon, the isotopic fractionation occurs that leads to the observed impoverishment of lipids in the ^{13}C isotope. (Let us note that, even in the way they posed the problem, DeNiro and Epstein based their work on the traditional idea of the existence of individual fractionation *barriers* on the pathways of biosynthesis.) To solve the problem, they used, as the original substance in transformation of *Escherichia coli*, one of the predecessors on the pathway of biosynthesis of lipids: glucose, pyruvate, or acetate. The results of the experiments are shown in Table 7.1.

As a comment on Table 7.1 let us mention that the original glucose, pyruvate, and acetate were independent chemical preparates. Here only the extent of difference between the isotopic composition of the lipids and of the source was significant. When glucose was the source of the carbon, the

TABLE 7.1

EFFECT OF SOURCE OF CARBON ON THE ISOTOPIC COMPOSITION OF THE LIPID FRACTION OF *E. coli*[a]

Source of carbon	$\delta^{13}C$ (‰) *Source*	*Lipids*
Glucose	−9.5	−15.7
		−16.3
Sodium pyruvate	−20.5	−28.9
		−28.9
Sodium acetate	−20.1	−21.5
		−20.7

[a] Two cultures were studied individually for each type of original carbon (after DeNiro and Epstein, 1977).

lipids were depleted by 6 to 7 ‰ in the ^{13}C isotope; when pyruvate was the source, by 8 %; and when acetate was the source, by 0.5 to 1.5 ‰. DeNiro and Epstein concluded that isotopic fractionation was related to the stages of decarboxylation and oxidation of the pyruvate to acetyl-CoA. In their opinion, there was no isotopic effect at the stage of formation of pyruvate from glucose because the lipids were almost equally impoverished in the ^{13}C isotope relative to both sources. The same was true for the stage of formation of lipids from acetate.

Later DeNiro and Epstein carried out an experiment *in vitro* to determine with which specific atom of pyruvate the supposed isotopic effect was associated (Table 7.2). The pertinent part of their experiment consisted of the following. The original pyruvate was introduced into a reaction with *pyruvate decarboxylase*. Then the isotopic composition of the acetaldehyde and CO_2 that formed was analyzed (the methyl and carbonyl carbon of the acetaldehyde were analyzed separately). The isotopic distribution was compared in two cases, when the enzymatic reaction was brought to full completion, total disappearance of the pyruvate (Experiments A and E) and in a very early stage of the reaction, with 0.5 to 1.9 % conversion of pyruvate into the product (Experiments B, C, D, F). The latter experiment was carried out at different temperatures.

It is obvious that in the first case, with 100 % output, the reaction product acquires an isotopic composition identical to the original, i.e., measurement of the intramolecular distribution in acetaldehyde and CO_2 shows the intramolecular distribution in the original pyruvate (Experiment E). In

TABLE 7.2

ISOTOPIC EFFECTS[a] OF PYRUVATE DECARBOXYLASE[b]

Experiment	Conditions of experiment: Temp. (°C)	Conditions of experiment: Reaction (%)	$\delta^{13}C$ (‰): Acetaldehyde		$\delta^{13}C$ (‰): Carbon dioxide
A	—	100	−20.6 ± 0.4		−22.5 ± 0.3
B	15	0.5	−26.5 ± 0.2		−28.1 ± 0.5
C	25	1.4	−28.5 ± 0.2		−28.8 ± 0.1
D	35	0.8	−30.2 ± 0.7		−30.2 ± 0.5
			Methyl carbon	*Carbonyl carbon*	
E	—	100	−20.1 ± 0.6	−19.9 ± 0.3	−22.3 ± 0.2
F	25	1.9	−21.1 ± 0.6	−34.0 ± 0.4	−28.8 ± 0.1

[a] DeNiro and Epstein (1977) called these effects kinetic.
[b] After DeNiro and Epstein (1977).

the second case, the isotopic effect characteristic of the reaction should be manifested in pure form. It must be noted that DeNiro and Epstein formerly called that effect *kinetic*, although their own work provided no grounds for that. Apparently it simply did not seem possible that biological isotopic effects of any other type could exist.

The results of the experiments presented in Table 7.2 were interpreted by DeNiro and Epstein as follows. Acetaldehyde formed in the 1.4% reaction at a temperature of 25°C (Experiment C) was 7.9‰ lighter than the original pyruvate (Experiment A), which had a $\delta^{13}C = -20.6$‰. Analysis of each of the three carbon atoms of the pyruvate (Experiment E) and products of the 1.9% reaction at 25°C (Experiment F) showed that the depletion of the aldehyde in the ^{13}C isotope was chiefly related to the carboxyl carbon atom.

On the basis of these observations, DeNiro and Epstein came to the following conclusions. First, in the course of conversion of glucose into pyruvate, the isotopic effect is small or absent. Second, oxidation of pyruvate to acetyl-CoA by pyruvate dehydrogenase decreases the ^{13}C content in acetyl relative to pyruvate. The amount of depletion in the ^{13}C isotope is determined by a temperature-dependent kinetic isotopic effect. Depletion of acetyl-CoA in the ^{13}C isotope is related to the carbonyl carbon atom, and the methyl group inherits the isotopic composition of the glucose. Third, inclusion of the acetyl group depleted in the ^{13}C isotope in the lipid component is responsible for the depletion of the lipid fraction in the ^{13}C isotope.

The difference in $\delta^{13}C$ between the methyl and carbonyl groups of acetyl is inherited by the carbon atoms of the lipid components originating from them.

These conclusions concerning important details of the biosynthesis of lipids are logical consequences of considering the results of the experiment from the traditional viewpoint that biological isotope effects are kinetic in nature. However, examination of these results from the viewpoint of the theory of biological fractionation of isotopes presented here leads to different conclusions.

The enzymatic process investigated by DeNiro and Epstein has the following scheme:

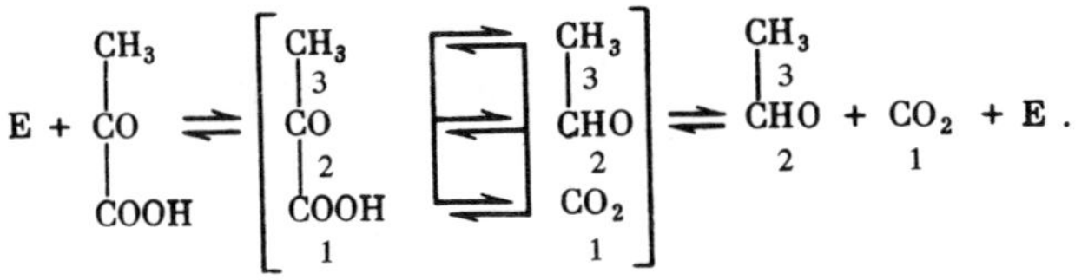

The conditions of the experiment in which the isotopes were fractionated between carbon atoms of pyruvate and acetaldehyde correspond to the model of an isolated substrate–product enzymatic system with a low degree (1.0–1.9%) of process completion ($f \ll 1$). There is no transfer effect. In principle, isotopic effects of thermodynamic and kinetic nature are possible. The fractionation of isotopes is described by the formula

$$\delta^{13}C_{P_i} = \delta^{13}C_{S_{0_i}} + \kappa\left(\frac{\beta_{P_i}}{\beta_{S_i}} - 1\right) \times 10^3 + \Delta_{\lambda_i}.$$

The isotopic compositions of each of the carbon atoms of the pyruvate ($\delta^{13}C_{S_{0_i}}$), and of acetaldehyde and CO_2 ($\delta^{13}C_{P_i}$) were measured. The coefficient κ characterizes the reaction as a whole; i.e., it is the same for the transformation of carbon in all three positions. The values of the thermodynamic isotopic factors β_{P_i} and β_{S_i}, calculated using the additive method, follow:

$$\beta_{COOH}(\text{pyruvate}) = L_{C-C} + L_{C-O} + L_{C=O} + l_{C=O} + l_{C-C} = 1.2019;$$

$$\beta_{CO}(\text{pyruvate}) = 2L_{C-C} + L_{C=O} + l_{C=O} + l_{C-O} = 1.1933;$$

$$\beta_{CH_3}(\text{pyruvate}) = 3L_{C-H} + L_{C-C} + l_{C=O} + l_{C-C} = 1.1357;$$

$$\beta_{CO_2} = 2L_{C=O} = 1.1916;$$

$$\beta_{CHO}(\text{acetaldehyde}) = L_{C-C} + L_{C=O} + L_{C-H} + 1.1706;$$

$$\beta_{CH_3}(\text{acetaldehyde}) = 3L_{C-H} + L_{C-C} + l_{C=O} = 1.1344.$$

A kinetic isotopic effect can occur when the bond between the carbonyl and carboxyl carbon of the pyruvate is broken. The probability of enrichment of the carbon of the product in the light isotope on both sides of the ruptured bond is the same in first approximation $\Delta_{(COOH)} \simeq \Delta_{(CO)} \simeq \Delta_\lambda$. Methyl carbon transforms without reorganization of the bond. Thus for the three carbon atoms we have the following system:

$$-28.8 = -22.3 + \kappa\left(\frac{1.1916}{1.2019} - 1\right) \times 10^3 + \Delta_\lambda,$$

$$-34.0 = -19.9 + \kappa\left(\frac{1.1706}{1.1933} - 1\right) \times 10^3 + \Delta_\lambda,$$

$$-21.2 = -20.1 + \kappa\left(\frac{1.1344}{1.1357} - 1\right) \times 10^3.$$

Solving the first two equations jointly, we find $\kappa = 0.73$ and $\Delta_\lambda = -0.3\permil$. Comparison of the calculated value of the kinetic isotopic effect ($-0.3\permil$) with the values of the observed isotopic shifts ($-6.5\permil$ in the conversion $COOH \rightarrow CO_2$ and $-14.1\permil$ in the conversion $CO \rightarrow CHO$) shows that the contribution of the kinetic component to the total value of the isotopic effect in this enzyme reaction is insignificant.

The quantity κ has a reasonable value—it is positive and falls within the limits $0 \leq \kappa \leq 1$ prescribed by the theory. Substituting the value obtained for κ in the third equation yields, within the limits of error of the experiment, the identity of the left- and right-hand parts: $-21.2 \simeq -20.9$. Therefore, $\kappa = 0.73$ independently satisfies the third equation.

Table 7.3 gives the experimental data obtained by DeNiro and Epstein and values of the β_i factors of the corresponding fragments. It shows that the change in isotopic composition of carbon in acetaldehyde and CO_2, when compared with the isotopic composition of the corresponding carbon in pyruvate, is directly correlated with the ratio of their β_i factors. If the isotopic composition of the carbon atoms of acetaldehyde and CO_2 is calculated under the assumption that the isotopic effect is purely thermodynamic in nature (with $\kappa = 0.73$), the values obtained are very close to those measured experimentally.

This quantitative agreement, in our opinion, is good evidence for the mechanism of biological fractionation of isotopes proposed above. In this interpretation, different conclusions are drawn from the experiments of DeNiro and Epstein. First, there is no basis for assuming that isotopic fractionation does not occur in the stage of transformation of glucose into pyruvate. The corresponding β_i factors are convincing evidence that isotopic shifts must occur between the carbon atoms of pyruvate and their

TABLE 7.3

INTRAMOLECULAR DISTRIBUTION OF CARBON ISOTOPES IN PYRUVATE AND IN ACETALDEHYDE AND CO_2 OBTAINED AS A RESULT OF THE ENZYMATIC REACTION OF PYRUVATE WITH PYRUVATE DEHYDROGENASE[a]

	Pyruvate		*Acetaldehyde and* CO_2		
Group	β_i	$\delta^{13}C_{exp}$	β_i	$\delta^{13}C_{exp}$	$\delta^{13}C_{calc}$
CH_3	1.1357	−20.1	1.1344	−21.1	−20.9
—CO	1.1933	−19.9	—	—	—
—COOH	1.2019	−22.3	—	—	—
—CHO	—	—	1.1706	−34.0	−33.9
CO_2	—	—	1.1917	−28.8	−28.5

[a] Experimental data of DeNiro and Epstein (1977); calculated values of $\delta^{13}C$ determined from ratio of β factors of corresponding carbons of product and predecessor with $\kappa = 0.73$.

predecessors in glucose (Table 7.4). However, the β_Σ factors of glucose and pyruvate virtually coincide. Therefore, despite intramolecular shifts, the isotopic composition of the molecules of glucose and pyruvate should be similar. This is observed experimentally and creates the semblance of the absence of fractionation at this stage.

Second, acetaldehyde and acetate have markedly lower β_Σ factor values (see Table 7.4). Therefore, the acetate carbon should be depleted in ^{13}C

TABLE 7.4

THERMODYNAMIC ISOTOPIC FACTORS OF COMPOUNDS PRECEDING LIPIDS ON THE PATHWAY OF BIOSYNTHESIS FROM GLUCOSE

Compound	β_i			β_Σ
Glucose	CH_2OH—	—$(CHOH)_4$—	—CHO	
	1.1611	1.1820	1.1735	1.1771
Pyruvate	CH_3—	—CO—	—COOH	
	1.1357	1.1933	1.2019	1.1769
Acetate	CH_3—	—COOH		
	1.1363	1.1973		1.1668
Acetaldehyde	CH_3—	—CHO		
	1.1344	1.1706		1.1525
Lipids	COOH—···	CH_2—CH_2—···	CH_3	
	1.1985	1.522 1.1522	1.1329	

relative to pyruvate carbon. This creates the impression that fractionation of isotopes occurs at this stage. As a matter of fact, the fractionation process in this stage is no different from that in the preceding stages.

The appreciable isotopic shift between the aldehyde carbon of acetaldehyde and carbonyl carbon of pyruvate obliged DeNiro and Epstein to presume that this effect occupies an unusual place in the process of isotopic fractionation in lipid biosynthesis. In our interpretation it is quite ordinary. Isotopic shifts of similar nature occur in other carbon positions (see Table 7.3). Its higher value is caused simply by the larger difference in the β_i factors of the carbonyl and aldehyde carbon compared with the difference in β_i factors in the other carbon positions.

Third, on the basis of the similarity of the isotopic composition of the carbon of acetate and lipids, DeNiro and Epstein concluded that in the stage of synthesis of lipid structures from acetyl-CoA there is no isotopic fractionation and, consequently, the carbon atoms of the lipids inherit the isotopic composition of the methyl and carboxyl carbon of the acetate from which they originate. However, the absence of appreciable differences between the isotopic composition of the carbon of lipids and acetate does not mean (as for glucose and pyruvate) that there is no isotopic fractionation. The β_Σ factor values of lipids lie in the same range of values as the β_Σ factors of acetaldehyde. Therefore, it should be expected that their carbon will have similar isotopic composition on the whole. However, in individual carbon positions appreciable isotopic shifts can be accomplished. The ratio of β_i factors is such that the CH_2 groups of lipids originating from the carboxyl carbon of acetate would be depleted in the ^{13}C isotope, and the CH_2 groups originating from the methyl carbon of acetate enriched in the ^{13}C isotope.

In conclusion, it must be pointed out that the results of the experiment of DeNiro and Epstein illustrate the theoretical point that, in an isolated enzymatic reaction, isotopic fractionation takes place only between product and predecessor carbon, and in that case a thermodynamically ordered distribution of isotopes within a molecule of the product may not be achieved. Thermodynamically ordered distribution of isotopes is a property of metabolic biosystems, the result of a higher metabolic level of isotopic fractionation. Actually, despite the fact that acetaldehyde and CO_2 are the products of an enzymatic reaction, the intramolecular distribution of carbon isotopes in acetaldehyde and between acetaldehyde and CO_2 does not correspond completely to the values of the β_i factors. Carbonyl carbon, which has a relatively high β_i factor value ($\beta_i = 1.1706$), is depleted in the ^{13}C isotope ($\delta^{13}C = -34.0‰$), while methyl carbon with a low value of the β_i factor ($\beta_i = 1.1344$) is enriched in the heavy isotope ($\delta^{13}C = 21.1‰$), relative to both carbonyl carbon and CO_2 carbon ($\delta^{13}C = -28.8‰$,

$\beta_i = 1.1916$). From this it must be concluded that there is a definite limitation of biochemical experiments *in vitro* as models of isotopic fractionation in living organisms.

II. MICROBIOLOGICAL OXIDATION OF ETHANOL TO ACETIC ACID

Rinaldi *et al.* (1974) investigated the isotopic composition of methyl and carboxyl carbon in acetic acid formed by microbiological oxidation of ethanol. The experiment proceeded as follows. A pure culture of *Acetobacter suboxydans* was grown on a 2% solution of synthetic ethanol at 30°C for 72 hr. Ethanol was the only source of energy for the microorganisms. As a result of the experiment, 0.0101 mol of acetic acid was formed from 0.0343 mol of the original ethanol; 0.0242 mol of ethanol remained unused.

The ethanol used in the experiment was obtained abiogenically by hydration of ethylene. Analysis of the intramolecular distribution of carbon isotopes showed that the hydroxyl carbon of the original ethanol was 4.6‰ poorer in the ^{13}C isotope than the methyl carbon, as shown in the following tabulation:

Ethanol	CH_3—	CH_2OH	Σ	*Acetic acid*	CH_3—	COOH	Σ
$\beta^{13}C$	1.1332	1.1582	1.1457	$\beta^{13}C$	1.1360	1.1972	1.1666
$\delta^{13}C_{exp}$	−25.5	−30.1	−27.8	$\delta^{13}C_{exp}$	−34.6	−30.8	−32.7

In the acetic acid obtained as a result of microbiological oxidation of ethanol, the intramolecular distribution of isotopes was the opposite—carboxyl carbon was 3.8‰ richer in the ^{13}C isotope than methyl carbon. The conclusion of the authors of the experiment was that the carboxyl carbon of the acetyl group of CoA has the ability to be enriched in the ^{13}C isotope. The mechanism of this process is not discussed in the work by Rinaldi *et al.*

Examination of the results presented above does not at first glance create the impression that the measured isotopic compositions of carbon have any relation to the values of the respective β factors. The carbons of the methyl groups of ethanol and acetic acid differ markedly in isotopic composition, although their thermodynamic isotopic factors are nearly the same. On the other hand, the isotopic composition of the carboxyl carbon of the acetate coincides with the isotopic composition of the hydroxyl carbon of ethanol, whereas the values of the corresponding β_i factors are substanially different. On the whole, acetic acid was depleted in the ^{13}C isotope, although it is characterized by a higher value of the β_Σ factor than ethanol. Nevertheless, as we shall see, the observed distribution of isotopes can be described by the theoretical considerations derived above.

The process of fractionation of isotopes in the given case is complicated by two circumstances. First, biochemical oxidation of ethanol in the living cell is accompanied by passage of matter through the cell walls. Thus isotopic transfer effects are possible. Second, measurement of the isotopic composition of the product refers to the moment when an appreciable part of the ethanol (about 30%) was transformed (the condition $f \ll 1$ cannot be taken and it is necessary to work with a change in isotopic composition of the metabolites during the reaction.)

In other respects the microbiological experiment of Rinaldi *et al.* presents favorable possibilities for analysis of the isotopic fractionation process. Acetic acid fermentation with *Acetobacter suboxydans* occurs without formation of any satellite products. The authors of the experiment confirm this by pointing out the existence of a precise isotopic balance of carbon between the original ethanol on the one hand and the unused ethanol and acetic acid on the other (the percentage of carbon going into an increase in biomass of the microorganisms can be neglected):

$$(-28.4)(0.0343) = (-26.6)(0.0242) + (0.0101)(-32.7).$$

Acetic acid is the final product. It is not subject to further oxidation. Its methyl carbon is derived from the methyl carbon of ethanol, and the carboxyl carbon from the hydroxyl carbon of ethanol.

The process of biochemical oxidation of ethanol can be described by an enzymatic reaction which is shown in Fig. 7.1. The model corresponding to

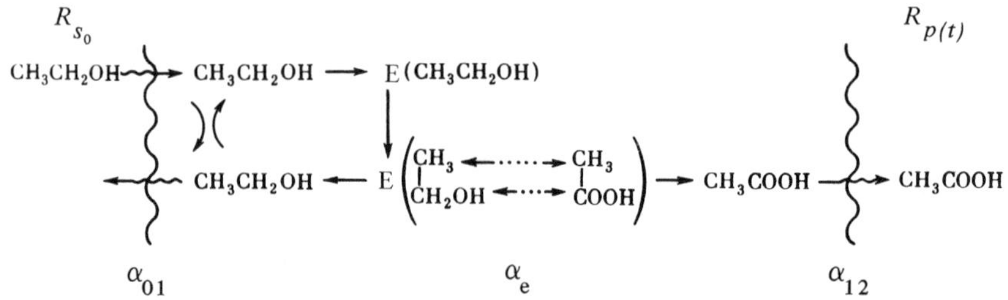

FIG. 7.1 Model of the process of biochemical oxidation of ethanol to acetic acid.

this case was considered in the relevant section V of Chapter 5. The isotopic composition of a substance which in addition to biochemical transformation is subjected to transfer through cell membranes is related to the isotopic composition of the original metabolite by Eq. (5.63). The fractionation coefficient in that case is determined by the relationship:

$$\frac{1}{\alpha_e \alpha_{TR}} = 1 + \kappa\left(\frac{\beta_P}{\beta_S} - 1\right) + (\Delta_\lambda + \Delta_{TR}) \times 10^{-3}. \tag{7.1}$$

Because 30% of the original amount of ethanol was converted in this experiment, it is necessary to use a model which takes into account the change in isotopic composition of the substrate as it is used up. Fractionation of isotopes in that model is described by the relationship [see Eq. (5.70)]

$$\frac{R_{P(t)}}{R_{S_0}} = \frac{1}{f}[1 - (1 - f)^{1/\alpha_e \alpha_{TR}}]. \tag{7.2}$$

Converting to $\delta^{13}C$ values and using Eq. (7.2), we obtain an expression which establishes the connection between the isotopic compositions of the product and substrate in the following form:

$$\delta^{13}C_{P_{i(t)}} = \delta^{13}C_{S_{0_i}} + \left\{\frac{1}{f}[1 - (1 - f)^{1 + \kappa(\beta_{P_i}/\beta_{S_i} - 1) + (\Delta\lambda_i + \Delta_{TR_i}) \times 10^{-3}}] - 1\right\} \times 10^3. \tag{7.3}$$

For the reaction depicted in Fig. 7.1, two equations of the type of (7.3) can be formulated separately: the first, for the methyl carbon of ethanol converted into methyl carbon of acetic acid, and the second for the hydroxyl carbon of ethanol converted into carboxyl carbon of acetic acid.

Both equations should contain the same value of κ. The same is true of the Δ_{TR} value, as during the transfer the kinetic preference of the isotopically light molecule over the molecule containing the ^{13}C isotope in one of the positions does not depend on which structural group contains the ^{13}C isotope. Things are less certain with respect to possible kinetic isotopic effects. Evidently, if a kinetic isotopic effect occurs, it would be greater for the hydroxyl carbon atom, for which the bond is reorganized, than for the methyl carbon atom. The values of the β_i factors of the isotopic forms taking part in the reaction have the following numerical values:

$$\beta^{13}C_{CH_3}(\text{ethanol}) = 3L_{C-H} + L_{C-C} + l_{C-O} = 1.1332,$$

$$\beta^{13}C_{CH_2OH}(\text{ethanol}) = 2L_{C-H} + L_{C-C} + L_{C-O} = 1.1582,$$

$$\beta^{13}C_{CH_3}(\text{acetic acid}) = 3L_{C-H} + L_{C-C} + l_{C-O} + l_{C=O} = 1.1360,$$

$$\beta^{13}C_{COOH}(\text{acetic acid}) = L_{C-C} + L_{C-O} + L_{C=O} = 1.1972.$$

The extent of completion of the reaction $f = 0.0101/0.0343 = 0.3$. Thus we have the system of equations:

$$-34.6 = -25.5 + \left\{\frac{1}{0.3}\left[1 - (1 - 0.3)^{1+\kappa(1.1360/1.1332-1)+(\Delta_{TR}+\Delta_{(2)\lambda})\times 10^{-3}}\right] - 1\right\} \times 10^3,$$

$$-30.8 = -30.1 + \left\{\frac{1}{0.3}\left[1 - (1 - 0.3)^{1+\kappa(1.1972/1.1582-1)+(\Delta_{TR}+\Delta_{(1)\lambda})\times 10^{-3}}\right] - 1\right\} \times 10^3.$$

After transformation, this system is reduced to the following two equations:

$$-11.0 = 2.5\kappa + \Delta_{TR} + \Delta_{(2)\lambda}$$

and

$$-0.8 = 33.7\kappa\, \Delta_{TR} + \Delta_{(1)\lambda}.$$

The presence of four unknowns makes the system indeterminate.

To calculate to what extent the observed isotopic effects agree with the theoretical model, let us simplify the problem. We shall assume equality of kinetic isotopic effects for the first and second atoms of carbon: $\Delta_{(1)\lambda} = \Delta_{(2)\lambda}$ (in a special case they may be equal to zero) and we shall consider the total nonthermodynamic isotopic effect Δ, which includes the kinetic isotopic effect and the transfer effect. We shall examine whether the values of the parameters of the fractionation process obtained by using these assumptions are valid. A solution of the system yields $\kappa = 0.33$ and $\Delta = -11.8$‰ in this case. The value of the coefficient κ is positive and falls between 0 and 1; it satisfies the requirements dictated by the meaning of the quantity κ. The negative value of Δ is also consistent with its meaning, as in the transfer process (and as a result of the kinetic isotopic effect) the product is depleted in the heavy isotope. Thus the experimental data seems to agree in principle with the theoretical model.

The fractionation process is interpreted as follows. As a result of the enzymatic reaction proper, values of $\delta^{13}C$ close to the following should be obtained†:

$$\delta^{13}C_{CH_3} = -25.5 + 0.33((1.1360/1.1322) - 1) \times 10^3 = -24.7‰,$$

$$\delta^{13}C_{COOH} = -30.1 + 0.33((1.1972/1.1582) - 1) \times 10^3 = -20‰.$$

† The significant figures of the values in these equations do not suggest corresponding precision in the calculations. They were chosen to preserve the formal arithmetical correspondence.

In this case the acetic acid would be enriched in the ^{13}C isotope ($\delta^{13}C_{CH_3COOH} = -22.3‰$) compared with the original ethanol, in agreement with the ratio of β_Σ factors of these compounds. However, as a result of the isotopic transfer effect and possibly of a certain percentage of kinetic isotopic effect, the values of $\delta^{13}C$ are shifted toward more negative ones:

$$\delta^{13}C_{CH_3} = -24.7 - 11.8 = -36.5‰,$$

$$\delta^{13}C_{COOH} = -20.10 - 11.8 = -31.8‰.$$

These values, and also the value $\delta^{13}C_{CH_3COOH} = -34.3‰$, would characterize the isotopic composition of the carbon of the acetic acid in its first batches (for $f \ll 1$). Because a considerable part of the substrate was converted in the course of microbiological oxidation of ethanol, the observed isotopic effects were somewhat reduced. As a result the $\delta^{13}C$ values of acetic acid given in the preceding tabulation were recorded.

REFERENCES

Abelson, P., and Hoering, T. (1961). Carbon isotope fractionation in amino acids by photosynthetic organisms. *Proc. Nat. Acad. Sci. USA* **47**(5), 623.

DeNiro, M., and Epstein, S, (1977). Mechanism of carbon isotope fractionation associated with lipid synthesis. *Science* **197**, 261.

Rinaldi, G., Meinschein, W. G., and Hayes, J. M. (1974). Carbon isotopic fractionations associated with acetic acid production by *acetobacter suboxydans*. *Biomed. Mass Spectrom.* **1**(6), 412.

CHAPTER 8

Fractionation of Isotopes in Biological Systems

I. THE CALVIN CYCLE

A. C_3 PLANTS

The Calvin cycle was considered previously in the discussion of the production mechanism of intramolecular isotopic effects. The reactions of this cycle are important in the biological fractionation of isotopes because in them the primary assimilation of inorganic carbon takes place and the difference between the isotopic composition of the carbon of autotrophs and the original CO_2 is established.

As mentioned in Chapter 2, terrestrial plants which accomplish photosynthesis by means of the Calvin cycle are characterized by an isotopic composition in the range of -22 to $-34‰$. In most cases the isotopic composition of the higher plants lies between $\delta^{13}C$ values of -24 to $-28‰$. The isotopic composition of atmospheric carbon dioxide averages $-7‰$ (Keeling, 1961), but in the ground-level layer of air in intracontinental areas it is closer to $-9‰$ (Galimov, 1966). Thus terrestrial plants of C_3 type are enriched in the ^{12}C isotope relative to the original carbon dioxide by 15 to 20‰. As is known, the term "C_3 plants" (in contrast to C_4 plants) means that the primary stable product of fixation of CO_2 in them is a tricarbon compound—phosphoglyceric acid (PGA).

Assimilation of external CO_2 includes two processes to which fractionation of isotopes may be related: the passage of CO_2 through the cell wall and solution in cytoplasm and the enzymatic conversion of CO_2 carbon into the carbon of organic compounds.

B. Enzymatic Carboxylation of Ribulose Diphosphate *in Vitro*

The first reaction of the Calvin cycle consists of carboxylation of ribulose 1,5-diphosphate, yielding two molecules of 3-phosphoglyceric acid (Bassham and Calvin, 1957). This reaction can be carried out *in vitro.*

The initial experiment for studying the fractionation of carbon isotopes by this reaction was carried out by Park and Epstein (1960). They used a system containing the enzyme ribulose carboxylase, ribulose 1,5-diphosphate (RbDP), and a bicarbonate buffer. During the experiment, complete transformation of ribulose diphosphate into phosphoglyceric acid (PGA) occurred and about 0.3% of the carbon of the bicarbonate buffer was used up. The isotopic composition of carbon of the reagents and products of enzymatic fixation of CO_2 is shown in the following tabulation:

	$\delta^{13}C$ (‰)	
Material	*Experiment* 1	*Experiment* 2
Ribulose 1,5-diphosphate	−13.7	−13.7
Total CO_2 dissolved in the reaction mixture	−5.4	−0.2
Triphosphoglyceric acid	−15.2	−13.9

The triphosphoglyceric acid was depleted in the ^{13}C isotope in both experiments that were conducted. Five-sixths of the carbon inherited its isotopic composition from ribulose diphosphate and one-sixth from CO_2. Calculation of the isotopic composition of the latter from the equation of the material balance yielded a value of $\delta^{13}C = -22.8$‰ for experiment 1 (17.4‰ depletion in the ^{13}C isotope) and $\delta^{13}C = -14.9$‰ for experiment 2 (14.7‰ depletion in the ^{13}C isotope). In addition, as Park and Epstein (1960) assumed, there is an isotopic effect in the process of passage of CO_2 through the cell wall. They investigated the isotopic composition of the carbon from carbon dioxide dissolved in the cell cytoplasm. It was enriched in the light isotope ($\delta^{13}C = -13.6$‰) relative to the carbon of atmospheric carbon dioxide ($\delta^{13}C = -7$‰). Summing up the isotopic effects at the stage of CO_2 arrival in the cell fluid (~7‰) and during enzymatic carboxylation (~17‰), Park and Epstein obtained a value sufficient to explain the observed enrichment of plants in the light isotope.

Based on this, Park and Epstein (1960, 1961) concluded that the main biological fractionation of isotopes occurs in the stage of formation of phosphoglyceric acid, which is the first stable product of photosynthesis.

For some time, experiments like those made by Park and Epstein (1960, 1961) were not reproduced. Thirteen years later, after the discovery of the differences between the isotopic compositions of plants of C_3 and C_4 type, Whelan *et al.* (1973) repeated the experiment with carboxylation of ribulose diphosphate *in vitro*. Using ribulose carboxylase isolated from sorghum (a plant of C_4 type), they obtained a fractionation value of -42.9‰ at 25°C and -26.1‰ at 18°C (relative to the carbon of HCO_3^-). Following this, experiments with enzymatic carboxylation of ribulose diphosphate were made by several other investigators. In the separate experiments, very different fractionation coefficients were obtained. The results obtained by Deleens *et al.* (1974) were surprising. In one of the experiments with ribulose carboxylase isolated from spinach they found isotopic fractionation reaching -98.7‰. In other cases their value of the fractionation coefficient varied from -48 to -83.6‰ (Table 8.1). Experiments made by Christeller *et al.* (1976) on carboxyl isolated from soy beans yielded fractionation values from -35.8 to -37.6‰ in the temperature range of 15 to 35°C (relative to HCO_3^-). In the experiments of Estep *et al.* (1978), who used preparates of ribulose diphosphate carboxylase iso-

TABLE 8.1

EXPERIMENTS WITH ENZYMATIC CARBOXYLATION OF RIBULOSE 1,5-DIPHOSPHATE *in vitro*[a,b]

	Plant from which enzyme was isolated		
Result	*Spinach*		*Maize*
Specific activity of preparate (μmol CO_2/mg·hr)	14		14
Isotopic composition of PGA (‰)	−19.4	−24.3	−23.6
	−20.2	−27.8	−24.0
Isotopic composition of fixed carbon (‰)	−53.4	−88.8	−78.8
	−58.2	−103.0	−81.0
Isotopic separation Δ (‰)	−48.8	−83.6	−73.4
	−52.8	−98.7	−75.8

[a] See Deleens *et al.* (1974).

[b] *Remarks*: The original ribulose 1,5-diphosphate (RbDP) had $\delta^{13}C = -12.6$‰. The isotopic composition of the bicarbonate buffer was $\delta^{13}C_{HCO_3^-} = -5.66$‰. The isotopic composition of fixed carbon was calculated from the equation $\frac{5}{6}\delta^{13}C_{RbDP} + \frac{1}{6}\delta^{13}C_{C_{fix}} = \delta^{13}C_{PGA}$. The isotopic separation was calculated from the formula $\Delta = (\delta^{13}C_{C_{fix}} - \delta^{13}C_{HCO_3^-})/(1 - \delta^{13}C_{HCO_3^-} \times 10^{-3})$. The two results under each entry pertain to two experiments with decarboxylation of the same preparate. For spinach, experiments were also carried out with two independently prepared preparates.

lated from various organisms, isotopic effects in the range of -34.3 to $-51.1‰$ (relative to bicarbonate) were established.

It should be mentioned that some uncertainty related to the calculation of the isotopic composition of the original carbon dioxide exists in the results of experiments of this type. In practice, the investigators used bicarbonate. Its carbon isotopic composition is the basis for calculating the amount of fractionation. At 25°C, 98% of the carbon dioxide is present in the form of the HCO_3^- ion. However, in the carboxylation of ribulose diphosphate, it is not the bicarbonate ion which takes part in the reaction but the CO_2 molecule. Both the thermodynamic and the kinetic isotopic effects related to this conversion deplete CO_2 in the ^{13}C isotope compared with bicarbonate. In the works of Park and Epstein (1960, 1961) and of several others, this fact was not taken into account. In later studies (Christeller *et al.*, 1976; Whelan *et al.*, 1973; Estep *et al.*, 1978) a correction was introduced which corresponded to the constant of isotopic equilibrium between HCO_3^- and CO_2. At equilibrium, dissolved CO_2 is depleted by 8‰ relative to HCO_3^- at 25°C (Mook *et al.*, 1974; Wendt, 1968). Therefore, in the previously mentioned works (Christeller *et al.*, 1976; Whelan *et al.*, 1973; Estep *et al.*, 1978), the measured isotopic effects are decreased by 8 to 9‰ (depending on the temperature at which the experiment was conducted). However, it must be noted that, although the exchange between CO_2 and HCO_3^- takes place fairly rapidly [the constant of the rate of hydration of CO_2 at 25°C is 0.0375 sec^{-1} (Gibbons and Edsall, 1963; Ho and Sturtevant, 1963)] and isotopic equilibrium is reached in less than 1 hr, there is no evidence that isotopic exchange equilibrium actually was reached under the conditions of the experiments cited. The existence of a directional flux of carbon dioxide produced by the reaction can lead to isotopic effects different from equilibrium effects. This must be remembered in accepting the calculations of fractionation coefficients made by Estep *et al.* (1978). They calculated the range of variation of the amount of isotopic fractionation in the carboxylation of RbDP (relative to CO_2) as being within -26.1 to $-42.9‰$. The average value was close to $-35.0‰$. Taking into account the results obtained by all investigators, the range of variation is from -5 to $-90‰$.

Thus experimental study of the process of carboxylation of ribulose diphosphate *in vitro* showed considerable enrichment of the assimilated carbon in the light isotope compared with the carbon of the inorganic substrate. The typical problems of the results obtained are as follows: (1) the value of the isotopic effect *in vitro* is too high, substantially exceeding the observed enrichment of the light isotope in plants and (2) the variation in the isotopic effect observed in the various experiments far exceeds the range of variation of the isotopic composition of carbon of plants in nature.

C. Theoretical Analysis of Isotopic Fractionation in the Calvin Cycle

The process of carboxylation takes place in several stages (Rubin and Gavrilenko, 1977). First, isomerization of ribulose diphosphate occurs. Second, the carboxylation reaction itself occurs, with the formation of the intermediate product 2-carboxy-3-ketoribitol 1,5-diphosphate. The final stage leads to the formation of two molecules of phosphoglyceric acid. If the intermediate stages are reversible, then relationships of (5.52) type, which directly relate the isotopic composition of the carbon atoms of tri-phosphoglyceric acid and the original ribulose 1,5-diphosphate, can be used. Analysis of these relationships shows that as a result of the reaction a difference should arise in the intramolecular distribution of isotopes between the carbon atoms of PGA and the predecessor atoms in the ribulose diphosphate. The thermodynamic isotopic effects should result in a small enrichment of the total carbon of PGA in the heavy isotope (by about 2‰ for $\kappa = 0.5$).

Experiments *in vitro* do not permit observation of the total picture of the isotopic shifts between the product carbon and the predecessor carbon because under the conditions of the experiment the ribulose-1,5-diphosphate is completely used up. In this case the isotopic composition of its carbon reproduces the carbon in PGA in the corresponding positions. The only isotopic effect that can be observed results from joining the CO_2 carbon to the second atom of isomerized ribulose diphosphate:

$$\delta^{13}C_{CO_2} = \delta^{13}C_{COOH} + \kappa(1.2017/1.1916 - 1) \times 10^3 + \Delta_\lambda.$$

The small thermodynamic component of this isotopic effect corresponds to a slight enrichment of carboxyl carbon in the ^{13}C isotope relative to the original CO_2. Consequently, the enrichment of assimilated carbon of PGA in the light isotope observed in the experiments is produced by a kinetic isotopic effect Δ_λ.

However, it is not known whether the isotopic fractionation occurring in organisms in the process of assimilating CO_2 is accomplished in the same way as it occurs in enzymatic carboxylation of ribulose diphosphate *in vitro*. Doubts are raised by the previously mentioned experimental results, namely, the high and widely varying measured fractionation values.

Even if the anomalously high values of fractionation (50–80‰) are not taken in the calculation and average values of $\Delta C_{org}–CO_2 = -35\%$ are the starting point (Christeller *et al.*, 1976; Estep *et al.*, 1978), it would be expected that the isotopic composition of C_3 plants would be about −42.0‰ on the average. However, this value is far beyond the limits of the observed range of $\delta^{13}C$ of C_3 plants. To the expected value must be added the possible isotopic effect in the introduction of CO_2 from the air into the cell fluid. Park and Epstein (1960) believed that this effect constitutes about $-7‰$. Whelan *et al.* (1973) doubted the data of Park and Epstein, citing work in which there was an absence of fractionation between gaseous and dissolved CO_2 (Deuser and Degens, 1967; Mook *et al.*, 1974; Wendt, 1968). The latter actually is only 1.0009 at 25°C, as was demonstrated by Thode *et al.* (1965). However, this is not a valid argument because in this case it was a matter of equilibrium of isotopic exchange in the CO_2(gas)–CO_2(solution) system. Meanwhile, in the process of intake of atmospheric CO_2 kinetic isotopic effects are possible, caused by preferential diffusion and sorption of isotopically light CO_2 (Baertschi, 1952). Thus only values in the range of -40 to $-50‰$, when observed in plants, can be considered in agreement with the results obtained in experiments *in vitro*.

Another important point is the exceptional variability of fractionation values observed in the experiments *in vitro*. The range of variation of the values of the isotopic effects is from -5 to $-90‰$(relative to calculated values of $\delta^{13}C_{CO_2}$). Even in the same series of experiments, which had as its purpose to obtain "reproducible results,... by careful control of the reaction conditions" (Estep *et al.*, 1978, p. 680), the spread of fractionation values was from -25.5 to $-42.3‰$ (with respect to calculated $\delta^{13}C_{CO_2}$). This exceeds the whole range of variations of isotopic composition of C_3 plants growing on land, although a vast number of other factors affect the latter including variations in the isotopic composition of the original carbon dioxide.

It is obvious that there is some essential difference between fractionation of isotopes when the process takes place in the living organism and when it takes place under laboratory conditions.

Estep *et al.* (1978) acknowledge that the isotopic effects of the reactions of carboxylation of RbDP *in vitro* are "much too high" and suggest several explanations. They note that in living organisms the isotopic effect may be lower because the concentration of CO_2 in the atmosphere is at a level that does not saturate ribulose diphosphate. However, under conditions of high concentration of CO_2, individuals would be encountered whose isotopic composition would correspond to maximum fractionation ($\delta^{13}C = -40$ to $-50‰$ or less). In nature, environments with a high concentration of CO_2 are fairly frequently encountered, for instance, in soils, but organisms with such isotopic composition have not been reported. In addition, numerous laboratory experiments are known in which organisms grew under conditions of an abundant supply of CO_2 at concentrations (5–10%) considerably exceeding the concentration of CO_2 in the atmosphere (Degens *et al.*, 1968). In these cases the organisms usually had an isotopic composition of -20 to $-25‰$.

Later these same authors quoted the work of Kelly (1976) on "the possibility of existence of other autocatalytic pathways of CO_2 fixation which might in part explain the rather great difference between isotopic fractionation *in vivo* and *in vitro*" (Estep *et al.*, 1978, p. 686). Actually, the existence of parallel pathways of CO_2 fixation with a smaller fractionation coefficient could reduce the enrichment of the organism as a whole in the light isotope. However, greater differences in the isotopic composition of individual compounds and biochemical fractions should then arise, caused by the existence of pathways resulting from forms of assimilated carbon that differ sharply in isotopic composition. Meanwhile, in investigations of intermolecular and intramolecular subjects, no such isotopic differences (of the order of -40 to $-50‰$) have been observed (see Chapter 4).

In general, any factor reducing the value of isotopic enrichment of plants would inevitably lead simultaneously to an increase in the range of variations of isotopic composition of plants (because of the inevitable variability of this factor under natural conditions). The difficulty is to reconcile the larger isotopic effects and their wide range *in vitro* with the much smaller values of enrichment and, at the same time, narrower range of variation of isotopic composition of plants in nature.

Possibly the most logical way to eliminate this contradiction, although it may seem paradoxical, is to assume that the isotopic effects in the reaction of decarboxylation of RbDP developed in an organism are small, i.e., to assume that in this case the *in vitro* experiments are not equivalent to the *in vivo* process. That conclusion can possibly be justified for a number of reasons. Bahr and Jensen (1974) recently showed that *in vivo* the enzyme ribulose diphosphate carboxylase is characterized by a $K(CO_2)$ constant essentially different from that for a purified preparate. It has been acknowledged (Estep *et al.*, 1978) that this might be a reason for the production of an equilibrium isotopic effect *in vivo*. As shown previously, the theoretical value of the thermodynamic isotopic effect of the carboxylation reaction is small. According to our calculations of the $\delta^{13}C$ value, there would be a small enrichment (2–4 ‰) of the carboxyl carbon in the ^{13}C isotope relative to CO_2. Not long ago, O'Leary and Yapp (1978) demonstrated experimentally that in an equilibrium decarboxylation reaction, where the decarboxylation of isocitric acid was investigated using the enzyme isocitrate dehydrogenase, carboxyl carbon was enriched by 2.7 ‰ in the ^{13}C isotope with respect to CO_2. We note that this confirms the validity of our method of calculating the thermodynamic isotopic effects in reactions of organic compounds. Mistakenly, the authors cited their results as contradictory to our point of view. They write:

> ...Galimov† recently proposed that isotope ratios in biological systems are the result of equilibrium rather than kinetic isotopic effects. The value of the equilibrium isotopic

† The reference is to Galimov (1973).

effect of carboxylation [i.e., +2.7‰] indicates that if isotopic equilibrium were achieved at the stage of carboxylation in photosynthesis, plants would be enriched in the ^{13}C isotope with respect to atmospheric CO_2, i.e., just the opposite of what is observed (O'Leary and Yapp, 1978, p. 159).

But we do not believe that enrichment of plant carbon is caused solely by an isotopic effect in the carboxylation stage. If one considers only the thermodynamic isotopic effect, then inclusion of CO_2 in the form of the carboxyl group of the primary acceptor results, as previously stated, in a small enrichment of the carboxyl carbon in the ^{13}C isotope. Enrichment of organic carbon in the light isotope would take place in the stage of reduction of carboxyl carbon.

The first product of photosynthesis, 3-phosphoglyceric acid, is transformed, after phosphorylization to 1,3-diphosphoglyceric acid, into glyceraldehyde-3-phosphate with the participation of NADF·H and the enzyme glyceraldehyde phosphate dehydrogenase (Rubin and Gavrilenko, 1977):

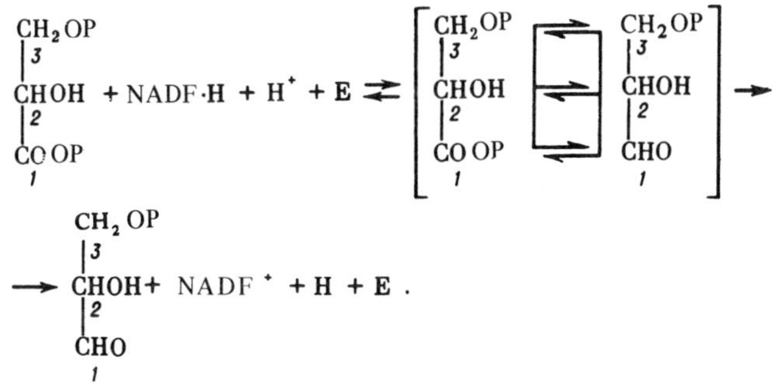

Reduction of carboxyl carbon occurs at this stage. The isotopic effect is expressed by the relationship

$$\delta^{13}C_{P_{(1)}} = \delta^{13}C_{S_{0_{(1)}}} + \kappa(1.1738/1.2004 - 1) \times 10^3 + \Delta_{\lambda_{(1)}},$$

or

$$\Delta_{CHO-COOH} = -22\kappa + \Delta_{\lambda_{(1)}}.$$

The thermodynamic isotopic effect of this reaction suggests enrichment of the assimilated carbon in the ^{12}C isotope commensurate with what is observed in plants. The realization of the thermodynamic isotopic effect requires fulfillment of the conditions formulated in Chapter 5. These conditions probably are fulfilled in the Calvin cycle, which functions in C_3 plants.

Thus enrichment of C_3 plants in the light isotope relative to the original CO_2 is explainable without the fact that in the first reaction of photosynthesis—carboxylation of ribulose diphosphate—there is enrichment of the assimilated carbon in the ^{12}C isotope. On the contrary, even some depletion of it in the light isotope (by 2 to 4‰) is presumed. Therefore, the question of the nature of isotopic effects at the stage of photosynthetic fixation of carbon must remain unanswered. The ratio of thermodynamic and kinetic isotopic effects in this process can be ascertained only by means of experiments which take into account all aspects of the problem.

II. THE HATCH–SLACK CYCLE

In the 1960s it was established that the first product of CO_2 fixation in photosynthesis is not in all cases the 3-carbon compound phosphoglyceric acid. In some plants exogenous carbon first enters into the composition of 4-carbon compounds such as oxaloacetic and malic acids (Karpilov, 1960; Kortschak, 1965). From this discovery plants with C_3- and C_4-types of photosynthesis were differentiated. Typical C_4 plants are maize, sorghum, and sugar cane. In succulents both types of photosynthesis operate and, depending on conditions such as the length of daylight, conversion occurs from one type of assimilation of CO_2 to the other (Lerman and Queiroz, 1974; Maruyama *et al.*, 1966).

There is a series of physical differences between C_3 and C_4 plants. Some time ago it was established that plants such as maize or sugar cane have a very low threshold concentration of CO_2—7 to 9 parts per million (ppm) compared with the usual 60–150 ppm for C_3 plants (Moss, 1962). Generally very efficient CO_2 assimilation is typical of C_4 plants.

A remarkable feature of C_4 plants is the absence or very low rate of photorespiration and, as a result, high productivity of photosynthesis (Jackson and Volk, 1970; Laing and Forde, 1971). The photorespiration coefficient for C_4 plants lies within 1.05 and 1.1, while for C_3 plants a ratio of 3.0 or more is typical (Zelitch and Day, 1973).

There are a number of special features in the structure of the chloroplasts and the anatomy of the chlorophyll cells of C_4-type plants (Laetsch, 1969; Tregunna *et al.*, 1970; Whelan *et al.*, 1973).

The physiological characteristics of plants of C_4 type are regarded as the result of adaptation to the conditions of the evolving environment (Kennedy and Laetsch, 1974; Smith and Epstein, 1971). Plants of C_4 type, which have economical gas and water exchange, are better adapted to the conditions of a hot, dry climate. Most C_4-type plants grow in tropical regions and in desert zones (Kennedy and Laetsch, 1974).

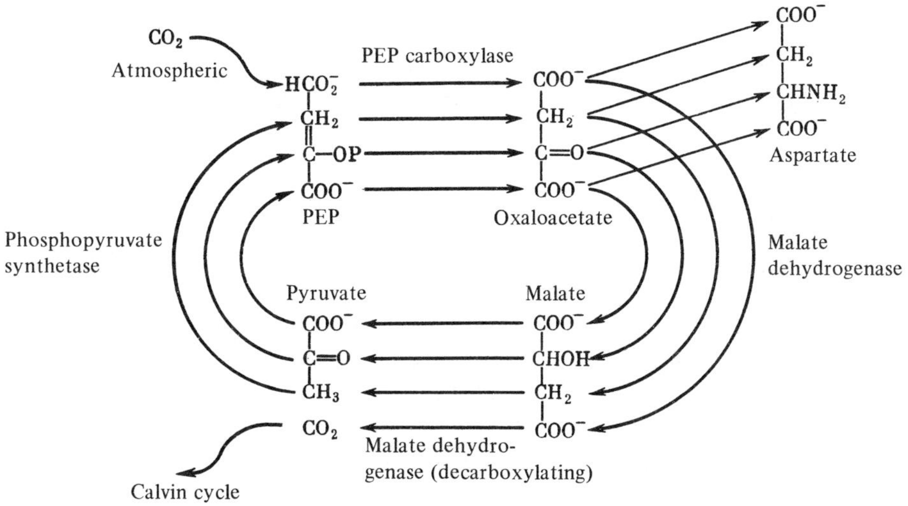

FIG. 8.1 The Hatch–Slack cycle.

Investigations have shown that, in addition to the Calvin cycle, a cycle which has been named the Hatch–Slack cycle functions in plants of C_4 type (Hatch and Slack, 1966, 1970). The first acceptor of CO_2 in this cycle is phosphoenol pyruvate (PEP) (Fig. 8.1). The enzyme phosphoenol pyruvate carboxylase catalyzes the attachment of CO_2. As a result, oxaloacetic acid is formed. It is reduced to malic acid. Transamination of the malic acid leads to formation of aspartic acid. These 4-carbon dicarboxylic acids in C_4-type plants contain about 90% of the tagged carbon in the first 6 sec after exposure to $^{14}CO_2$. In the same exposure time more than 90% of the ^{14}C in C_3 plants is localized in phosphoglyceric acid and sugar phosphates (Chen *et al.*, 1971; Kortschak *et al.*, 1965). Sometimes the most common metabolite among the products of primary fixation of CO_2 is alanine (Kennedy and Laetsch, 1974).

The Hatch–Slack cycle is completed by decarboxylation of organic acids and regeneration of the phosphoenol pyruvate. The liberated CO_2 is refixed in the Calvin cycle.

A. CARBON ISOTOPIC COMPOSITION OF C_4 PLANTS AND SUCCULENTS

The distinguishing feature of C_4 plants proved to be their sharp difference from C_3 plants in carbon isotopic composition. Whereas terrestrial plants of C_3 type are enriched in the ^{12}C isotope by 15 to 20‰ relative to atmospheric carbon dioxide, in most cases C_4 plants are enriched by only 2 to 8‰.

Bender (1968) apparently was the first to suggest that the relative depletion of some plants (e.g., maize and sugar cane) in the light isotope is related to the functioning of the Hatch–Slack cycle in those plants. Smith and Epstein (1971), who investigated a substantial collection of terrestrial plants, showed that their isotopic compositions form two distinct ranges of $\delta^{13}C$ values. Plants of C_3 type belong to the category of those highly enriched in the ^{12}C isotope, with $\delta^{13}C$ from -23.2 to -34.3‰. Plants of C_4 type constitute the category of plants slightly enriched in the light isotope, with $\delta^{13}C$ from 5.6 to 18.6‰. Lesser enrichment in the ^{12}C isotope proved to be typical of CO_2 in soils on which C_4-type plants grow (Lerman, 1972) [compared with soil CO_2 from a region where predominately C_3 plants are cultivated; Galimov (1966)]. Low enrichment of carbon in the ^{12}C isotope, as in modern C_4 plants, has also been established in fossil forms of plants of that type (Troughton *et al.*, 1974). Plants of the same family and even the same species may belong to either C_3 or C_4 types. The isotopic composition of plants of the same species is essentially different in these cases, depending on which type of photosynthetic fixation of carbon takes place in them (Smith and Epstein, 1971).

Succulents have isotopic compositions intermediate between C_3 and C_4 plants (Bender *et al.*, 1973; Lerman, 1974a; Osmond *et al.*, 1973). Their isotopic composition may vary depending on the surrounding environment (Black, 1973) and length of the photoperiod (Lerman and Queiroz, 1974). Laboratory experiments have shown that succulents growing during long daylight, when the Calvin cycle is active, are similar to C_3 plants in isotopic composition. In a short photoperiod, when the Hatch–Slack cycle is active, succulents approach C_4 plants in isotopic composition (Lerman and Queiroz, 1974). Lerman *et al.* (1974), investigated an acid extract chiefly containing the products of primary fixation of carbon in the Hatch–Slack cycle. It had an isotopic composition of -10.8 to -13.5‰, while the insoluble fraction (chiefly cellulose) had a $\delta^{13}C$ from -16.5 to -18.5‰. The data pertain to succulent plants. The authors mention that the isotopic composition of the fraction varies as a function of the relative intensity of the carbon fixation processes by way of carboxylation of RbDP and PEP in the plant investigated. Whelan *et al.* (1973) made a comparative analysis of several components and intermediate metabolites from a C_4 plant (sorghum) and a C_3 plant (cotton). All the investigated components of the former were isotopically heavier. The isotopic composition of sorghum leaves averaged -10.9‰ and of cotton leaves -27.0‰. In sorghum the isotopically heaviest intermediate metabolite was malate at -7.8‰. Of other intermediate metabolites, aspartic acid had $-9.9\,‰$, glutamic acid $-11.5\,‰$, alanine $-9.3\,‰$, and succinic acid $-9.6\,‰$. Glucose in sorghum had $-9.9\,‰$ compared with $-24.6\,‰$ in cotton; protein as a whole had $-10.2\,‰$ in sorghum and $-27.9\,‰$ in cotton. The lipid fraction in sorghum was relatively enriched in the ^{12}C isotope with a $\delta^{13}C = -16.1\,‰$.

Thus in plants of C_4 type the products of primary fixation of carbon in the cycle of four-carbon acids are depleted in the ^{12}C isotope. To a lesser extent, but still substantial compared with C_3 plants, the products of refixation of carbon dioxide in the Calvin cycle are also depleted. This determines the relative depletion in the ^{12}C isotope of carbon entering into all the metabolic pathways of C_4 plants. However, in the biosynthesis of lipids they are enriched in the ^{12}C isotope, relative to carbohydrates, to about the same extent as in C_3 plants. Evidently the intermolecular distribution of isotopes in those biochemical fractions whose biosynthesis is the same for C_3 and C_4 plants must be the same. In our laboratory the isotopic composition of the components of the lipid fraction of C_4 plants was studied (Galimov *et al.*, 1977). The lipids were extracted from mature corncobs. The results of isotopic analysis are given in Table 8.2.

TABLE 8.2

ISOTOPIC COMPOSITION OF CARBON OF THE LIPID FRACTION COMPONENTS OF MAIZE AND THE $\beta^{13}C$ VALUES OF THE RESPECTIVE COMPOUNDS[a]

Component	*Number of fraction on chromatogram*	$\delta^{13}C$ (‰)	$\beta^{13}C$
Corncobs as a whole	—	−11.9	—
Total lipids	—	−16.3	—
Residue after extraction	—	−11.1	—
Polar lipids	1	−17.0	—
Unidentified	2	−19.7	—
Monoglycerides	3	−18.2	1.1562
Diglycerides	4	−18.9	1.1562
Sterols	5	−17.0	1.1590
Fatty acids	6	−21.5	1.1540
Unidentified	7	−18.1	—
Triglycerides	8	−21.0	1.1548
Unidentified	9	−25.4	—
Waxes	10	−25.0	1.1527
Carbohydrates	11	−25.7	1.1520

[a] Details of the procedure of separating the components of the lipid fraction and calculation of $\beta^{13}C$ are given in Chapter 4 and in Galimov and Shirinskiy (1975).

The ratio of isotopic compositions of the individual components reflects a tendency toward thermodynamically ordered distribution of isotopes that is precisely the same as that which occurs in the lipid fraction of C_3 plants (Galimov and Shirinskiy, 1975).

Thus the question of the characteristics of the isotopic composition of C_4 plants focuses on examination of the characteristics of fractionation of isotopes in the reaction cycles in which carbon dioxide is assimilated.

B. Isotopic Fractionation in the Process of Enzymatic Carboxylation of Phosphoenol Pyruvate *in Vitro*

As previously mentioned, most investigators relate biological fractionation of isotopes to the first reaction of inclusion of external CO_2 in the biological cycle. In the Calvin cycle that is the reaction of carboxylation of ribulose 1,5-diphosphate; in the Hatch–Slack cycle, carboxylation of phosphoenol pyruvate (PEP). A number of experiments have been made to study the fractionation of carbon isotopes in enzymatic carboxylation of PEP.

Whelan *et al.* (1973) carboxylized PEP using phosphoenol pyruvate carboxylase isolated from sorghum leaves. The temperature was kept at 24°C, the pH at 8.5. The reaction mixture also contained malate dehydrogenase and HCO_3^-. The PEP was completely converted in the experiment. The percentage of bicarbonate used was 0.2%. Similar experiments with phosphoenol pyruvate carboxylase were made on C_4 plants by Deleens *et al.* (1974) and by Reibach and Benedict (1977). The isotopic composition of the fixed carbon was calculated from the equation

$$\delta^{13}C_{C_{fix}} = 4\delta^{13}C_{malate} - 3\delta^{13}C_{PEP}.$$

In contrast to ribulose 1,5-diphosphate, the active molecule in enzymatic carboxylation of PEP is not CO_2 but bicarbonate (Maruyama *et al.*, 1966). Therefore, the enzymatic fractionation is determined directly from the equation

$$\Delta = \delta^{13}C_{C_{fix}} - \delta^{13}C_{HCO_3^-}.$$

The results of the experiment, presented in Table 8.3, show the following: the value of the isotopic effect in carboxylation of PEP *in vitro*, averaging $-2.6‰$, is incomparably smaller than in carboxylation of RbDP, and the scatter of the data obtained in various experiments by different investigators is not more than $\pm 2‰$ (this value is within the limits of calculated error of the experiment which is $\pm 4.4‰$ (Whelan *et al.*, 1970). Therefore, in contrast to the experiments with RbDP the results are readily reproduced.

TABLE 8.3

FRACTIONATION OF CARBON ISOTOPES IN ENZYMATIC CARBOXYLATION OF PEP *in Vitro*

Plant from which the enzyme PEP carboxylase was isolated	$\delta^{13}C$ (‰)				*Reference*
	HCO_3^-	*PEP*	*Malate*	$\Delta C_{fix} - HCO_3^-$	
Maize	−31.25	−19.17	−22.31	−0.51	Reibach and Benedict (1977)
	−31.10	−19.20	−23.31	−3.55	Reibach and Benedict (1977)
	−5.7	−18.7	−16.6	−4.6	Deleens *et al.* (1974)
	−5.7	−18.7	−15.8	−1.8	Deleens *et al.* (1974)
Sorghum	−29.1	−21.2	−23.9	−2.9	Whelan *et al.* (1973)
	−29.1	−19.1	−22.2	−2.5	Whelan *et al.* (1973)
Average				−2.6	

Let us calculate the theoretical value of the enzymatic isotopic effect in the enzymatic reaction of carboxylation of PEP:

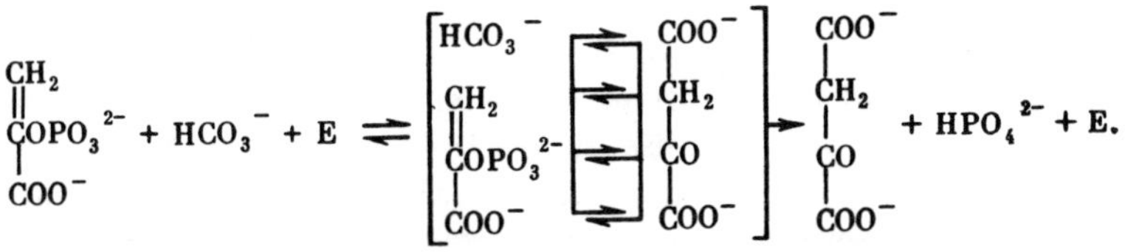

Because the pyruvate is completely converted, an isotopic effect *in vitro* is possible only for the carboxyl carbon derived from CO_2, i.e.,

$$\delta^{13}C_{COOH} = \delta^{13}C_{HCO_3^-} + \kappa(1.2000/1.2058 - 1) \times 10^3 + \Delta_\lambda$$

or

$$\Delta_{COOH-HCO_3^-} = -4.8\kappa + \Delta_\lambda.$$

If $\kappa \simeq 0.5$ is taken, the thermodynamic isotopic effect of the reaction is −2.4‰. This corresponds almost exactly to the values of the isotopic effects observed experimentally.

C. REASONS FOR THE DIFFERENCE IN CARBON ISOTOPIC COMPOSITION OF C_3 AND C_4 PLANTS

It is easiest to consider the differences in isotopic composition of C_3 and C_4 plants to be the result of a difference in fractionation of isotopes in the first photosynthetic reactions. Actually, in the carboxylation of RbDP, which is the first reaction of photosynthesis of C_3 plants, a very substantial enrichment of the fixed carbon in the light isotope has been established *in vitro*, while in C_4 plants substantially less enrichment in the light isotope corresponds to the first reaction of carboxylation of PEP, which is accompanied by a small isotopic effect. In C_4 plants the Hatch–Slack cycle is coupled with the Calvin cycle so the carbon dioxide liberated in decarboxylation of the first 4-carbon organic acids in the Hatch–Slack cycle is refixed in the Calvin cycle by carboxylation of ribulose diphosphate. It is known that ribulose diphosphate carboxylase operating in plants of C_4 type produces isotopic effects in the carboxylation of RbDP *in vitro* which are of approximately the same amount and range of variations as the analogous enzyme liberated from C_3 plants (Deleens *et al.*, 1974; Whelan *et al.*, 1973) (for instance, see the data for maize and spinach in Table 8.1). However, the reasoning here is different. In the first reaction of carboxylation of PEP the isotopic effect is about $-3‰$. Therefore, the first carbon fixed has an isotopic composition of about $-10‰$ (for an isotopic composition of atmospheric CO_2 of $-7‰$).† This isotopic composition is inherited by the CO_2 formed in decarboxylation of organic acids in the Hatch–Slack cycle. The liberated carbon dioxide is refixed in the Calvin cycle with its considerable isotopic effect inherent in the reaction of carboxylation of RbDP. But because the pool of this carbon dioxide is small and it is nearly all used up, resulting enrichment of the reaction products in the ^{12}C isotope does not take place and the plant, as a whole, remains depleted in the light isotope. Such an explanation has been proposed by a number of investigators (Lerman, 1974b; Reibach and Benedict, 1977; Whelan *et al.*, 1973) in one version or another.

However, another explanation is possible if the alternative suggested in Chapter 5 is developed to explain the isotopic effect in the Calvin cycle. Enrichment of fixed carbon may be related in principle not to the kinetic isotopic effect of carboxylation of RbDP but to thermodynamic isotopic

† The following details should be noted. In carboxylation of PEP the active form is HCO_3^- (Maruyama *et al.*, 1966). Carbon dioxide comes into the cytoplasm from the surrounding atmosphere. Therefore, possible isotopic fractionation in the CO_2–HCO_3^- system must be kept in mind. If it is assumed, as was done in the case of the experiments with RbDP *in vitro*, that an equilibrium distribution of isotopes exists in the system in question, then the isotopic composition of HCO_3^- at 25°C should be shifted by 8.2‰ toward positive values of $\delta^{13}C$ relative to CO_2. Then the carbon fixed by C_4 plants should have $\delta^{13}C = -1$ to $-3‰$, i.e., it should be isotopically heavier than the carbon of atmospheric carbon dioxide. Obviously this is not so. But then it is doubtful that a correction of 8.2‰, which decreases the large discrepancy between the natural fractionation of isotopes and that observed in this reaction *in vitro*, can be used in the case of experiments with carboxylation of RbDP.

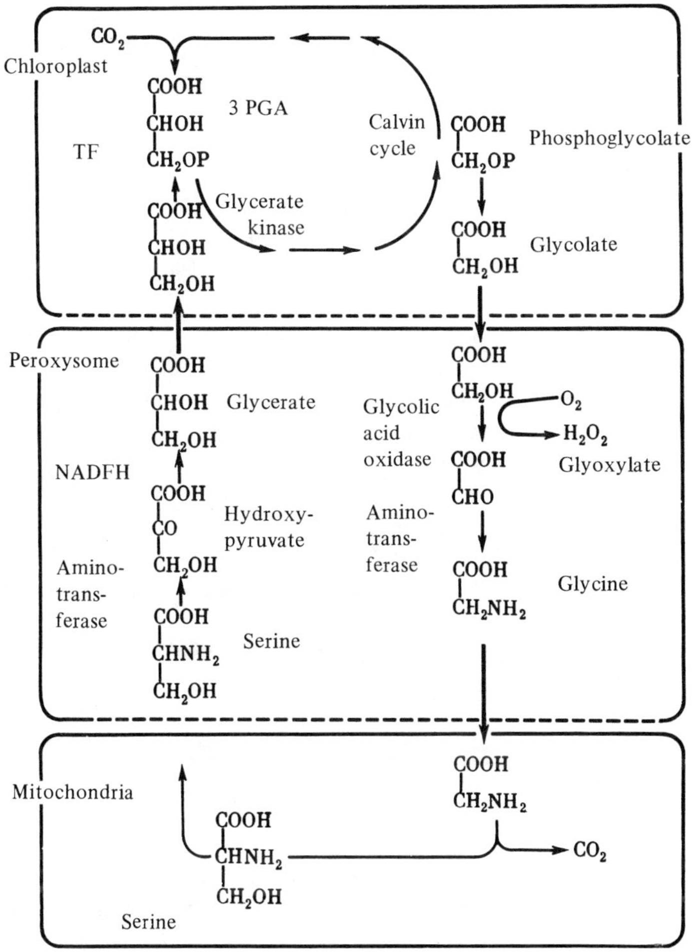

FIG. 8.2 Scheme of transformation of glycolate. From Rubin and Gavrilenko, 1977.

effects at the reduction stage of the assimilated carbon. For this effect to be realized, it is necessary that a reverse reaction pathway exist (as shown in Chapter 5) from product to predecessor. The key difference in the character of fractionation of isotopes in plants of C_3 and C_4 type may lie in precisely this point, as there is such a pathway in C_3 plants while in C_4 plants either there is no such pathway or it is weakly expressed. The reverse pathway in question is photorespiration.

The substrate of photorespiration is the first products of photosynthesis produced in the Calvin cycle (Rubin and Gavrilenko, 1977). The so-called glycolate pathway of oxidation to CO_2 of reduced primary products of photosynthesis is illustrated in Fig. 8.2. As previously mentioned, little or no photorespiration is a specific feature of C_4 plants (Laing and Forde, 1971). Therefore, in C_3 plants an isotopic effect accompanying reduction of carbon in the Calvin cycle is realized which leads to enrichment of biological carbon as a whole in the light isotope, whereas no analogous thermodynamic isotopic effect can be realized at analogous stages in C_4 plants. This divergence can explain the difference in isotopic composition of C_3 and C_4 plants in the framework of the general principles which are being developed in this work.

The questions of whether such assumptions are valid or whether the kinetic isotopic effect plays the role assigned to it evidently can be answered by experiments on C_3 plants under conditions of inhibited photorespiration. If enrichment of plants in the ^{12}C isotope is caused by an isotopic effect during primary fixation of CO_2, then inhibiting photorespiration theoretically should have no effect on the isotopic composition of a plant. But if the assumption of enrichment of carbon in the ^{12}C isotope in the stage of reduction is valid, then inhibition of photorespiration should lead to depletion of C_3 plants in the light isotope to the level typical of C_4 plants. Such an experiment probably is capable of answering the knotty question of the causes of the isotopic effect between biological carbon and the carbon of external CO_2.

III. THE KREBS CYCLE

As is known, the Krebs (tricarboxylic acid) cycle is the basis of the process of respiration. Acetyl groups formed in catabolism of lipids, proteins, and carbohydrates are oxidized in this cycle (also called the citric acid cycle) to CO_2. The liberated energy is stored in the form of ATP (Lehninger, 1974).

Acetate enters the reaction of the cycle, condensing with oxaloacetic acid (oxaloacetate). The original product is citric acid (citrate). Further transformations in the cycle lead to splitting off of two carbon atoms in the form of CO_2 and regeneration of the oxaloacetate.

In some organisms which use acetate not only as a source of energy but also as the initial metabolite for the synthesis of biochemical components, the glyoxylate cycle is present (Lehninger, 1974).

Figures 8.3 and 8.4 depict the Krebs and glyoxylate cycles in an illustration convenient for following the pathways of the carbon atoms. Carboxyl carbon of acetate (dashed line) passes from one metabolite to another, invariably remaining in carboxyl form. Ultimately this carbon is

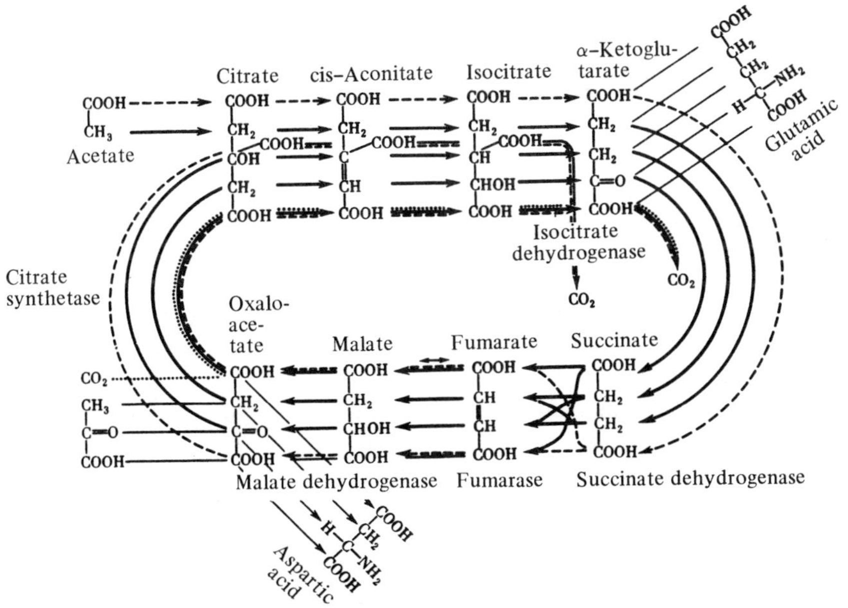

FIG. 8.3 The Krebs cycle.

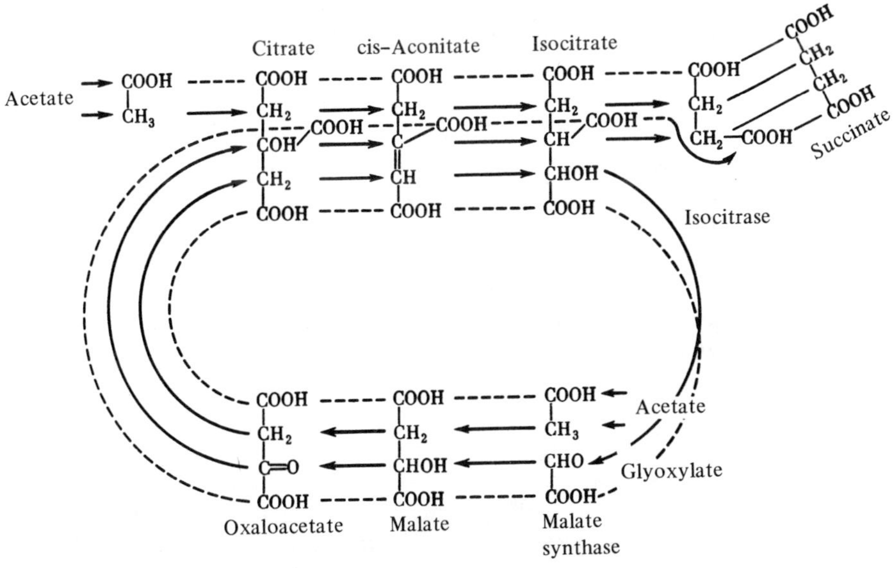

FIG. 8.4 The glyoxylate cycle.

removed in the form of CO_2 as a result of decarboxylation of isocitrate and α-ketoglutarate.

The carbon of the methyl group of acetate takes a much more capricious pathway. Each turn of the cycle is the whorl of a spiral for it, moving this carbon to a position adjacent to the one it occupied in the compound in the previous cycle. Ultimately this carbon, like the carboxyl carbon of acetate, is released in the form of CO_2 via the carboxyl groups of isocitric and α-ketoglutaric acids.

Examination of the tricarboxylic acid cycle again brings us to Eq. (5.97). From it, it follows that, if the thermodynamic isotopic factors of comparable compounds or carbon atoms are the same, i.e., $\beta_1 = \beta_k$, then the thermodynamic component of the isotopic effect is absent (the first term in the right-hand part is equal to zero). However, in this case the isotopic differences of thermodynamic origin between $\delta^{13}C_1$ and $\delta^{13}C_k$ may exist if the carbon in question is assimilated into groups other than the first and kth positions in intermediate metabolites. This is because in that case $\beta_i \neq \beta_{i+1}$ and, under the condition $\Delta\kappa_i \neq 0$, the second term containing the β factors of the intermediate metabolites is not equal to zero. An example is the carbon atoms of the CH_2 groups of α-ketoglutaric acid. They are related by a chain of transformations in the cycle: carbon of CH_2 group adjacent to carboxyl travels the path CH_2 (succinate) to CH (fumarate) to CHOH (malate) to CO (oxaloacetate) to HCO (citrate) to C-(*cis*-aconitate) to HC (isocitrate) and is returned to another CH_2 group, α-ketoglutarate, adjacent to the original one.

If the chemical surroundings of the carbon are the same not only in comparable compounds but in intermediate metabolites, then in formula (5.97) not only does $\beta_1 = \beta_{k_1}$, but also $\beta_i = \beta_{i+1}$, and both terms, which depend on thermodynamic isotopic effects, become zero. Only isotopic effects related to transport of matter and kinetic isotopic effects are possible. An example is the pathway of the carboxyl carbon of acetate, which remains in carboxyl form in all metabolites of the cycle until it is expelled from the cycle in the form of CO_2.

The compounds of the citric acid cycle thus present an excellent opportunity for calculating the relative value of isotopic effects of a different nature. Unfortunately the isotopic distribution in this system has not yet been studied experimentally.

As the carbon originating from the methyl group of acetate successively makes the rounds of all carbon positions in the compounds of the cycle except the carboxyl line, and the latter, in turn, is the pathway of the carboxyl group of acetate, it appears that the conditions for establishment of thermodynamically ordered distribution of isotopes in the compounds of the cycle would be fulfilled (to the extent that it exists in the original acetate). However, the condition of a closed stage must also be fulfilled, which is a problem for the Krebs cycle.

The fractionation of isotopes in the cycle can be calculated indirectly, taking into account that the metabolites of the citric acid cycle are the predecessors of the amino acid series, the isotopic composition of which is known (see Table 8.4).

TABLE 8.4

ISOTOPIC COMPOSITION OF GLUTAMIC AND ASPARTIC ACIDS IN A NUMBER OF ORGANISMS[a]

Organism	*Amino acid fraction analyzed*	*Glutamic*	*Aspartic*	*Difference*
Chlorella pyrenoidosa	Total	−18.7	−6.6	−12.1
	Carboxyl	−8.8	2.2	−11.0
	Residual	−21.1	−15.6	−5.5
Anacystis nidulans	Total	−11.1	−9.7	−1.4
	Carboxyl	11.7	2.8	8.9
	Residual	−16.7	−22.1	5.4
Scenedesmus quadricauda	Total	−22.6	−12.3	−10.3
	Carboxyl	−5.6	−0.8	−4.8
	Residual	−26.6	−24.3	−2.3
Chromatium sp.	Total	−15.6	−21.8	6.2
	Carboxyl	2.2	−15.9	18.1
	Residual	−20.1	−27.6	7.5
Euglena gracilis	Total	−17.3	−9.6	−7.7
	Carboxyl	−0.2	1.0	−1.2
	Residual	−21.5	−20.3	−1.2
Gracilaria sp.	Total	−17.2	−14.4	−2.8
	Carboxyl	−8.4	−5.3	−3.1
	Residual	−19.3	−18.4	−0.9

[a] Data of Abelson and Hoering (1961).

Glutamic acid is formed by direct amination of α-ketoglutaric acid. All five carbon atoms of α-ketoglutaric acid are converted to corresponding atoms of glutamic acid. The carboxyl groups of glutamic acid originate from the carboxyl groups of α-ketoglutaric acid.

The carbon of one of the carboxyl groups of oxaloacetic acid, which may be a predecessor of aspartic acid via malic acid, originates from the carbon of the carboxyl group of α-ketoglutaric acid. The carbon of the other carboxyl group of oxaloacetic acid is the predecessor of the carbon of the second carboxyl group of α-ketoglutaric acid in the course of the cycle.

It would seem that the carboxyl carbons of aspartic and glutamic acids should be very similar in isotopic composition. However, as noted in Table 8.4, the differences between them within each organism are substantial. In the given case the thermodynamic component of the isotopic effects in the intermediate enzymatic reactions is close to zero (the difference in β factors may not exceed the value of the isotopic bond numbers of l_{C-X} type). Consequently, the observed deviations are not related to any uncertainty in the calculations or character of thermodynamic isotopic effects. On the contrary, they reflect the scale of the uncertainties which may be introduced into a thermodynamically ordered distribution of isotopes.

One other observation is of interest. After the loss of the carboxyl group the remaining four carbon atoms of α-ketoglutaric acid are the predecessors of the corresponding carbon atoms of oxaloacetic acid. In the process of decarboxylation a kinetic isotopic effect may take place. Its result would be enrichment of the liberated carbon dioxide and the four-carbon fragment in the light isotope relative to the corresponding carbons of the α-ketoglutaric acid, which remains unsplit and branches off in the formation of glutamic acid. If a kinetic isotopic effect of this type plays the predominant role and thermodynamic isotopic effects are insignificant, then the oxaloacetic acid and later the aspartic acid would be enriched in the light isotope relative to glutamic acid. On the other hand, if the enzymatic reactions of the cycle are accompanied by thermodynamic isotopic effects related to oxidation of the CH_2 and carboxyl carbon of α-ketoglutaric acid, then the aspartic acid would be isotopically heavier than the glutamic, as the β factor values of the carboxyl and carbonyl carbon are greater than those of the corresponding carbonyl and CH_2 carbons.

From Table 8.4 it is seen that in five out of the six organisms investigated the aspartic acid is isotopically heavier than the glutamic. Thus the thermodynamic nature of enzymatic isotopic effects is manifested, but they are markedly attenuated. There may be several reasons for this. One obviously is the presence of branching and, correspondingly, the presence of anaplerotic reactions compensating for the loss of intermediate metabolites of the cycle. The Wood–Workman reaction is such a reaction; it introduces carbon dioxide directly into the Krebs cycle by condensing it with pyruvate. The existence of this pathway is not *per se* the reason for enrichment of carboxyl carbon in the heavy isotope, as suggested earlier by Abelson and Hoering (see Chapter 7), but it can cause variations in $\delta^{13}C$ which depend on the extent of functioning of this pathway and of the current isotopic composition of CO_2 in the cytoplasm.

REFERENCES

Baertschi, P. (1952). The fractionation of carbon isotopes in the absorption of carbon dioxide. *Helv. Chim. Acta* **35**, 1030–1036.

Bahr, J. T., and Jensen, R. G. (1974). Ribulose diphosphate carboxylase from freshly ruptured spinach cloroplasts having an *in vivo* $K_m(CO_2)$. *Plant. Physiol.* **53**, 39–44.

Bassham, J. A., and Calvin, M. (1957). "The Path of Carbon in Photosynthesis." Prentice–Hall, Englewood Cliffs, New Jersey.

Bender, M. M. (1968). Mass-spectrometric studies of carbon 13 variations in corn and other grasses. *Am. J. Sci. Radiocarbon. Suppl.* **10**, 468–472.

Bender, M. M., Rohuani, I., Vines, H. M., and Black, C. C. (1973). $^{13}C/^{12}C$ ratio changes in crassulacean acid metabolism plants. *Plant. Physiol.* **52**, 427–430.

Black, C. C. (1973). Photosynthetic carbon fixation in relation to net CO_2 uptake. *Annu. Rev. Plant Physiol.* **24**, 280.

Chen, T. M., Brown, R. H., and Black, C. C. (1971). Photosynthetic $^{14}CO_2$ fixation products and activities of enzymes related to photosynthesis in Bermuda grass and other plants. *Plant. Physiol.* **47**, 199.

Christeller, J. T., Lang, W. A., and Troughton, J. H. (1976). Isotope discrimination by ribulose-1,5-diphosphate carboxylase. *Plant. Physiol.* **57**, 580–582.

Craig, H. (1953). The geochemistry of the stable carbon isotopes. *Geochim. Cosmochim. Acta* **3**, 53–92.

Degens, E. T., Guillard, R. R. L., Sackett, W. M., and Hellebust, J. A. (1968). Metabolic fractionation of carbon isotopes in marine plankton. I. Temperature and respiration experiments. *Deep-Sea Res.* **15**, 1–9.

Deleens, V., Lerman, J. C., Nato, A., and Moyse, A. (1974). Carbon isotope discrimination by carboxylating reactions in C_3, C_4 and CAM plants. "Proc. of the Third Intern. Congr. on Photosynthesis" (M. Avron, ed.), pp. 1267–1276. Elsevier, Amsterdam.

Deuser, W. G., and Degens, E. T. (1967). Carbon isotope fractionation in the system CO_2 (gas)—CO_2 (aqueous)—HCO_3 (aqueous). *Nature* **215**, 1033–1035.

Estep, M. F., Tabita, F. R., Parker, P. L., and van Baalen, C. (1978). Carbon isotope fractionation by ribulose-1-5-biphosphate carboxylase from various organisms. *Plant. Physiol.* **61**(4), 680.

Galimov, E. M. (1966). Carbon isotopes of soil CO_2. *Geochem. Intern.* **3**(5), 889–897.

Galimov, E. M. (1973). Organic geochemistry of carbon isotopes. *In* "Advances in Organic Geochemistry" (B. Tissot, ed.), pp. 439–452. Technip, Paris.

Galimov, E. M., and Shirinskiy, V. G. (1975). Ordered distribution of carbon isotopes in individual compounds and components of the lipid fraction of organisms. *Geokhimiya*(4), 503–528.

Galimov, E. M., *et al.* (1977). Investigation of the distribution of carbon isotopes in biogenic compounds. *In* "Trudy VIII Int. Congr. on Organic Geochemistry" (A. V. Sidorenko, ed.), Vol. 3. Nauka, Moscow.

Gibbons, B. H., and Edsall, J. T. (1963). Rate of hydration of carbon dioxide and dehydration of carbonic acid at 25°. *J. Biol. Chem.* **238**(10), 3502–3507.

Hatch, M. D., and Slack, C. R. (1966). Photosynthesis by sugarcane leaves: A new carboxylation reaction and pathway of sugar formation. *Biochem. J.* **101**(1), 103.

Hatch, M. D., and Slack, C. R. (1970). Photosynthetic CO_2-fixation pathways. *Annu. Rev. Plant. Physiol.* **21**, 141–162.

Ho, C., and Sturtevant, J. M. (1963). The kinetics of the hydration of carbon dioxide at 25°. *J. Biol. Chem.* **238**(10), 3499–3501.

Jackson, W. H., and Volk, R. J. (1970). Photorespiration. *Annu. Rev. Plant. Physiol.* **221**, 385.

Karpilov, Yu. S. (1960). Distribution of radioactive carbon ^{14}C among the products of photosynthesis in maize. *Trudy Kazakhs. S.-Kh. Inst.* **41**(1).

Keeling, C. (1961). The concentration and isotopic abundances of carbon dioxide in rural and marine air. *Geochim. Cosmochim. Acta.* **24**, 277–298.

Kelly, G. J., Latzko, E., and Gibbs, M. (1976). Regulatory aspects of photosynthetic carbon metabolism. *Annu. Rev. Plant. Physiol.* **21**, 181–205.

Kennedy, R. A., and Laetsch, W. M. (1974). Plant species intermediate for C_3, C_4 photosynthesis. *Science* **184**(4141), 1087–1088.

Kortschak, H. P., Hartt, C. E., and Burr, G. O. (1965). Carbon dioxide fixation in sugarcane leaves. *Plant. Physiol.* **40**(2), 209–213.

Laetsch, W. M. (1969). Relationship between chloroplast structure and photosynthetic carbon fixation pathways. *Sci. Prog.* **47**, 323–351.

Laing, W. H., and Forde, B. J. (1971). Comparative photorespiration in amaranthus, soybean and corn. *Planta (Berlin)* **98, 221.**

Lehninger, A. (1974). "Biochemistry." Mir, Moscow (Russian translation).

Lerman, J. C. (1972). Soil CO_2 and groundwater: Carbon isotope composition. In "Proc. 85th Intern. Conf. Radiocarbon Dating" (T. A. Rafter and T. Grant-Taylor, eds.). R. Soc. N.Z., Wellington.

Lerman, J. C. (1974a). Variation in the carbon isotope composition of a plant with crassulacean acid metabolism. *Plant. Physiol.* **53**, 581–584.

Lerman, J. C. (1974b). How to interpret variations in the carbon isotope ratio of plants. *In* "Environmental and Biological Control of Photosynthesis. Proceedings of Conference Held at Limbergs Universitair Centrun, Belgium," (T. Marcelle, ed.), pp. 323–333. W. Junk, the Hague.

Lerman, J. C., and Queiroz, O. (1974). Carbon fixation and isotope discrimination by a crassulacean plant: dependence on the photoperiod. *Science* **183**, 1207.

Lerman, J. C., Delleens, E., Nato, A., and Moyse, A. (1974). Variations in the carbon isotope composition of a plant with crassulacean acid metabolism. *Plant. Physiol.* **53**, 581–584.

Maruyama, H., Easterday, R. L., Chang, H. C., and Lane, M. D. (1966). The enzymatic carboxylation of phosphoenolpyruvate. *J. Biol. Chem.* **241**(10), 2405–2412.

Mook, W. G., Bommerson, J. C., and Staverman, W. H. (1974). Carbon isotope fractionation between dissolved bicarbonate and gaseous carbon dioxide. *Earth Planet. Sci. Lett.* **22**, 169–176.

Moss, D. M. (1962). The limiting carbon dioxide concentration for photosynthesis. *Nature* **193**, 587.

O'Leary, M. H., and Yapp, C. J. (1978). Equilibrium carbon isotope effect on a decarboxylation reaction. *Biochem. Biophys. Res. Commun.* **80**(1), 155–160.

Osmond, C. B., *et al.* (1973). Carbon isotope discrimination in photosynthesis of CAM plants. *Nature* **246**, 41.

Park, R., and Epstein, S. (1960). Carbon isotope fractionation during photosynthesis. *Geochim. Cosmochim. Acta* **21**, 110–126.

Park, R., and Epstein, S. (1961). Metabolic fractionation of ^{13}C and ^{12}C in *Plants. Plant. Physiol.* **36**, 133–138.

Reibach, P. H., and Benedict, C. R. (1977). Fractionation of stable carbon isotopes by phosphoenolpyruvate carboxylase from C_4 plants. *Plant Physiol.* **59**(4), 564–568.

Rubin, B. A., and Gavrilenko, V. F. (1977). "Biochemistry and Physiology of Photosynthesis." Izd-vo Moskov. Univ., Moscow.

Smith, B. N., and Epsteins, S. (1971). Two categories of $^{13}C/^{12}C$ ratios for higher plants. *Plant. Phys.* **47**, 380–384.

Thode, H. G., Shima, M., Rees, C. E., and Krishnamurty, K. V. (1965). Carbon-13 isotope effects in systems containing carbon dioxide bicarbonate, carbonate and metal ions. *Can. J. Chem.* **43**(3), 582–595.

Tregunna, E. B., Smith, B. N., Berry, J. A., and Downton, W. J. S. (1970). Some methods for studying the photosynthetic taxonomy of the angiosperms. *Can. J. Bot.* **48**, 1209–1214.

Troughton, J. H., Wells, P. V., and Mooney, H. A. (1974). Photosynthetic mechanisms and paleoecology from carbon isotope ratios in ancient specimens of C_4 and CAM plants. *Science* **185**(4151), 610.

Wendt, I. (1968). Fractionation of carbon isotopes and its temperature dependence in the system CO_2 gas–CO_2 in solution and HCO_3^-–CO_2 in solution. *Earth Planet. Sci. Lett.* **4**(1), 64–68.

Whelan, T., Sackett, W. M., and Benedict, C. R. (1970). Carbon isotope discrimination in a plant possessing the C_4 dicarboxylic acid pathway. *Biochem. Biophys. Res. Commun.* **41**, 1205–1210.

Whelan, T. Sackett, W. M., and Benedict, C. R. (1973). Enzymatic fractionation of carbon isotopes by phosphoenolpyruvate carboxylas. *Plant. Physiol.* **51**, 1051–1054.

Woo, K. C., *et al.* (1970). Deficient photosystem II in agranal bundle sheath chloroplast of C_4 plants. *Proc. Nat. Acad. Sci. USA* **67**, 18–25.

Zelitch. L., and Day, P. R. (1973). The effect of net photosynthesis of pedigree selection for low and high rates of photorespiration in tobacco. *Plant. Physiol.* **52**, 33–37.

CHAPTER 9

Biological Fractionation of Oxygen Isotopes

I. $^{18}O/^{16}O$ OF BIOLOGICAL OXYGEN

The principles of biological fractionation of isotopes previously considered are obviously applicable not only to carbon isotopes but also to the isotopes of other elements. However, the relationships between the different kinds of isotopic effects vary for different elements, depending on their chemical properties.

Oxygen is the only element present in all the components of the overall equation of photosynthesis, e.g.,

$$CO_2 + H_2O = \{CH_2O\} + O_2.$$

An idea of the isotopic compositions of these components can be obtained from the available geochemical data and the results of a few specialized biological experiments.

The isotopic composition of the oxygen of atmospheric carbon dioxide averages $\delta^{18}O = 41‰$ (Bottinga and Craig, 1969). In an aqueous medium the oxygen of carbon dioxide is in isotopic equilibrium with the oxygen of the water. The coefficient of fractionation of oxygen isotopes in the system $CO_2 \rightleftarrows H_2O$ is 1.0412 at 25°C (O'Neil *et al.*, 1975). In other words, aqueous carbon dioxide is enriched in the ^{18}O isotope by approximately 40‰ relative to the oxygen of the water.

The isotopic composition of seawater is close to $\delta^{18}O = 0.0‰$ (SMOW standard, see Chapter 1) and relatively constant. The isotopic composition of intracontinental waters ranges from +30‰ in closed waterbodies of arid zones to −50‰ in circumpolar areas (Ferronskiy *et al.*, 1975). The dependence of the isotopic composition of aqueous carbon dioxide on the isotopic composition of the water, because of oxygen isotope exchange between them, determines a correlation between the isotopic composition of the water and the oxygen of aquatic plants living in it (Epstein *et al.*, 1977). The isotopic composition of investigated water plants ranges from

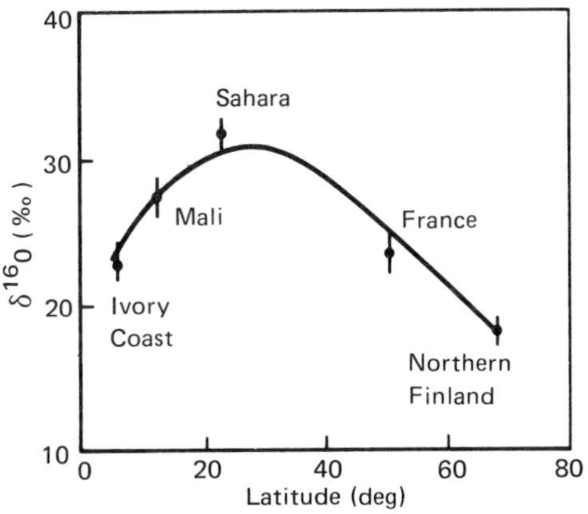

FIG. 9.1 Dependence of $\delta^{18}O$ of Plants on Geographic Latitude. (From Long *et al.*, 1978.)

$\delta^{18}O = 14.2‰$ to $28.3‰$ (Epstein *et al.*, 1977). The difference between the isotopic composition of water and of the oxygen of water plants is fairly stable and amounts to about 28‰, as Epstein *et al.* (1977) have shown.

Water evaporating from the surface of the ocean is depleted in the ^{18}O isotope. Therefore, atmospheric moisture and meteoric and groundwaters, i.e., ultimately all the water supplied to terrestrial plants, are isotopically lighter than ocean water. Meteoric waters are increasingly depleted in the ^{18}O isotope as the altitude of the locality and geographic latitude increase. In the tropics, rains have a $\delta^{18}O$ from 0 to $-5‰$, in the temperate zone from -5 to $-15‰$, and in polar latitudes less than $-15‰$ (Craig and Gordon, 1965). On the other hand, due to evaporation of moisture from the surface of leaves, the intracellular water in plants is enriched in the heavy oxygen isotope (Craig *et al.*, 1963; Gonfiantini *et al.*, 1965). The more active the evaporation (transpiration), the more substantial the enrichment. This explains the dependence of the $\delta^{18}O$ of plants on humidity observed by several investigators (Ferhi *et al.*, 1977; Long *et al.*, 1978) ($\sim 2‰$ per 10% change in humidity) and temperature (Gray and Thomson, 1976; Libby *et al.*, 1976). It is interesting to depict the dependence of the average value of $\delta^{18}O$ of plants on geographic latitude (Long *et al.*, 1978) (Fig. 9.1). This reflects the competition of the two previously mentioned processes. In high latitudes samples are depleted in the ^{18}O isotope because both the efficiency of transpiration is low and the precipitation is isotopically light. In middle latitudes relatively high efficiency of transpiration is coupled with moderate depletion of precipitation in the ^{18}O isotope. In tropical regions, due to the high humidity, the enrichment of plants in the ^{18}O isotope is less pronounced even though the precipitation there is relatively heavy isotopically. On the whole, terrestrial plants are characterized by an isotopic composition of organic oxygen in the range of $\delta^{18}O = 15$ to 35‰ (Epstein *et al.*, 1977; Gray and Thomson, 1976; Libby *et al.*, 1976; Long *et al.*, 1978).

Oxygen of the atmosphere is enriched in the ^{18}O isotope by 23.5‰ (Kroopnick, 1972) compared with SMOW (the Dole effect). However, this value cannot be identified with the value of the corresponding biological isotopic effect. The isotopic composition of atmospheric oxygen is determined by the balance of the oxygen liberated in photosynthesis and that consumed in respiration and oxidation processes (Dole, 1955; Lane and Dole, 1956). It is known that oxygen in contact with plants is enriched in the ^{18}O isotope. Dole and Jenks (1944) found that oxygen liberated in photosynthesis has an isotopic composition corresponding to the isotopic equilibrium between molecular oxygen and that of water (i.e., it is enriched in ^{18}O relative to water by about 8‰). It has been suggested that the oxygen of photosynthesis coincides in isotopic composition with water, and the observed enrichment of biogenic oxygen in the ^{18}O isotope is caused by the preferential consumption of isotopically light oxygen in the process of respiration (Kutyurin, 1972; Vinogradov and Kutyurin, 1962). Lane and Dole (1956), who investigated the fractionation of oxygen isotopes in the respiration process, established that the oxygen isotopes are fractionated by various organisms with a factor ranging from 1.006 to 1.029, averaging 1.016. Molecular oxygen occupies an intermediate position between water and organic oxygen in isotopic composition.

DeNiro and Epstein (1979)† investigated the distribution of oxygen isotopes in water, CO_2, and cellulose of a plant (wheat) grown in a closed chamber under controlled conditions. A series of experiments was made. In one of them carbon dioxide with normal isotopic composition was used, another used carbon dioxide sharply enriched in the ^{18}O isotope (Table 9.1). In the experiment modeling natural conditions, the following was established: cellulose oxygen is enriched in the ^{18}O isotope relative to water and depleted relative to respired carbon dioxide, the oxygen of water given off by plants is enriched in the ^{18}O isotope relative to ground water, and respired carbon dioxide is enriched in the ^{18}O isotope relative to assimilated carbon dioxide.

The enrichment of intracellular water in the heavy isotopes of oxygen and hydrogen observed in the experiment illustrates the isotopic effect caused by transpiration.

† I thank DeNiro and Epstein (1979) for acquainting me with their work before its publication.

TABLE 9.1

FRACTIONATION OF ISOTOPES DURING EXPERIMENTAL GROWTH OF A WHEAT PLANT IN A CLOSED CHAMBER[a,b]

CO_2	δD_{SMOW} (‰)	$\delta^{18}O_{SMOW}$ (‰)	$\delta^{13}C_{PDB}$ (‰)
With normal isotopic composition			
CO_2			
Inlet		6.5	−39.6
Outlet		33.3	−39.4
Water			
Inlet	−52	−8.1	
Outlet	−52	−8.4	
Distilled from plant	−2	−0.8	
Cellulose	−13	26.8	−57.2[a]
Enriched in ^{18}O			
CO_2			
Inlet		985	−37.3
Outlet		160	N.D.[c]
Water			
Inlet	−52	−8.0	
Outlet	−51	−8.3	
Distilled from plant	−7	13.2	
Cellulose	−16	35.4	−53.6

[a] From DeNiro and Epstein (1979).

[b] The unusual isotopic composition of carbon of the plants is explained by enrichment of absorbed carbon dioxide in the light isotope. If instead of $\delta^{13}C = -39.6$‰ this carbon dioxide had a value of $\delta^{13}C = -7$‰, typical of atmospheric CO_2, the plant would have a value of $\delta^{13}C = -24$‰, typical of terrestrial plants.

[c] N.D. = not determined.

Cellulose oxygen is enriched in the ^{18}O isotope by 28‰ relative to intracellular water, i.e., the coefficient of fractionation of isotopes between organic matter and water for the investigated terrestrial plant was the same as for water plants.

In the experiment with enriched CO_2 ($\delta^{18}O$ ranging from 160 to 985‰) the isotopic composition of cellulose changed by a relatively small value (8.6‰). The isotopic composition of intracellular water changed by about the same amount (14‰). In other words, the difference in enrichment of absorbed carbon dioxide in the ^{18}O isotope had little effect on the isotopic composition of the plant. From this it was concluded that "oxygen derived from carbon dioxide equilibrates with the oxygen of water in the plant during the synthesis of cellulose" (DeNiro and Epstein, 1979, p. 52). In the opinion of the authors of the cited work, the exchange occurs even before the CO_2 is included in 3-phosphoglyceric acid. However, they admit that it might occur during biosynthesis of cellulose in the D-glyceraldehyde phosphate stage.

Molecular oxygen was not investigated in this experiment, but as already mentioned, it has an isotopic composition intermediate between that of the oxygen of water and of the biomass of the plant. Thus the components of the overall reaction of photosynthesis are arranged in order of increasing content of the ^{18}O isotope: intracellular water, molecular oxygen, organic oxygen, and carbon dioxide.

II. NATURE OF OXYGEN ISOTOPIC EFFECTS IN BIOLOGICAL SYSTEMS

The main processes determining the pathway of oxygen in photosynthesis and respiration are known (see Lehninger, 1974). Under the stimulation of light, photooxidation of water and liberation of free oxygen takes place:

$$2H_2O + 2NADF^+ \xrightarrow[(ADP \to ATP)]{h\nu} 2NADF{\cdot}H + 2H^+ + O_2. \tag{9.1}$$

The reduction potential and energy of ATP, stored in this reaction, are later expended in the stage of reduction of the CO_2 carbon:

$$CO_2 + 2HADF{\cdot}H + 2H^+ \xrightarrow[ATP \to ADP]{} \{CH_2O\} + H_2O + 2NADF. \tag{9.2}$$

The overall equation of photosynthesis thus has the form

$$CO_2^* + 2H_2O^{\#} \longrightarrow \{CH_2O^*\} + O_2^{\#} + H_2O^*, \tag{9.3}$$

where like signs indicate the genetic relationship of oxygen atoms.

In the respiration process oxidation of organic compounds occurs. In the Krebs cycle, complete dehydrogenation of acetic acid is accomplished (Lehninger, 1974):

$$(CH_2O^*)_2 + 2H_2O^{\#} \longrightarrow 2CO_2^{*\#} + 8H, \tag{9.4}$$

and in reactions of oxidational phosphorylation accompanying respiration, electron transfer to molecular oxygen takes place according to the reaction

$$NAD{\cdot}H + H^+ + \tfrac{1}{2}O_2^{\#} \xrightarrow[ADP \to ATP]{} NAD^+ + H_2O^{\#}. \tag{9.5}$$

Therefore, the overall reaction of respiration is

$$\{CH_2O^*\} + O_2^{\#} \longrightarrow CO_2^{*\#} + H_2O^{\#}. \tag{9.6}$$

If there were no isotopic effects accompanying these processes, it would be expected that the isotopic composition of organic oxygen ($\delta^{18}O_{org}$)

would, as stated in Eq. (9.3), correspond to the isotopic composition of CO_2 oxygen. Due to the partial refixing of respired carbon dioxide [Eq. (9.4)] the value of $\delta^{18}O_{org}$ should be shifted somewhat toward the isotopic composition of water oxygen. However, in the $CO_2 \rightleftarrows H_2O$ system a rapid isotopic exchange of oxygen occurs (Cohn and Urey, 1938). As the experiment described above (DeNiro and Epstein, 1979) showed, isotopic equilibrium between CO_2 and H_2O manages to set in before CO_2 is involved in biosynthesis. Since the amount of water far exceeds the amount of all other components taking part in metabolism, including CO_2, the isotopic composition of the latter will be determined by the isotopic composition of water plus the coefficient of fractionation between H_2O and CO_2 at a given temperature. In other words, the isotopic composition of the original CO_2 for biosynthesis would not depend on the isotopic composition of external CO_2, in particular on the addition of refixed respiratory CO_2. Nevertheless, the isotopic composition of organic oxygen does not coincide with the isotopic composition of CO_2 oxygen. It differs from the latter by $-14‰$ and from the isotopic composition of aqueous oxygen by 28‰ (carbon dioxide at 25°C is enriched by 42‰ in the ^{18}O isotope relative to water). Epstein and coauthors (1977; DeNiro and Epstein, 1979) suggest that the 28‰ value can be explained by assuming that cellulose oxygen is made up of H_2O and CO_2 oxygen in proportions of one to two. An alternative explanation is that cellulose oxygen is derived from CO_2 but isotopic fractionation takes place between CO_2 and organic oxygen.

Fractionation of isotopes between organic oxygen and free molecular oxygen must occur precisely as stated in the overall process of photosynthesis plus respiration. Otherwise, according to Eq. (9.3) and (9.6) the isotopic composition of molecular oxygen would coincide with the isotopic composition of the oxygen in water.

The experimental data available today are insufficient to examine the problem of the nature of biological isotopic effects of oxygen in any detail. Below we shall limit ourselves to the evidence that the observed distribution of oxygen isotopes in the components of the overall equation of photosynthesis agrees in principle with the idea of the predominantly thermodynamic nature of isotopic effects in biological systems.

Let us consider possible thermodynamic isotopic effects of oxygen in organic compounds, on the basis of the values of the isotopic bond numbers of $\beta^{18}O$ derived in Chapter 3.

Organic compounds of biological components include the oxygen atom in the following structural positions. Proteins contain oxygen in the peptide bond. With carbon the oxygen atom forms the bond $(O{=}C)\sigma\pi_{sp_2-sp_2}$, characterized by an isotopic bond number of 0.105. In addition, two oxygen atoms enter into the composition of the carboxyl groups of amino acids if they are contained in the amino acid residuals (residuals of dibasic amino acids). One of these atoms is characterized by $\beta^{18}O = 1.105$, the other by $\beta^{18}O = 1 + L_{O-C} + L_{O-H} = 1.097$. On average the value of $\beta_{\Sigma}{}^{18}O$ of protein molecules should be close to 1.100.

In sugar molecules the oxygen atom is present in a state of sp_3 hybridization of orbitals and forms a σ bond with an sp_3-hybridized carbon atom and a hydrogen atom. The $\beta^{18}O$ value in this case is 1.097. One of the six atoms of oxygen (in hexoses) forms a $(O{=}C)\sigma\pi_{sp_2-sp_2}$ bond in aldoses ($\beta^{18}O = 1.105$) or a $(C{-}O{-}C)\sigma\pi_{sp_3-sp_2}$ bond in pyranoses ($\beta^{18}O = 1.130$). Thus the $\beta_{\Sigma}{}^{18}O$ factors of sugars (analogous to proteins) have values close to or greater than 1.100.

In lipids oxygen is present in the carboxyl, keto, and alcohol groups. The $\beta^{18}O$ values are 1.098–1.105. Oxygen of the ester groups is characterized by higher values of the β factor (1.130).

In lignin a substantial part of the oxygen atoms is found in methoxyl groups. The value of $\beta^{18}O$ of the oxygen of the $-OCH_3$ group is 1.130. Therefore, lignin probably would be somewhat enriched in the heavy isotope compared with the other biochemical components. On the whole, the β factors of the compounds entering into the composition of the lipid fraction and lignin are characterized by values of 1.100 to 1.110.

It is simple to calculate the $\beta^{18}O$ value of cellulose, for which the data on isotopic composition were discussed previously. This value is 1.1086 (disregarding the effect of the distant surroundings).

Let us compare the values of the β factors characterizing the components of the CO_2–H_2O–$\{CH_2O\}$–O_2 system. From Table 9.2, the $\beta^{18}O$ values increase in the following order: water, molecular oxygen, organic oxygen, and carbon dioxide, i.e., precisely the direction in which their observed enrichment in the ^{18}O isotope increases. The importance of this coincidence probably should not be overestimated; there are other factors which might control the distribution of oxygen isotopes in the system in question. However, the fact that the relationships of the thermodynamic isotopic factors of organic and molecular oxygen, water, and CO_2 correspond to those relationships of their isotopic compositions which actually characterize them in biosystems is significant.

Study of intermolecular and intramolecular isotopic effects of oxygen in organic compounds will be of decisive importance for assessing the role of thermodynamic isotopic effects in biological fractionation of oxygen isotopes. The $\beta^{18}O$ values of various biochemical components differ very little, as the examination made above shows. Therefore their oxygen isotopic compositions probably are relatively similar, but intramolecular isotopic effects would be greater. As seen from the following tabulation, the $\beta_i{}^{18}O$ factors of hydroxyl, carboxyl oxygen, and oxygen forming the ester bond

TABLE 9.2

$\beta^{18}O$ FACTORS OF COMPOUNDS FORMING THE $CO_2 + H_2O \rightarrow \{CH_2O\} + O_2$ SYSTEM (AT 300 K)[a]

Compound	$\beta^{18}O$
H_2O	1.0734[b]
O_2	1.0818
$\{CH_2O\}$-cellulose	1.1086
CO_2	1.1175

[a] See Tables 3.18 and 3.19.

[b] The $\delta^{18}O$ value for liquid water is indicated, according to O'Neil and Adami (1969), taking into account revision made later for fractionation in the CO_2–H_2O system (O'Neil *et al.*, 1975).

differ by values which would specify a definite and substantial difference in their isotopic composition:

Structural group	*Bond*	$\beta_i{}^{18}O$
C—O—H	Hydroxyl	1.097
C=O	Carboxyl	1.105
C—O—C	Ester	1.130

REFERENCES

Bottinga, Y., and Craig, H. (1969). Oxygen isotope fractionation between CO_2 and water, and the isotopic composition of marine atmospheric CO_2. *Earth Planet Sci. Lett.* **5**(N5), 285–295.

Cohn, M., and Urey, H. C. (1938). Oxygen exchange reactions of organic compounds and water. *J. Am. Chem. Soc.* **60**(3), 679–687.

Craig, H., and Gordon, L. (1965). Deuterium and oxygen-18 variations in the ocean and the marine atmosphere. *In* "Stable isotopes in Oceanographic Studies and Paleotemperatures, *Spoleto* 1965, "(E. Tongiorgi, ed.), pp. 9–130. Consiglio Nazionale delle Richerche, Laboratoria di Geologia Nucleare, Pisa.

Craig, H., Gordon, L. I., and Horibe, Y. (1963). Isotopic exchange effects in the evaporation of water. 1. Low-temperature experimental results. *J. Geophys. Res.* **68**(17), 5079–5087.

DeNiro, M. J., and Epstein, S. (1979). $^{18}O/^{16}O$ ratios of terrestrial plant cellulose are independent of the $^{18}O/^{16}O$ ratio of atmospheric carbon dioxide. *Science* **204**, 51–53.

Dole, M. (1955). The oxygen isotope cycle in nature. *Proc. 2nd Conf. Nucl. Processes Geol. Settings*, pp. 13–19.

Dole, M., and Jenks, J. (1944). Isotopic composition of photosyntheric oxygen. *Science* **100**, 409.

Epstein, S., Thomson, P., and Yapp, C. J. (1977). Oxygen and hydrogen isotopic ratios in plant cellulose. *Science* **198**, 1209–1215.

Ferhi, A. M., Long, A., and Lerman, J. C. (1977). Stable isotopes of oxygen in plants—A possible paleohygrometer. *Hydrology Water resources Arizona Southwest* **7**, 191–198.

Ferronskiy, V. I., Dubinchuk, B. T., and Polyakov, V. A. (1975). "Natural Isotopes in the Hydrosphere." Nedra, Moscow.

Gonfiantini, R., Gratzien, S., and Tongiorgi, E. (1965). "Oxygen Isotopic Composition of Water in Leaves: Isotopes and Radiation in Soil–Plant Nutrition Studies," pp. 405–410. IAEA, Vienna.

Gray, J., and Thomson, P. (1976). Climatic information from $^{18}O/^{16}O$ ratios of cellulose in tree rings. *Nature* **262**, 481–482.

Kroopnick, P., and Craig, H. (1972). Atmospheric oxygen: isotopic composition and solubility fractionation. *Science* **175**, 54.

Kutyurin, V. M. (1972). Water origin of oxygen of photosynthesis and variability of the isotopic composition of oxygen given off by plants. *In* "Outlines of Modern Geochemistry and Analytical Chemistry." Nauka, Moscow.

Lane, G. A., and Dole, M. (1956). Fractionation of oxygen isotopes during respiration. *Science* **123**, 574.

Lehninger, A. (1974). "Biochemistry." Mir, Moscow (Russian translation).

Libby, L. M., Pandolfi, L. I., Payton, P. H., Marshall, J., III, Becker, B., and Giertz-Sienbenlist, V. (1976). Isotopic tree thermometers. *Nature* **261**, 284–288.

Long, A., Lerman, J. C., and Ferhi, A. (1978). Oxygen-18 in tree rings: paleothermometers of paleohygrometers? *4th Intern. Conf. Geochronology, Cosmochronology, Isotope Geology*, Snowmass-at-Aspen, Aug. 20–25, pp. 253–254.

O'Neil, J. R., and Adami, L. H. (1969). The oxygen isotope partition function ratio of water and the structure of liquid water. *J. Phys. Chem.* **73**(6), 1553–1558.

O'Neil, J. R., Adami, L. H., and Epstein, S. (1975). Revised value for the ^{18}O fractionation between CO_2 and H_2O at 25°C. *J. Res. U.S. Geol. Surv.* **3**(5), 623–624.

Vinogradov, A. P., and Kutyurin, V. M. (1962). On the question of the mechanism of dehydration of water in the process of photosynthesis. *In* V. Internat. Biochem. Cong. Mechanism of Photosynthesis, *Symposium VI*. Izd-vo AN SSSR, Moscow.

CHAPTER 10

New Isotopic Criteria in the Solution of Problems of the Genesis of Organic Compounds in Nature

I. EXPERIMENTAL INVESTIGATION OF THE INTRAMOLECULAR DISTRIBUTION OF ISOTOPES IN ABIOGENIC ANALOGS OF SOME BIOGENIC COMPOUNDS

Many organic compounds present in biological systems can be synthesized chemically without the participation of organisms. Both in the laboratory and in nature various hydroxy acids, amino acids, and other more complex compounds can be synthesized abiogenically. Although chemically they are indistinguishable from their biogenic analogs, their isotopic composition may be different.

Their isotopic composition is determined by the isotopic composition of the precursor and by fractionation of isotopes during formation of the compound. The isotopic composition of abiogenically synthesized compounds may vary widely depending on the character and origin of their precursor. High-molecular-weight organic compounds of biological origin generally vary in carbon isotopic composition within the -15 to $-35‰$ range, but some biological compounds can be found whose isotopic composition deviates substantially from this range. Therefore, the isotopic composition of the total carbon of a compound cannot be a reliable criterion for recognizing its biological or nonbiological origin.

The preceding chapters provide grounds for expecting, however, that in systems of organic compounds (between individual compounds) or within one compound (between the individual structural groups) the distribution of isotopes in biogenic and abiogenic substances would be characteristically different. The biological specificity of the mechanism leading to thermodynamically ordered distribution of carbon isotopes in biomolecules suggests that no ordered distribution should be present in abiogenic organic compounds.

To test this point experimentally, the intramolecular distribution of carbon isotopes was investigated in a number of biological compounds and their abiogenically synthesized analogs (Vinogradov *et al.*, 1976). Several compounds isolated from living organisms, all containing the methoxyl ($—OCH_3$) group, were investigated. They were *trans*-anethole, isoeugenol, vanillin, and anisic acid. The same compounds were synthesized under abiotic conditions. In so doing, the methoxyl group was incorporated by treating the appropriate phenol predecessor with dimethyl sulfate obtained from methyl alcohol.

The methoxyl group was split off from each compound and both its isotopic composition and the isotopic composition of the carbon of the compound as a whole were measured. The results of this experiment are given in Table 10.1. In biogenic compounds a small and fairly uniform enrichment of the OCH_3 group in the light carbon isotope was observed. It corresponds to a somewhat lower value of the β_i factor of the OCH_3 group ($\beta_i = 1.141$) compared with the β_Σ factors of the corresponding compounds, which are approximately 1.155–1.160. In the synthetic analogs the carbon of the methoxyl group is sharply enriched in the ^{12}C isotope in individual compounds, with no relationship to β-factor values. The carbon of the methyl alcohol that was the starting point for the carbon of the methoxyl groups of the abiogenic compounds had a $\delta^{13}C = -34.5‰$, and these methoxyl groups were more enriched in ^{12}C than in methyl alcohol. This enrichment is expected on the basis of kinetic consideration during abiotic synthesis, whereas in biomolecules the carbon of any structural group may be either enriched or depleted in the light isotope depending on the ratio of values of thermodynamic isotopic factors (β_i factors) of the carbon structures that form these molecules.

This difference is clear when biogenic and industrial acetic acid (CH_3COOH) are compared. The value of the β_i factor of the methyl group (1.131) is much lower than that of the carboxyl group (1.197). Therefore, in accordance with the principle of thermodynamic ordering of isotopes, in biogenic acetic acid the carboxyl group should be isotopically heavier than the methyl. The experiment confirmed this prediction. As noted in Chapter 4, in biogenic acetic acid (isolated from apple cider; Meinschein *et al.*, 1974) the carboxyl group is about 20‰ richer in the ^{13}C isotope than the methyl. This corresponds to a higher value of the β_i factor of carboxyl carbon (1.197) compared with the β_i factor of methyl carbon (1.131).

The intramolecular distribution of isotopes in industrial acetic acid

TABLE 10.1

INTRAMOLECULAR ISOTOPIC EFFECT FOR CARBON IN SOME BIOGENIC COMPOUNDS AND THEIR SYNTHETIC ANALOGS

		Synthetic[a]		*Biogenic*[a]	
Compound	*Structure*	*1*	*2*	*1*	*2*
Anethole	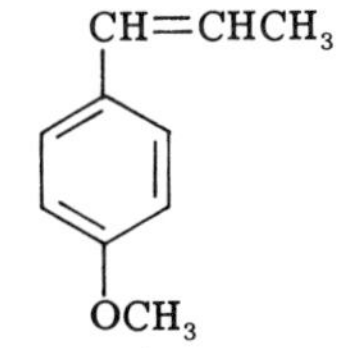	−32.2		−28.0	
OCH_3 group		−68.7	−36.5	−30.9	−2.9
Eugenol	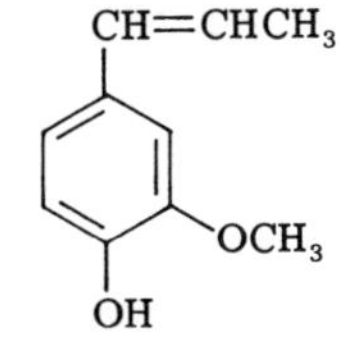	−27.7		−30.8	
OCH_3 group		−49.7	−22.0	−34.0	−3.2
Vanillin	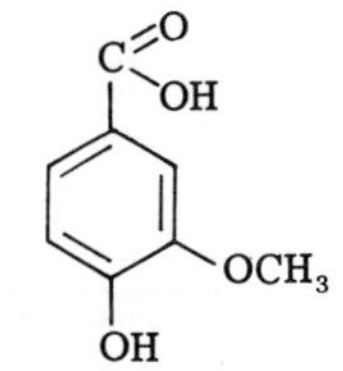	−29.0		−29.9	
OCH_3 group		−44.4	−15.4	−32.7	−2.8
Anisic acid	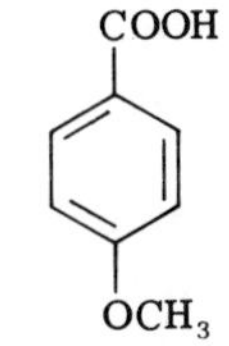	−30.7			
OCH_3 group		−93.3	−62.6		

[a] *1*, $\delta^{13}C$ of the total carbon of the compound (in per mil); *2*, $\delta^{13}C_{OCH_3} - \delta^{13}C_{total}$.

turned out to be the opposite (Table 10.2). It was obtained by oxidation of hydrocarbons, in particular, ethylene. In this case there was less of the ^{13}C isotope in the carboxyl group than in the methyl. Also, the extent of depletion was different, depending on the means of obtaining the acetic acid. It varied from 1 to 20‰. The depletion of the carboxyl group of industrially synthesized acetic acid in the ^{13}C isotope agrees with the expected kinetic isotopic effect in the oxidation process.

TABLE 10.2

INTRAMOLECULAR DISTRIBUTION OF CARBON ISOTOPES IN SYNTHETIC ACETIC ACID

Means of obtaining acetic acid	$\delta^{13}C$‰		$\delta^{13}C_{CH_3} - \delta^{13}C_{COOH}$	*Reference*
	CH_3	COOH		
Ethylene–ethanol–acetaldehyde–acetic acid	−39.5	−40.4	0.8	Yankwich and Promislov, 1953
Oxidation of propane + butane	−35.4	−40.6	10.2	Yankwich and Promislov, 1953
Ethane–ethylene–ethyl sulfate–diethyl sulfate–ethanol–acetaldehyde–acetic acid	−15.1	−44.4	29.2	Meinschein *et al.*, 1974

It was important to check whether any intramolecular separation of isotopes occurs in abiogenic compounds if the kinetic isotopic effect is suppressed. For this purpose the amino acid glycine (CH_3NH_2COOH) was synthesized abiogenically from a monocarbon compound (CH_3NO_2), so that both the carboxyl and methyl groups of the glycine would have the same carbon source (Vinogradov *et al.*, 1976). The pathway of synthesis was chosen to minimize kinetic isotopic effects during the synthesis each of the carbon structural groups in glycine. When the amino acid ultimately obtained was analyzed, it was discovered that the isotopic composition of the carbon of both its structural groups was identical despite the markedly different values of the corresponding β_i factors. There was no intramolecular isotopic exchange during the synthesis of this abiogenic compound.

Thus the difference in the intramolecular distribution of carbon isotopes in biological and abiogenically synthesized organic compounds is seen. In the first case there is a tendency toward ordered distribution of isotopes corresponding to the values of the thermodynamic isotopic factors. In the second case the isotopic effects are not ordered in value or sign but depend on special conditions and the method of synthesis of the given compound.

II. THERMODYNAMIC ISOTOPIC EFFECTS IN NONBIOLOGICAL SYSTEMS

The differences in isotopic composition of biogenic and artificially synthesized compounds established above can serve as the basis for formulating a new method of recognizing the biological or nonbiological nature of organic compounds. Of course thermodynamic isotopic effects are inherent not only in biological systems. In nature there are many processes in which isotopic exchange takes place and thermodynamic isotopic effects are manifested.

An example of a reaction in which isotopic equilibrium is established rather rapidly is the $CO_2 \rightleftarrows H_2O$ system. At 25°C, equilibrium of isotopic exchange for oxygen is achieved in less than 1 hr (Cohn and Urey, 1938). For the slow precipitation of carbonates, equilibrium is achieved in the $CaCO_3 \rightleftarrows H_2O$ system. The isotopic exchange equilibrium constant and the temperature dependence of this system have not only been calculated theoretically but have also been well studied experimentally (Friedman and O'Neil, 1977). Therefore, the isotopic composition of carbonates can be used for precise determination of the temperature of the aqueous environment in which they were precipitated, in particular the temperature of ancient sedimentary basins. This is called the paleotemperature method.

Isotopic equilibrium for carbon is also reached rather rapidly in the $CaCO_3 \rightleftarrows CO_2$ system (McCrea, 1950). This system is important in geochemical respects. Due to the definite value of the thermodynamic isotopic effect between atmospheric carbon dioxide and sedimentary carbonates, the carbon isotopic composition of the latter reflects that of atmospheric carbon dioxide in previous geologic epochs. Investigation of the variations in isotopic composition of the carbonate carbon through geologic times allows the recreation of variations of the isotopic composition of atmospheric CO_2—that very important member of the geochemical cycle of carbon in the geologic history of the earth (Galimov *et al.*, 1975).

However, in the absence of catalysts rapid isotopic exchange is normally the exception rather than the rule. Even in such a simple system as CO_2—CH_4, isotopic equilibrium for carbon is not reached at 25°C during prolonged geologic time. For example, analysis of carbon dioxide and methane from gaseous inclusions in a mineral of an igneous rock evolved near the surface in Devonian time (about 400 million years ago) shows that the difference in carbon isotopic compositions of these compounds still did not correspond to the temperature conditions of emplacement of this rock (Galimov and Petersil'ye, 1967). Even at 500°C under laboratory conditions, equilibrium was not observed in the $CO_2 \rightleftarrows CH_4$ system (Sackett and Chang, 1979).

The atoms belonging to the carbon skeleton of molecules of organic compounds are exchanged with still more difficulty. Investigations made with very simple carbon compounds (Galimov *et al.*, 1972) showed their isotopic composition to be nearly unchanged to temperatures of the order of 300°C (Table 10.3). Only at a temperature approaching 500°C is an appreciable change in carbon isotopic composition finally observed.

TABLE 10.3

VARIATION IN ISOTOPIC COMPOSITION IN A CH_4–C_2H_6–C_3H_8–C_4H_{10} MIXTURE KEPT IN A CLOSED SYSTEM FOR 4 HR AT DIFFERENT TEMPERATURES[a]

	$\delta^{13}C$ (‰)			
Hydrocarbons	20°C	150°C	300°C	500°C
Experiment 1[b]				
CH_4	−41.5	−41.4	−40.7	−36.8
C_2H_6	−32.3	−32.2	−32.0	−29.5
C_3H_8	−26.6	−26.8	−26.2	−19.9
C_4H_{10}	−26.3	−26.4	−26.4	−21.0
Experiment 2[c]				
CH_4	−41.5	−40.5	41.2	−36.3
	—	−41.4	−40.8	−35.6
C_2H_6	−32.3	−32.0	−31.6	−31.0
	—	−31.9	−31.2	—
C_3H_8	−26.6	−26.1	−27.4	−26.7
	−26.1	−26.1	−27.5	−26.9
C_4H_{10}	−26.2	−26.4	−26.7	−24.1
	−26.7	−26.0	−27.1	−24.3
Total	−32.4	−32.6	−32.4	−27.8
hydrocarbons	−32.6	−32.2	−31.5	−28.6

[a] From Galimov *et al.* (1972).

[b] Experiment 1: hydrocarbons dissolved in petroleum at a pressure of 200 atm.

[c] Experiment 2: hydrocarbons formed from a gaseous mixture in the presence of water vapor, with an initial pressure of 1 atm. Two values in each column correspond to parallel experiments.

However, it apparently is not caused by isotopic exchange but by incipient reactions of the formation of olefins at these temperatures, with the subsequent formation of a polymerized carbonaceous substance which evidently is accompanied by a kinetic isotopic effect. As a result, all the components of the gaseous mixture are enriched in the ^{13}C isotope.

A slow rate of isotopic exchange of carbon in organic compounds is a typical property of abiogenic systems. As a rule, the more complex the organic compound, the more difficult the isotopic exchange. It follows, first, that thermodynamic isotopic effects are not characteristic of abiogenic systems of organic compounds. This is valid in any case for homogeneous systems. Second, the distribution of isotopes which is established can be retained in organic compounds at fairly low temperatures (200–250°C) during prolonged geologic time.

III. ISOTOPIC CRITERIA FOR THE RECOGNITION OF BIOGENIC AND ABIOGENIC COMPOUNDS

Keeping in mind what has been said concerning isotopic effects in biological and nonbiological systems, three levels can be indicated at which isotopic analysis can be used for identification of the biological or nonbiological origin of a substance.

The first is determination of the total isotopic composition of the carbon in a substance and comparison of it with the typical range of variations for substances of biological origin. Although occasionally that assessment is effective, it is generally ambiguous, as different processes can lead to the formation of substances with approximately the same isotopic composition. For instance, even in early works on the geochemistry of carbon isotopes it was established that the isotopic composition of petroleum carbon is similar to the isotopic composition of the carbon of living organisms. It would seem that decisive evidence of the origin of the petroleum from fossil organic matter was obtained. However, it was ascertained later that abiogenically synthesized carbonaceous matter chemically similar to petroleum, in particular bituminoids, occurring in igneous rocks has nearly the same isotopic composition as petroleum (Galimov and Petersil'ye, 1968). Therefore, the isotopic criteria used for comparison of total carbon of a substance actually provided no solution for the problem of the origin of petroleum. Below we shall discuss this problem to show another approach to its solution.

Discovery of thermodynamically ordered distribution of isotopes in biological systems opened up the possibility of formulating a more delicate, two-level isotopic criterion for identification of the nature of organic compounds. The following criterion is based on the possibility that the thermodynamic isotopic effects in enzymatic reactions produce a thermodynamically ordered distribution of isotopes between biocomponents under certain conditions. In a system of organic compounds the systematic relationship of their isotopic compositions to the values of the corresponding β_{Σ} factors, therefore, is indicative of the biological origin of these compounds. The more compounds examined and the closer the relationship, the more significant this conclusion will be (Galimov *et al.*, 1977).

However, it cannot be ruled out that, in some processes of heterogeneous catalysis, isotopic effects of the same type as in enzymatic reactions could arise. In the classification of Boreskov (1977), enzymatic reactions are assigned to catalytic processes in which intermediate interaction of the reagents with the catalyst takes place concertedly. Not only enzymatic but also many fluid-phase reactions catalyzed by metal complexes take place by a concerted mechanism. In some cases it is possible that the conditions guaranteeing an intracomplex thermodynamic isotopic effect analogous to the thermodynamic isotopic effect of the enzymatic process will be satisfied.

Although the relationship between the isotopic composition and values of the β factors in a system of organic compounds is a very strong sign of biological origin of a substance, it must be remembered that in some cases a similar relationship can be established in the process of heterogeneous catalysis.

Apparently, a superior criterion for recognition of biological and nonbiological compounds is the character of intramolecular distribution of isotopes.

As discussed in previous chapters, an ordered distribution of isotopes in biosystems is accomplished on two levels. The first level corresponds to the distribution of isotopes in an isolated enzymatic process. At this level a ratio of isotopes is established between the immediate predecessor and product of a reaction that corresponds to the β-factor values (e.g., between the carbonyl carbon of pyruvate and the aldehyde carbon derived from it by the decarboxylation of pyruvate). But an intramolecular thermodynamically ordered distribution of isotopes in the acetaldehyde (between aldehyde and methyl carbon) cannot be established solely as the result of this reaction. To establish that, the presence of an interrelated system of enzymatic reactions with direct and inverse pathways of transformation is necessary. In other words, intramolecular ordering of isotopes is realized in a system of higher order than an isolated enzymatic reaction, namely, a metabolizing system, i.e., in an organism.

A compound having a thermodynamically ordered distribution of isotopes must have been formed not simply as a result of some catalytic reaction, including a biochemical reaction, but as a participant in the biological cycles of an organism. This gives the intramolecular isotopic criterion a special value for application to problems of the origin of life. As a matter of fact, any reaction considered individually, including one catalyzed by an enzyme, is not a biological process. On the other hand, a system of enzymatic reactions characterized by the criteria mentioned above is a very simple model of a biological process. Obviously, for such a metabolic system to arise in biological evolution, i.e., in the course of natural selection, it must have been phase isolated, as Oparin (1977) points out. In essence, it corresponds to the minimum requirements which would satisfy the concept of an organism. Apparently, the formation of this system denotes the boundary which separates the living from the nonliving. With that definition we can immediately specify the criterion for finding such systems—intramolecular thermodynamically ordered distribution of isotopes.

In contrast to other diagnostic criteria of biogenic matter (e.g., optical activity, or typomorphic nature of structures), ordered intramolecular distribution of isotopes may be inherent in any biomolecule regardless of its composition, stereochemistry, etc. Therefore, the intramolecular distribution of isotopes evidently is the most important criterion which can serve as the basis for recognizing biogenic and abiogenic structures.

On the other hand, in biological systems the establishment of thermodynamically ordered intramolecular distribution of isotopes is associated with a number of limiting conditions, as was shown in Chapter 5; these are reversibility of transformations on the pathways connecting the biosynthesis of a given compound, steady state of the system of corresponding reactions, and absence of appreciable kinetic isotopic effects and transfer effects in the chains of transformation. Therefore, in biomolecules considered individually, intramolecular thermodynamically ordered distribution of isotopes can often be disrupted. Consequently, a practical method should be based on the study of the intramolecular isotopic effect for a multitude of molecules. Because direct investigation of intramolecular isotopic distribution in an organic compound is a complicated analytical problem, simplified versions of the method based on the study of the isotopic composition of typical structural groups can be used.

For instance, according to the principle of thermodynamically ordered distribution of isotopes, the carbon of the carboxyl groups of biomolecules should be enriched in the heavy isotope. Although departures from this rule are possible in individual compounds, on average it must be fulfilled for a biomass. In those cases where the formation of carboxyl groups was the result of abiogenic oxidation, carboxyl carbon would not show enrichment in the ^{13}C isotope. On the contrary, due to the kinetic isotopic effect, carboxyl carbon would be enriched in the light isotope. Therefore, the decarboxylation and the study of the relationship of isotopic compositions of carboxyl and residual organic carbon obviously can be used to ascertain the processes of biogenic oxidation of organic matter and, under certain conditions, to recognize the biogenic and abiogenic forms of organic compounds in general.

IV. DISTRIBUTION OF ISOTOPES IN PRODUCTS OF THE FISCHER–TROPSCH REACTION

At the beginning of this chapter a comparative analysis was made of biogenic and abiogenic chemical analogs in connection with the question of the details of their intramolecular distribution of isotopes. We shall now consider several more interpretations of the isotopic characteristics of substances which possibly took part in the formation of prebiological systems.

The Fischer–Tropsch process is generally considered to be the most probable chemical model of abiogenic synthesis of organic compounds in nature. This concerns the possible formation of both organic compounds under cosmic conditions, for instance, those present in carbonaceous chondrites, and hydrocarbons and bitumens found in igneous rocks.

Fractionation of carbon isotopes in the Fischer–Tropsch reaction was investigated by Lancet and Anders (1970). The experiment was carried out at a temperature of 400 K with a pressure of 1 kg/cm^2 and with equimolar volumes of CO and H_2 in the presence of a cobalt catalyst.

During the reaction both oxidized and reduced products were formed. Most of the oxygen of the CO went into H_2O, but some went into carbon dioxide, which comprised a few tenths of a percent of the reaction products. Methane predominated among the reduced products (69–88% by the end of the reaction). The content of C_{2+} hydrocarbons that were volatile at 400 K reached about 3% by the end of the experiment. Some of the polymerized organic compounds (about 1.5%) which were not distilled off at 400 K remained on the catalyst as coke. The results of isotopic analysis of the carbon of the reaction products are given in Fig. 10.1.

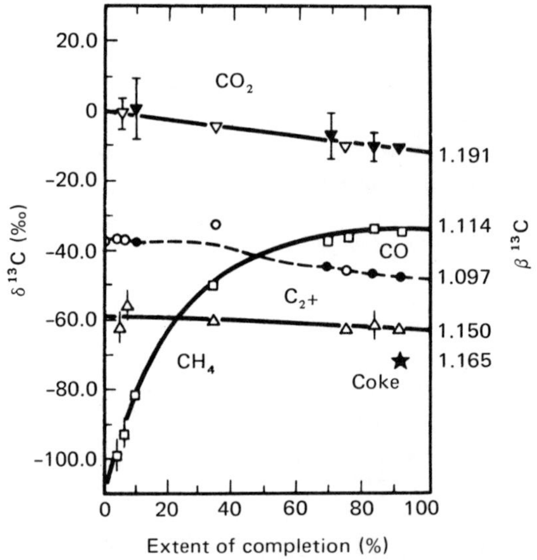

FIG. 10.1 Distribution of isotopes in products of the Fischer–Tropsch reaction.

Carbon dioxide formed in the Fischer–Tropsch process is enriched by about 35.5‰ in the ^{13}C isotope relative to the original CO. This difference remains fairly constant during the reaction. The carbon of the C_{2+} fraction, on the other hand, is enriched by about 19.9‰ in the light isotope. The high-molecular product remaining on the catalyst is characterized by a greater enrichment in the light isotope—by 33.3‰ relative to the CO carbon. The carbon of methane formed at the beginning of the reaction is maximally enriched in the light isotope, when the difference between the isotopic composition of CO_2 and CH_4 carbon reaches 100‰, and between CO and CH_4 carbon, 60‰. As the reaction develops, the methane is rapidly depleted in the light isotope. Because methane is the predominant product of the reaction, its isotopic composition begins to approach that of the original CO.

On the right side of Fig. 10.1, the β-factor values of the compounds participating in the process are shown. A β-factor value of 1.150, characterizing low-molecular hydrocarbons on the average, is taken for the C_{2+} fraction. Coke can be characterized by a β-factor value corresponding to polymerized carbonaceous substances, for which $\bar{\beta} = 1.165$ (Galimov, 1973, p. 129).

Comparison of the $\delta^{13}C$ values and β-factor values characterizing the products of the Fischer–Tropsch reaction shows that there is no correspondence between them, which was to be expected for products of abiogenic synthesis.

V. DISTRIBUTION OF ISOTOPES IN PRODUCTS OF SYNTHESIS IN AN ELECTRICAL DISCHARGE

In the discharge of a high-voltage electric arc, amino acids are formed from a mixture of CH_4, NH_3, and H_2O. As is widely known, S. Miller first performed experiments of this type. In DeMarais's laboratory (NASA Ames Research Center; Moffett Field, California), the isotopic composition of carbon of amino acids synthesized in this manner were analyzed. The carbon isotopic composition of the total amino acids and of their carboxyl carbons was determined. These analyses, as furnished by S. Chang, are summarized in Table 10.4. As seen from these data, the carbon of the amino acids was enriched in the light isotope with respect to the original methane carbon. A thermodynamic isotopic effect in this system, on the contrary, should have led to enrichment of the amino acids in the heavy isotope, because the β factor of methane is characterized by a value of 1.114 and of

TABLE 10.4

ISOTOPIC COMPOSITION OF CARBON OF PRODUCTS OF ABIOGENIC SYNTHESIS IN AN ELECTRICAL DISCHARGE[a]

			$\delta^{13}C_{sample} - \delta^{13}C_{reagent}$	
Reagents	*Type of discharge*	*Product*	*From*	*To*
CH_4	Low voltage	C_2–C_5 hydrocarbons	−3.2	−0.9
	High voltage	C_2–C_5 hydrocarbons	−8.8	−6.3
CH_4, NH_3, H_2O	Low voltage	Amino acids		
		Total carbon	−4.5	−2.1
		Carboxyl carbon	−9.6	−2.7

[a] Data of S. Chang, personal communication (from D. J. DeMarais's laboratory).

amino acids, on the average, by a value of 1.175. Correspondingly, biogenic methane is always isotopically lighter than biogenic amino acids.

The contrast between the isotopic characteristics of abiogenically synthesized amino acids and their biogenic analogs is also emphasized by the opposite relationships of isotopic compositions of total and carboxyl carbon. The carboxyl carbon of amino acids synthesized in the electrical discharge is enriched in the light isotope, i.e., the isotopic effect has a sign opposite to that which corresponds to the ratio of β factors ($\beta^{13}C = 1.175$ for total carbon of the amino acids and $\beta^{13}C = 1.197$ for carboxyl carbon). In biogenic amino acids, carboxyl carbon as a rule is isotopically heavy in correspondence with the ratio of β factors.

An electrical discharge in an environment of pure methane produces higher-molecular-weight hydrocarbons. As seen from Table 10.4, the synthesized C_2 to C_5 hydrocarbons are enriched in the light isotope.

Meanwhile, the thermodynamic isotopic factors of C_2 to C_5 hydrocarbons are higher (1.130–1.150) than those in methane, i.e., in the case of a thermodynamic isotopic effect, enrichment of the high-molecular-weight hydrocarbons in the ^{13}C isotope would be expected. That is precisely what is observed in the distribution of isotopes in hydrocarbons originating from biological sources. In the gases of sedimentary rocks there is a progressive increase in content of the ^{13}C isotope from methane to the higher hydrocarbons in the series CH_4–C_2H_6–C_3H_8–C_4H_{10} (Galimov, 1973) (Fig. 10.2).

It is interesting to note that the inverse abiogenic character of distribution of isotopes in hydrocarbons is also encountered under natural conditions. An inverse distribution was first found in gaseous inclusions in minerals of igneous rocks of the Kola Peninsula (Galimov, 1973; Galimov and Petersil'ye, 1967).

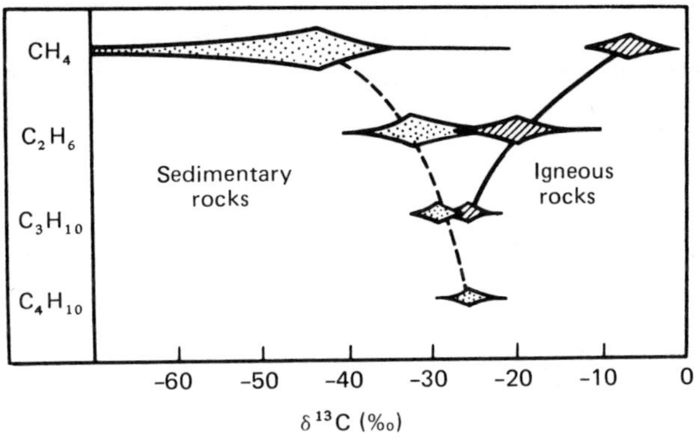

FIG. 10.2 Isotopic composition of the carbon of natural hydrocarbons, both originating from a biological source and synthesized abiogenically in igneous rocks.

VI. ISOTOPIC COMPOSITION OF THE CARBON COMPONENTS OF CARBONACEOUS CHONDRITES

The isotopic composition of the carbon of carbonaceous chondrites is interesting in two respects. First, in contrast to other types of meteorites, carbonaceous chondrites are rich in volatiles. They contain from tenths of a percent (type III) to a few weight percents of carbon and water (type I), while the carbon content in other types of meteorites does not exceed 0.02–0.04%, and the water content is still less (Mason, 1965). Their unique feature is the presence of various organic compounds, including amino acids, aromatic polymers, several hydroxyacids porphyrins, and others (Boato, 1954; Folsome *et al.*, 1973; Gelpi and Oro 1970; Hayes, 1970; Lawless *et al.*, 1971, 1974). Therefore, the origin of these cosmic bodies and the nature of the organic compounds in them is of considerable scientific interest.

Second, on the basis of a whole series of parameters the composition of the substance of carbonaceous chondrites is closer to the composition of the earth's crust than to the composition of meteorites of any other type. This enabled the author to suggest that in the last stages of its formation the earth consisted of material of the type of carbonaceous chondrites (Galimov, 1967). Such a possibility in one or another modification has been discussed by many investigators subsequently (Anders and Owen, 1977; Ringwood, 1970; Sackett and Chang, 1979). This hypothesis could be developed and refined on the basis of new data (Bannikova and Galimov, 1978),

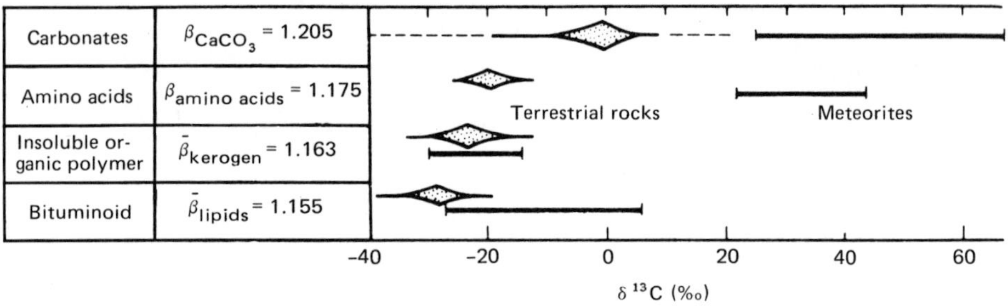

FIG. 10.3 Ranges of variation of the isotopic composition of the carbon components of carbonaceous chondrites compared with the range of variations of corresponding terrestrial substances (average values of their β^{13}C factors are given at the left).

which among other things, suggests that on the earth's surface there originally existed a rather high concentration of reduced forms of carbon, including many organic compounds which could be of prebiological importance. This possibility lends special interest to the study of the character of isotopic distribution in the organic matter of carbonaceous chondrites.

As early as 1954, Boato (1954) showed that carbonaceous chondrites have an unusual isotopic composition of carbon and hydrogen, different from other types of meteorites. And previously noted, these compositions correspond best to estimates of the average isotopic composition of those elements in the outer shell of the earth (Galimov, 1968). At present, there is a fairly voluminous literature dealing with investigation of the substance of carbonaceous chondrites, including the isotopic composition of the various forms of carbon in them (Belsky and Kaplan, 1970; Chang *et al.*, 1978; Clayton, 1963; Krouse and Modzeleski, 1970; Vinogradov *et al.*, 1967). Figure 10.3 compares the range of variations of δ^{13}C for several forms of carbon in carbonaceous chondrites with the ranges of variation of corresponding terrestrial forms of carbon.

From this comparison it follows that (1) the variations in δ^{13}C in each form and the isotopic differences between individual forms are greater in the carbonaceous chondrites than in the terrestrial forms; (2) carbonates are anomalously enriched in the ^{13}C isotope (under terrestrial conditions carbonates of such isotopic composition are not encountered); (3) amino acids are enriched in the ^{13}C isotope, which distinguishes them from terrestrial amino acids and indicates their undoubtedly extraterrestrial origin; and (4) the correlation with β-factor values is not clear (in contrast to terrestrial biogenic forms given on the same graph), but it is not completely lacking, as was shown previously for abiogenic forms of carbon.

Recently American investigators carried out an extremely interesting and delicate investigation of the isotopic composition of amino acids in car-

TABLE 10.5

CARBON OF AMINO ACIDS ISOLATED FROM THE MURCHISON CARBONACEOUS CHONDRITE[a]

	$\delta^{13}C$ (‰)		
Fraction[b]	*Measured value*	*Background*	*Corrected value*
Valine	21.2	−15.5	22.6
Alanine	27.3	−15.7	30.4
Glycine	35.6	−26.6	43.7
Glutamic acid	27.2	−24.6	38.4

[a] Data from S. Chang, personal communication.
[b] The fractions are eluates of broad chromatographic bands, the centers of which correspond to the R_f of the amino acids in question.

bonaceous meteorites. These data have partially been published (Chang *et al.*, 1978).

Table 10.5 gives data on the isotopic composition of the individual amino acids from the Murchison meteorite (data furnished by Dr. Chang).

Comparison of the measured values of $\delta^{13}C$ with the $\beta^{13}C$ factors of the corresponding compounds brings out a remarkable relationship between these parameters—a relationship analogous to that of biological compounds (Fig. 10.4).

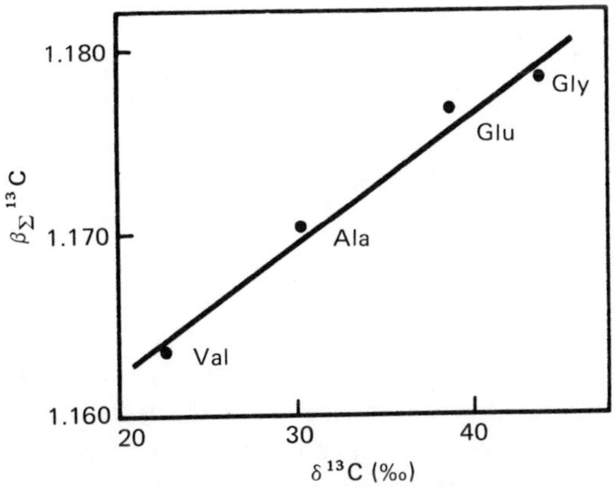

FIG. 10.4 Comparison of the carbon isotopic composition and $\beta^{13}C$ factors of individual amino acids isolated from the Murchison carbonaceous chondrite.

The anomalous enrichment of carbonates in the heavy isotopes mentioned above, which is strikingly different from terrestrial carbonates, is not difficult to explain if it is kept in mind that the dominant form of carbon in the carbonaceous chondrites is reduced carbon and the dominant form under the conditions of the earth's crust is the oxidized form (CO_2 and carbonate). Therefore, the difference in isotopic composition between these oxidized and reduced forms for an average value of the isotopic composition of total carbon ($\delta^{13}C = -7‰$) in carbonaceous chondrites shows up as a shift of the $\delta^{13}C$ values characterizing carbonates to above the average. Under terrestrial conditions there is a shift to below the average of the $\delta^{13}C$ values, characterizing a reduced form of carbon, for instance methane. We note that this reasoning suggests the thermodynamic nature of the isotopic effect. In the case of kinetic isotopic effects, the distribution of isotopes depends on the mechanism of the process rather than on the chemical form of the compounds. In this case the products of oxidation may be isotopically lighter than the initial reduced form.

With the thermodynamic isotopic effect one can also explain the anomalous enrichment of the amino acids in carbonaceous chondrites in the heavy isotope ($\delta^{13}C$ from 22 to 45‰). It would seem that amino acids synthesized abiogenically should be enriched in the light isotope as a result of the kinetic isotopic effect, as was observed in the experiments described previously. However, if the β-factor values of the amino acids are kept in mind ($\bar{\beta}^{13}C = 1.175$), it turns out that, with respect to the isotopic composition of other forms of carbon, especially carbonates, their isotopic composition is in approximately the same relationship as their β-factor values compared with the β factors of other forms. This again suggests the thermodynamic character of the isotopic effects accompanying the synthesis of the amino acids.

However, the profound differentiation of the various forms of carbon in isotopic composition, mentioned above, in the case of the thermodynamic character of isotopic effects is acceptable only for temperatures not exceeding a few tens of degrees Celsius.

In any case, in the carbonaceous chondrites we find a very specific distribution of isotopes which indicates tendencies toward a thermodynamically ordered distribution of isotopes, i.e., tendencies characterizing biological carbon. Possibly the manifestation of isotopic effects of thermodynamic nature is related to some very efficient forms of low-temperature heterogeneous catalysis. Further investigations of the isotopic composition of the individual compounds of carbonaceous chondrites, in particular investigation of the intramolecular distribution of isotopes, apparently could be vitally important for learning the mechanism of formation of organic compounds in carbonaceous chondrites.

REFERENCES

Anders, E., and Owen, T. (1977). Mars and Earth: origin and abundance of volatiles. *Science* **198**, 453.

Bannikova, L. A., and Galimov, E. M. (1978). On the character of the original material forming the outer shell of the Earth. *In* Abstracts of papers at the VII All-Union Symposium on Stable Isotopes in Geochemistry. GEOKhI AN SSSR, Moscow.

Belsky T., and Kaplan, I. R. (1970). Light hydrocarbon gases ^{13}C and origin of organic matter in carbonaceous chondrites. *Geochim. Cosmochim. Acta* **34**, 257–278.

Boato, G. (1954). The isotopic compositions of hydrogen and carbon in the carbonaceous chondrites. *Geochim. Cosmochim. Acta* **6**, 201.

Boreskov, G. K. (1977). Development of ideas on the nature of heterogeneous catalysis. *Kinetika i Kataliz* **8**(5).

Chang, S., Mack, R., and Lemon, K. (1978). Carbon chemistry of separated phases of Murchison and Allende meteorites. *In* "Lunar and Planetary Science IX." Pt. 1, p. 157. Conf. Sponsored NASA.

Clayton, R. N. (1963). Carbon isotope abundance in meteorite carbonates. *Science* **140**, 192.

Cohn, M., and Urey, H. C. (1938). Oxygen exchange reactions of organic compounds and water. *J. Am. Chem. Soc.* **60**(3), 679–687.

Folsome, C. E., Lawless, J. G., Romiez, M., and Ponnamperuna, C. (1973). Heterocyclic compounds recovered from carbonaceous chondrites. *Geochim. Cosmochim. Acta* **37**, 455–465.

Freidman, I., and O'Neil, J. R. (1977). Data of geochemistry. "Compilation of Stable Isotope Fractionation Factors of Geochemical Interest" Geol. Surv. Prof. Pap. 440 KK), Ch. KK. U.S. Gov. Printing Office, Washington, D.C.

Galimov, E. M. (1967). Two lines of evolution of the carbon of the Earth. *Geokhimiya.* (5).

Galimov, E. M. (1968). "Geochemistry of stable carbon isotopes." Nedra, Moscow.

Galimov, E. M. (1973). "Carbon isotopes in petroleum and gas geology." Nauka, Moscow.

Galimov, E. M. (1977). Carbon isotopes in the living and nonliving. *In* "Science and Man: Internat. Annual." Znaniye, Moscow.

Galimov, E. M., and Petersil'ye, I. A. (1967). Isotopic composition of the carbon of methane isolated in the pores and cavities of some igneous minerals. *Doklady AN SSSR* **176**(4).

Galimov, E. M., and Petersil'ye, I. A. (1968). Isotopic composition of the carbon of bituminoids of igneous and metamorphic rocks. *Doklady AN SSSR* **182**(1).

Galimov, E. M., Posyagin, V. I., and Prokhorov, V. S. (1972). Experimental study of the fractionation of carbon isotopes in the system CH_4–C_2H_6–C_3H_8–C_4H_{10} at different temperatures. *Geokhimiya* (8).

Galimov, E. M., Migdisov, A. A., and Ronov, A. B. (1975). Variations in the isotopic composition of carbonate and organic carbon of sedimentary rocks in the history of the Earth. *Geokhimiya* (3).

Galimov, E. M., Kodina, L. A., Generalova, V. N., Bogacheva, M. P., and Shirinskiy, V. G. (1977). Investigation of the distribution of carbon isotopes in biogenic compounds. *In* "Trudy VIII Internat. Congress on Organic Geochemistry." Nauka, Moscow.

Gelpi, E., and Oro, J. (1970). Organic compounds in meteorites. IV. Gas chromatographic–mass-spectrometric studies on the isopenoids and other isometric alkanes in carbonaceous chondrites. *Geochim. Cosmochim. Acta* **34**(9), 995.

Hayes, J. M. (1970). Organic consituents in meteorites: a review. *Geochim. Cosmochim. Acta* **31**(12), 1395.

Krouse, H. R., and Modzeleski, V. E. (1970). $^{13}C/^{12}C$ abundance in components of carbonacenous meteorites and terrestrial samples. *Geochim. Cosmochim. Acta* **34**(4), 459.

Lancet, M. S., and Anders, E. (1970). Carbon isotope fractionation in the Fischer–Tropsch synthesis and in meteorites. *Science*, **170**.

Lawless, J. G., Kvenvolden, K. A., Peterson, E., and Ponnamperuna C. (1971). Amino acids indigenous to the Murray metorite. *Science* **173**, 626.

Lawless, J. G., Zeitman, B., and Pereira, W. E., *et al.* (1974). Dicarboxylic acids in the Murchison meteorite. *Nature* **251**, 40–42.

McCrea, M. J. (1960). On the isotopic chemistry of carbonates and a paleotemperature scale. *J. Chem. Phys.* **18**(6), 849–857.

Mason, B. (1965). "Meteorites." Mir, Moscow (Russian translation).

Meinschein, W. G., Rinaldi, G. G., Hayes, J. M., and Schoeller, D. A. (1974). Intramolecular isotopic order in biologically produced acetic acid. *Biomed. Mass-Spectr.* **1**(3), 172.

Oparin, A. I. (1977). "Matter, Life, Intellect." Nauka, Moscow.

Ringwood, A. E. (1970). Origin of the Moon: the precipitation hypothesis. *Earth Planet. Sci. Lett.* **8**(2).

Sackett, W. M., and Chang, H. M. (1979). Experimental confirmation of the lack of carbon isotope exchange between methane and carbon dioxides at high temperature. *Geochim. Cosmochim. Acta* **43**, 273–276.

Turekian, K. K., and Clark, S. P. (1969). Inhomogeneous accumulation of the Earth from the primitive solar nebula. *Earth Planet. Sci. Lett.* **6**(5) 346–348.

Vinogradov, A. P., Galimov, E. M., Kodina, L. A., and Generalova, V. N. (1976). Intramolecular distribution of carbon isotopes as a criterion of biological or nonbiological origin of organic compounds. *Geokhimiya* (7).

Vinogradov, A. P., Kropotova, O. I., Vdovykin, G. P., and Grinenko, V. A. (1967). Isotopic composition of various phases of carbon in carbon-rich carbonaceous chondrites. *Geokhimiya* (3).

Yankwich, P. E., and Promislov, A. L. (1953). Carbon isotope constitution of some acetic acids. *J. Am. Chem. Soc.* **75**(9), 4881.

CHAPTER 11

Geologic Aspects of Thermodynamically Ordered Isotopic Distributions in Biological Products

Whereas biologically the most interesting possibilities of using the phenomenon of thermodynamically ordered distribution of isotopes in biomolecules are related to the solution of the problem of the causes and mechanism of this effect, geochemically the fact that ordered distribution of isotopes in biological material exists is important. This means that biogenic compounds bear a unique isotopic stamp—a typical distribution of carbon isotopes. This distribution is also retained when the organism dies and the biogenic molecules deposited in sediments undergo chemical changes during geologic time that gradually alter their structural features. For example, some hydrocarbon molecules in petroleum differ from the biomolecules that were their precursors. Consequently, geochemists find it difficult to determine how a given organic compound found in rocks originated. One of the main problems in organic geochemistry is to identify what the precursor compounds are and the chemical evolutionary pathways of specific organic compounds in rocks, but using the aforementioned stamp, this identification is possible in principle.

Another important point should be mentioned. As stressed previously, the distribution of isotopes may depart from a thermodynamically ordered one in individual biomolecules and biological systems owing to characteristics of their functioning or of the life activity of the specific organism. However, in geochemical examination of organic matter we are dealing with the product of transformation of a multitude of individuals. Individual differences are extinguished and the distribution becomes systematic. Therefore, it is fruitful to use the established principles of biological fractionation of isotopes to investigate the general regularities of behavior of organic matter under natural conditions.

I. CHEMISTRY OF THE TRANSFORMATION OF LIVING MATTER TO FOSSIL FORMS

With the death of an organism, the operation of the mechanism maintaining thermodynamically ordered distribution of isotopes in biomolecules ceases. The distribution of isotopes inherited from biological systems is metastably preserved in undisturbed structures of buried organic compounds. Where disruption occurs and new bonds are formed, isotopic fractionation may occur, caused chiefly by kinetic isotopic effects, i.e., because $^{12}C—^{12}C$ bonds are more labile than $^{12}C—^{13}C$ bonds. The more chemical reorganization the original biomolecules undergo, the weaker becomes the connection between the isotopic composition and the β factor of the compound.

In sedimentary rocks, organic matter is present mainly in the form of kerogen, a complex irregular polymer which is insoluble in water, organic solvents, acids, and alkalis. A small part of organic matter (from fractions of a percent to several percent) consists of bituminoid extracted from the rock by organic solvents. In young marine sediments and soils, humic substances are present as dark brown polymeric material which can be extracted from soils and sediments with dilute alkaline solutions. It is customary to divide this alkaline extract into two parts: humic acids, which are soluble in alkalis but insoluble in acids, and fulvic acids, which are soluble in both bases and acids.

It has long been known that the insoluble organic matter in rocks, i.e., kerogen, contains less of the ^{13}C isotope than living organisms. For instance, kerogen in marine sedimentary facies has a $\delta^{13}C$ from -23 to $-26‰$, while biological carbon in corresponding biocenoses has a $\delta^{13}C$ from -18 to $-22‰$.

It was assumed that this phenomenon was related to the fact that, during microbiological decomposition of dead organisms, proteins and carbohydrates, the fractions relatively depleted in the ^{12}C isotope, are eliminated, i.e., they decompose to the simple components CH_4, NH_3, CO_2, H_2S, etc. At the same time, lipids and other isotopically light components more resistant to microbiological attack are preserved. These produce the observed depletion of kerogen in the ^{13}C isotope relative to the original biological material as a whole.

Based on the regularities of distribution of isotopes in biological systems, a novel idea concerning the chemical process of early (diagenetic) transformation of biological material to the fossil form was successfully substantiated.

From the preceding chapters, it is clear that not all the carbon of proteins and carbohydrates is depleted in the light isotope. Proteins and carbohydrates are depleted as a whole in the light isotope because they contain a substantial number of functional groups in which the carbon is bound to heteroatoms and, as a result, is enriched in the ^{13}C isotope according to the rule of thermodynamically ordered distribution of isotopes.

Earlier it was thought that removal of functional groups could not lead to enrichment of the residual organic matter in the light isotope, as this process is accompanied by a kinetic isotopic effect. As a result, the low-molecular-weight fraction is enriched in the light isotope, and conversely, the residual substance is enriched in the heavy isotope. This was supported by the results of experiments on decarboxylation of organic acids (Melander, 1964). However, the intramolecular isotopic inhomogeneity of biological compounds was not taken into account.

In reality, there are two competing tendencies: one related to the specifics of intramolecular distribution of isotopes, and the other caused by the kinetic isotopic effect. Generally, the first tendency is stronger, so the resulting effect is liberation of CO_2 enriched in the ^{13}C isotope and, correspondingly, enrichment of the residual organic matter in the ^{12}C isotope.

From this point of view it is not necessary to relate the enrichment of kerogen in the light isotope to the process of elimination of proteins and carbohydrates. It is accepted that the process of their polymerization, melanoidin formation, takes place, accompanied by removal of functional groups. The melanoidin reaction consists of the interaction of compounds containing aldehyde and amine groups (carbohydrates and proteins) and leads to the formation of cinnamon-brown polymeric substances similar in their properties to the humic acids of marine sediments. Manskaya and Drozdova (1964) were the first to point out the geochemical significance of the melanoidin reaction.

With Drozdova, we conducted an experimental study of the fractionation of isotopes in the process of melanoidin formation. Glucosamine was heated with a small amount of protein hydrolyzate on a boiling water bath for 12, 18, and 30 hr. As a result of heating, a brown solution with a dark precipitate was formed. The melanoidin character of this reaction was demonstrated earlier (Manskaya and Drozdova, 1964). Table 11.1 summarizes the conditions of the experiment and the data obtained. The results show the following: the polymeric substances formed are enriched in the light isotope relative to the carbon of the original biological compound and there is a definite regularity in the relationships of the isotopic compositions of the products of the reaction, representing successive stages of polymerization. The reaction products soluble in alkalis and acids, i.e., products analogous to natural fulvic acids, are slightly enriched in the ^{12}C isotope relative to the original substance. The content of the light isotope

TABLE 11.1

PRODUCTS OF THE MELANOIDIN REACTION[a]

Heating time (hr)	Reaction product: Brown solution of fulvic acids[b]		Black precipitate of humic acids[c]		Black precipitate of kerogen	
	δ ^{13}C (‰)	Yield (%)	δ ^{13}C (‰)	Yield (%)	δ ^{13}C (‰)	Yield (%)
12	−27.00	11.4	−24.47	0.6	—	Trace
18	−26.68	10.7	−29.19	2.0	0	0.33
30	−26.32	3.6	—	Trace	−36.42	6.1

[a] Glucosamine, 1 g; protein hydrolyzate, 1.018 g in experiment 1 and 0.12 in experiments 2 and 3, H_2O, 15 ml, pH 8.0. Isotopic composition of original substances: glucosamine, −26.14‰; protein hydrolyzate, −19.24‰.

[b] No precipitate, high-molecular-weight substances removed by dialysis.

[c] Soluble in 1N NaOH, precipitates from 3N HCl.

[d] Insoluble melanoidins.

becomes higher in the more polymerized reaction products, soluble only in alkalis (products corresponding to natural humic acids), and while the enrichment of the latter in the ^{12}C isotope increases as the experiment continues, the enrichment of the former decreases. The product of the melanoidin reaction that was insoluble in alkalis and acids (which can be considered the analog of natural kerogen) was enriched in the light isotope to a still greater extent.

Comparing these results with the distribution of isotopes in natural substances, we can state that they are completely similar. In fact, investigation of the isotopic composition of the carbon of humic substances shows that the general rule is enrichment of humic acids in the ^{12}C isotope relative to fulvic acids. In turn, the carbon of kerogen is enriched in the ^{12}C isotope relative to humic acids (Nissenbaum, 1974; Nissenbaum and Kaplan, 1972; Nissenbaum *et al.*, 1971).

Investigation of the physicochemical properties of fulvic acids, humic acids, and kerogen shows that they form a sequential genetic series. Fulvic acids have a lower molecular weight than humic acids. They contain more native functional groups. There are more heteroatoms in their chemical composition than in humic acids. In turn, humic acids are less altered than kerogen. In the precipitates, the contents of humic acids and kerogen are usually mutually complementary. With an increasing degree of diagenetic alteration, the amount of fulvic acids decreases due to an increase in amount of humic acids, which in turn decreases as the kerogen content increases (Brown *et al.*, 1972; Nissenbaum, 1974; Nissenbaum and Kaplan, 1972).

On passing from fulvic to humic acids and as the molecular weight of the humic acids increases, a decrease in the content of the functional groups occurs, as spectral investigations show (Raschid and King, 1970; Stevenson and Goh, 1971).

The agreement between the distribution of carbon isotopes in polymeric compounds obtained in the experiment and the distribution of isotopes in corresponding natural polymers makes it possible to consider melanoidin formation as an adequate model of the process of diagenetic transformation of organic matter. The observed depletion of fulvic acids in the ^{12}C isotope as humic acids are enriched relative to it suggests that fulvic acids actually are intermediate compounds between the original protein–carbohydrate components and humic acids.

Enrichment of polymeric compounds in the ^{12}C isotope is related to the splitting off of functional groups (carboxyl, formyl, keto, etc.) which are enriched in the ^{13}C isotope. Another possible cause of isotopic fractionation during diagenesis is polymerization reactions *per se*.

Investigation of isotopic effects in polymerization reactions showed that a polymer may be enriched by several per mil compared with the carbon of a monomer at normal temperatures, when the yield of polymer is 20 to 25% (Berman *et al.*, 1977, 1978).

The chemistry of diagenetic transformation of organic matter (more accurately, it is a consequence of the activity of microorganisms) is represented as follows (Galimov, 1980) (Fig. 11.1). Originally, biopolymers break down into simpler fragments and monomers. Successive repolymerization of these fragments into polymers of another type, more resistant in the geologic setting, follows. This process is accompanied by splitting off of active functional groups.

Both the loss of isotopically heavy carbon bound to the carboxyl and other functional groups and the isotopic effect of polymerization produce the observed enrichment of the geopolymers in the ^{12}C isotope.

Transformation of biomolecules begins in the water layer before deposition. Here the first polymeric neogenic formations arise and the original enrichment of carbon in the light isotope relative to the carbon of plankton occurs. The organic matter dissolved in water is similar to fulvic acids (Nissenbaum *et al.*, 1972). The polymerization process continues in the sediment, and as a result, the content of humic acids increases at the expense of fulvic acids and the content of kerogen increases at the expense of humic acids. Ultimately, all the organic matter is converted to the stable form of kerogen.

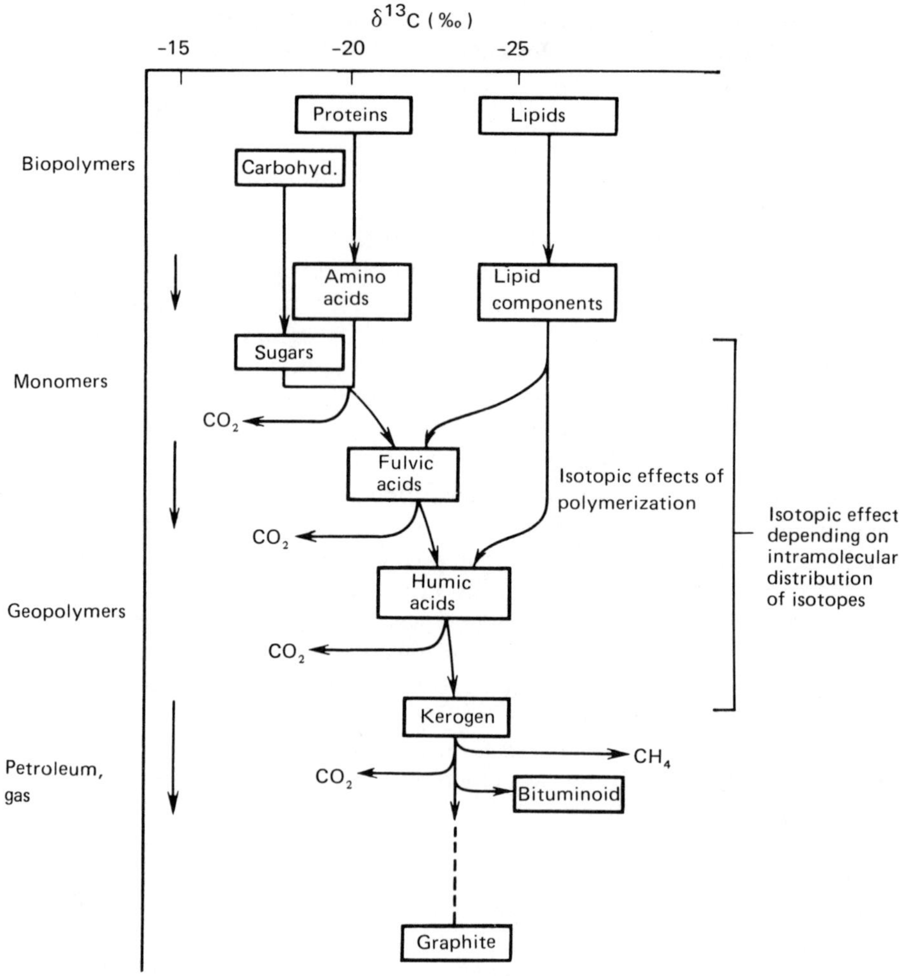

FIG. 11.1 The change of carbon isotopic composition in the course of transformation of biopolymers into fossil forms of organic matter.

Humic acids tend to form complexes with lyophobic compounds such as alkanes, fatty acids, pigments, and other lipid components (Philp and Calvin, 1976; Schitzer and Neyroud, 1975), which forms the basis of the future bituminoid fraction of organic matter. The formation of kerogen (and therefore the disappearance of humic acids) is the final stage of the diagenesis of organic matter and the beginning of the catagenetic phase of its transformation. In this stage kerogen is transformed under the influence of the following lithogenetic factors: increasing temperature and pressure and processes of recrystallization of the mineral part of the sediment. This process, on the one hand, proceeds along the pathway of further cyclization and condensation of the kerogen structure, ultimately leading to the formation of graphite; and, on the other, is accompanied by the splitting off of several mobile derivatives, including gases, liquid hydrocarbons, tetrapyrroles and other stable structures. The lipid-like fraction which is extracted from kerogen by organic solvents is called *bituminoid*. As a rule, the bituminoid content increases with the extent of catagenetic transformation of kerogen, and the content of low-polarity fractions, chiefly hydrocarbons, increases in the bituminoid fraction. Under certain conditions and at a certain stage of the transformation of organic matter this process reaches maximum development. This stage has been called the main phase of petroleum formation. The detection of the main phase of petroleum formation and identification of deposits going through this stage in the geologic column is an important practical problem.

From the regularities of thermodynamically ordered distribution of isotopes it follows that bituminoid should be enriched in the ^{12}C isotope relative to kerogen. The bituminoid may be separated into fractions of different polarity. The most polar fractions are asphaltenes, the least polar, hydrocarbons. To whatever extent the bituminoid inherits the structure of its biological predecessors, the polar fractions should be enriched in the ^{13}C isotope relative to the low-polarity fractions, as the organic matter in the polar fraction contains more heteroatoms and thus is characterized by higher values of $\beta^{13}C$.

In our laboratory a study was made of organic matter from unique samples of rocks obtained during deep-sea drilling in the Gulf of California by the drilling vessel Glomar Challenger (Galimov *et al.*, 1980). The Gulf of California is geologically interesting, having places with normal and very high heat flow next to one another. Therefore, the same organic matter can be studied in a wide range of temperature conditions.

Table 11.2 shows the variation in the studied characteristics of the samples organic matter with increasing exposure to heat. First, we see that the curves characterizing the relative bitumen content and concentration of hydrocarbons have a characteristic maximum (sample 477–7). The samples subjected to higher temperatures are already in a less productive stage of conversion of organic matter for petroleum formation. As is seen from Table 11.2, the tetrapyrrole compounds disappear at this stage, the content of isoprenoid hydrocarbons is sharply reduced, and the ratio of the isoprenoid phytane to the normal C_{18} alkane becomes much less than unity. As is

TABLE 11.2

VARIATION IN CHARACTER OF THE RELATIONSHIP $\delta^{13}C$–$\delta_{\Sigma}{}^{13}C$ AND OTHER CHARACTERISTICS OF ORGANIC MATTER WITH INCREASING DEGREE OF ITS THERMAL METAMORPHISM (SEDIMENTARY DEPOSITS OF THE GULF OF CALIFORNIA)

Sample no.		Relative content of bituminoid (1) and hydrocarbons (2)	$\frac{Phy}{n\text{-}C_{18}}$
476-6			1.1
479-43			6.5
474-40			0.9
477-7			3.0
477-16			0.37
481-13			0.27

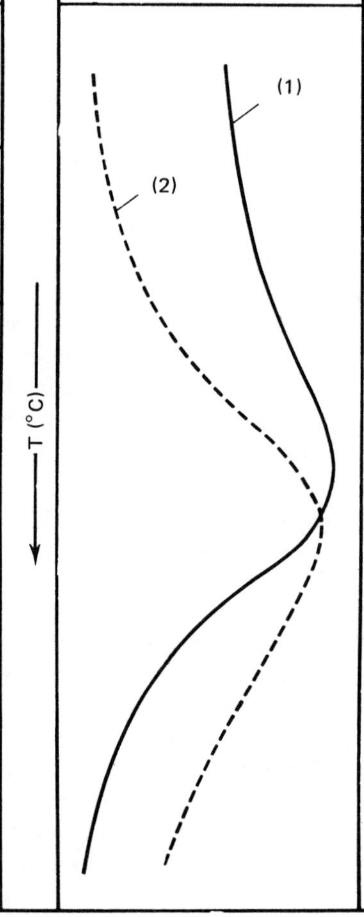

TABLE 11.2 *Continued*

VARIATION IN CHARACTER OF THE RELATIONSHIP $\delta^{13}C$–$\delta_{\Sigma}{}^{13}C$ AND OTHER CHARACTERISTICS OF ORGANIC MATTER WITH INCREASING DEGREE OF ITS THERMAL METAMORPHISM (SEDIMENTARY DEPOSITS OF THE GULF OF CALIFORNIA)

$\frac{\Sigma_{odd} C_{25}-C_{29}}{\Sigma_{even} C_{26}-C_{30}}$	Tetrapyrrole compounds	Bituminoid fractions 1 2 3 4 5
2.03	Chlorins	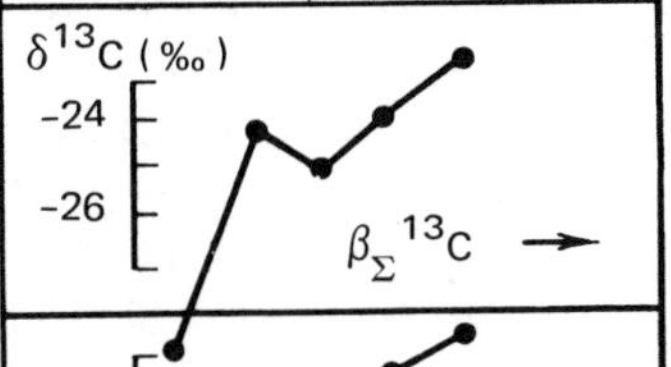
2.01	Ni-porphyrins	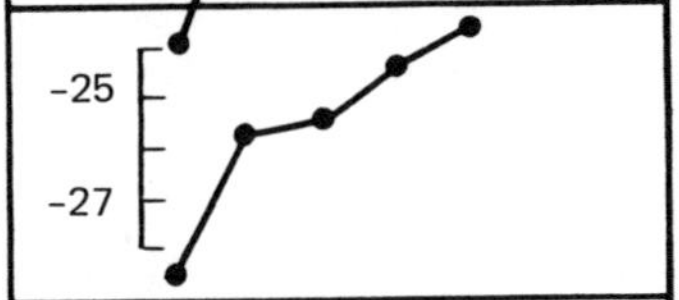
2.17	Ni-porphyrins	
1.52	Ni-porphyrins VO-porphyrins	
1.15	Absent	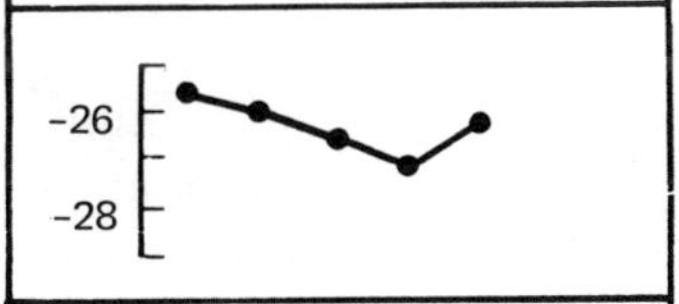
0.97	Absent	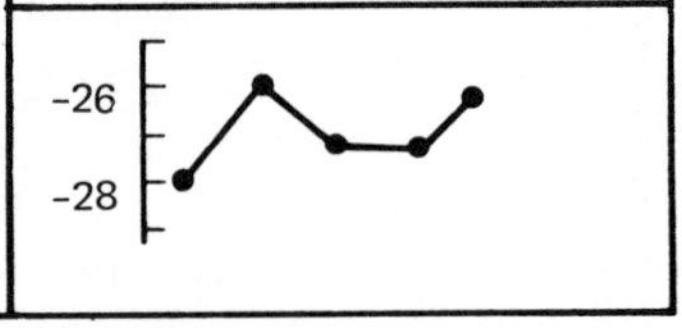

[a] Bituminoid fractions: hydrocarbons (1), hexane–benzene resins (2), benzene resins (3), alcohol–benzene resins (4), asphaltenes (5).

known, in biological material, fatty acids with an even number of carbon atoms predominate. In fossil organic matter, *n*-alkanes with an odd number of carbons are most common. In thermally transformed samples the content of even- and odd-number alkanes is equalized. Thus in the stage of transformation of organic matter following passage through the main stage of petroleum formation, organic matter loses most of the signs of its biological origin.

The isotopic composition of the bituminoid fraction in samples from areas of relatively slight thermal activity shows a distinct dependence on the degree of polarity—essentially a relationship of the $\delta^{13}C$–$\beta_{\Sigma}{}^{13}C$ type. In samples 477–16 and 481–13 there is no such relationship. In other words, along with other signs of biological origin, organic matter subjected to thermal metamorphism loses the thermodynamic ordering of isotopic distribution inherent in biological objects, although that ordering may be retained at the intramolecular level.

Thus the results of isotopic investigations, interpreted on the basis of established regularities of the biological behavior of isotopes, makes it possible to trace the postbiological history of organic matter.

II. THE ORIGIN OF PETROLEUM

The problem of the origin of petroleum traditionally includes the question whether petroleum is the result of transformation of organic remains or whether the petroleum hydrocarbons were synthesized under endogenetic conditions (the inorganic hypothesis of the origin of petroleum). If petroleum is of organic origin, then the question consists of whether the different classes of hydrocarbons (paraffin, naphthene, or aromatic) were formed from a common original material or had different biochemical components as their predecessors (inherited character of hydrocarbon structures). To understand the conditions of petroleum formation it is important to know the relationship of inherited and neogenic structures in different petroleums.

For this, the following approach, based on the rule formulated above, is used: correspondence between the isotopic composition of the petroleum hydrocarbons and the $\bar{\beta}$-factor values characterizing those hydrocarbons is evidence of a biological origin and of the inheritance of the given hydrocarbon structures from corresponding structural forms of biomolecules. If there is no such correspondence, then the given hydrocarbon structures should be considered of neogenic origin, possibly including abiogenic.

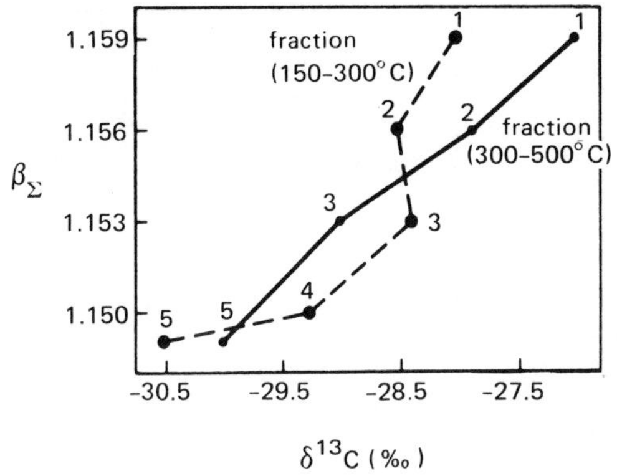

FIG. 11.2 $\delta^{13}C$–$\beta^{13}C$ relationship in hydrocarbons of natural origin (Krasnokamsk field of Perm district). Hydrocarbons: 1, polyaromatic; 2, biaromatic; 3, monoaromatic; 4, naphthene; 5, *n*-paraffin.

This approach was the basis of experiments carried out in our laboratory (Galimov, 1975b). A series of hydrocarbon fractions of different molecular weight (i.e., boiling in different temperature ranges) was isolated from a sample of natural petroleum from an oil field in the Volga–Ural district (the Krasnokamsk). From these fractions normal-paraffin, isoparaffin, naphthene, and aromatic hydrocarbons of different degree of cyclicity were extracted. The isotopic composition of their carbon was measured in a mass spectrometer; then the measured $\delta^{13}C$ values were compared with the calculated values of the β factors (Fig. 11.2).

For comparison the isotopic composition of the carbon of hydrocarbon fractions obtained synthetically by thermocatalysis at 200°C (Galimov, 1975b) was studied. The source of the hydrocarbons in this case was a single original compound—oleic acid (Fig. 11.3).

On the whole, a correlation between the $\delta^{13}C$ and $\beta^{13}C$ values of the same type as in biomolecules was observed for hydrocarbons of natural petroleum. In the synthetic hydrocarbons there was no such relationship. This indicates the biological origin of the hydrocarbons of the given petroleum. Later a dependence of the $\delta^{13}C$–$\beta^{13}C$ type was established for many petroleums (Shirinskiy, 1978).

The relationship between $\delta^{13}C$ and $\beta^{13}C$ values will become weaker as the substance is transformed from the biological forms to the corresponding hydrocarbon fraction. From Fig. 11.2 it is seen that the $\delta^{13}C$–$\beta^{13}C$ relationship is more clearly expressed for the higher-molecular fractions. This shows that the process of petroleum formation is directed toward successive

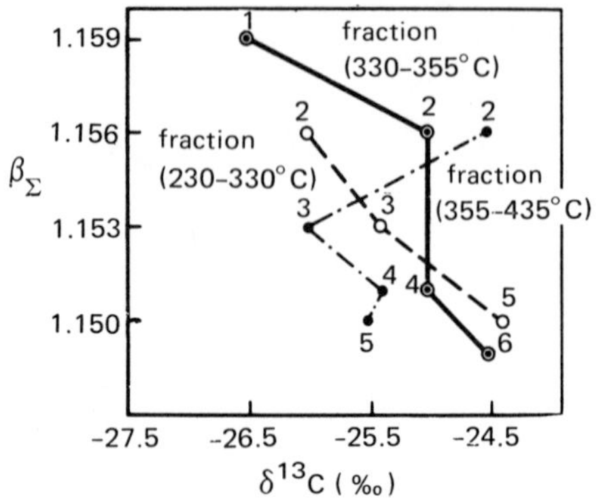

FIG. 11.3 Graph showing the lack of relationship of $\delta^{13}C$–$\beta^{13}C$ type in hydrocarbons synthesized by thermocatalysis of oleic acid. Hydrocarbons: 1, polyaromatic; 2, biaromatic; 3, monoaromatic; 4, naphthene (bicyclic); 5, naphthene (monocyclic); 6, *n*-paraffin.

reduction in size of the molecules. In other words, the low-molecular benzine fractions of petroleum are farther from the original biomolecules in genetic respects than the high-molecular-weight fractions.

This last is confirmed by the results of study of individual normal alkanes in petroleums. As a rule the low-molecular-weight paraffin hydrocarbons of the C_7 to C_{10} benzine fraction are depleted in the ^{12}C isotope relative to the higher-molecular-weight alkanes. They are the product of fragmentation of the original high-molecular-weight alkanes and in turn are the source of gaseous C_1 to C_4 hydrocarbons. This latter process, accompanied by a kinetic isotopic effect, results in the emergence of methane and its homologs enriched in the light isotope and a corresponding depletion of the hydrocarbons of the benzine fraction in the light isotope. For the same reason, the highly transformed so-called petroleum condensates are relatively depleted in the ^{12}C isotope.

Compounds for which definite biological predecessors can be indicated occur in petroleum. Thus it is suggested that normal alkanes are derived from fatty acids, isoprenoid hydrocarbons are the product of transformation of phytol, and porphyrins are formed from the porphyrin nucleus of chlorophyll. However, adherents of the inorganic origin of petroleum claim that the main classes of petroleum hydrocarbons, including normal alkanes, form as a result of abiogenic synthesis in the mantle or lower levels of the earth's crust, while porphyrins and isoprenoids (which are easily destroyed at high temperatures) actually are biogenic, but are introduced into the petroleum from the surrounding rocks, extracted by the flux of hydrocarbons arriving from depth. In other words, they are not genetically related to

TABLE 11.3

HYDROCARBON FRACTIONS OF PETROLEUM[a]

Hydrocarbon fraction	$\delta^{13}C$ (‰)
282–350°C fraction as a whole	−29.48
Branched-chain alkanes and naphthenes	−30.41
Aromatics	−28.21
n-Alkanes	−30.99
Napthenes	−29.90
ip-Alkanes (isoprenoids)	−30.30
i-Alkanes (nonisoprenoid isoalkanes)	−29.84

[a] Kokuy field, fraction boiling in the 282–350°C range; $\delta^{13}C$ of Petroleum as a whole is −29.23‰.

the main mass of petroleum, in particular to normal alkanes. Thus a solution of the problem of the presence or absence of a genetic relationship between the aforementioned compounds in petroleum takes on theoretical importance.

In our work (Galimov, 1975b, p. 371) we mentioned that the problem might be solved if a comparative analysis were made of the isotopic composition of isoprenoids and normal alkanes in petroleum. The fact is that the β factors of the phytol chain of chlorophyll and long-chain fatty acids are similar and have relatively low values. Therefore the isoprenoids should be similar in isotopic composition to normal alkanes and, like the latter, relatively enriched in the light isotope, if they are derived from a common biological source.

Experimental investigation (Bogacheva *et al.*, 1980) has shown that the isotopic composition of isoprenoids (ipC_{16} to ipC_{20}) actually is very close to that of normal alkanes (nC_{16} to nC_{21}) from the same petroleum (Table 11.3). On the other hand, the β factor of the porphyrin nucleus of chlorophyll has a higher value than the β factor of phytol. In Chapter 5 we saw that, in agreement with this and in correspondence with the principle of thermodynamically ordered distribution of isotopes, the carbon of the porphyrin nucleus of chlorophyll is depleted in the ^{12}C isotope. Consequently, the porphyrins in petroleum should be relatively depleted in the ^{12}C isotope. Experimental investigation showed that this is true (Table 11.4). In

TABLE 11.4

CARBON OF PETROLEUM PORPHYRINS IN SOME PETROLEUMS OF THE PERKAM'YE AREA

	$\delta^{13}C$(‰)	
Field	*Petroleum*	*VO-porphyrins*
Kuyedinskoye	−28.9	−27.1
	−28.2	−27.4
	−27.3	−26.5
	−27.7	−25.7
	−28.0	−26.0
Durinskoye	−29.4	−26.9
Al'nyazhskoye	−28.4	−28.0
	−28.0	−25.2
Kiyengopskoye	−27.7	−26.0
Andreyevskoye	−28.1	−26.8

addition, a direct correlation is observed between the isotopic composition of various petroleums and that of the porphyrins they contain.

Thus a genetic relationship of biological markers (the porphyrins and isoprenoids) to the hydrocarbons constituting the main mass of petroleum, and thus the organic origin of the petroleum itself, is demonstrated.

III. DETERMINATION OF THE GENESIS AND SOURCE OF THE HYDROCARBONS FORMING A GIVEN GAS FIELD

Methane is formed at all stages of transformation of organic matter. At the earliest stage it usually is the product of microbiological reduction of carbon dioxide, $CO_2 + 4H_2 = CH_4 + 2H_2O$, and enzymatic splitting of acetic acid, $CH_3COOH = CH_4 + CO_2$. Methane of microbiological origin is highly enriched in the light isotope ($\delta^{13}C$ from -50 to -95‰). It is isotopically one of the lightest compounds in nature. The enrichment of methane in the light isotope is usually explained by the kinetic isotopic effect. However, keeping in mind the extensive occurrence of thermodynamic isotopic effects in biochemical processes, it is impossible to ignore the fact that methane is characterized by a very low β-factor value. The $\beta^{13}C$ value of methane at 25°C is 1.114, while that of $\beta^{13}C_{CO^2} = 1.190$, i.e., the thermodynamic isotopic effect amounts to 65‰. This corresponds approximately to the observed difference in isotopic composition of CH_4 and CO_2 in natural conditions.

In later stages, when organic matter is buried in sediments at increasingly greater depths, the microbiological processes die out and methane is formed due to chemical fragmentation of organic compounds which constitute fossil organic matter—kerogen.

The rate of transformation of organic matter into petroleum depends on various factors. But, on the whole, the more deeply the organic matter is buried in sedimentary rocks, the more it is transformed. It has long been established that the carbon of methane in sedimentary deposits is isotopically much lighter than the carbon of organic matter. This difference decreases with depth, i.e., the deeper the pool, as a rule, the isotopically heavier the methane it contains (Galimov, 1968b).

This relationship was explained by the kinetic isotopic effect which decreases with increasing temperature, but the temperature in the bowels of the earth increases with depth. Therefore, with increasing depth the difference in isotopic composition between methane and the original organic matter, caused by the kinetic isotopic effect, decreases. That explanation was given by Galimov (1968a) and simultaneously by American investigator Sackett (1968). Later it was ascertained that, in those cases where methane is present in association with very highly transformed organic matter (anthracite coal) or at great depth, its isotopic composition may be the same as that of the carbon of associated organic matter or even heavier (Stahl, 1968; Teichmüller *et al.*, 1970). This cannot be explained by the kinetic isotopic effect, as that always leads to enrichment of the product in the light isotope.

An attempt to explain this by isotopic exchange effects, in particular between CO_2 and CH_4 (Prasolov and Lobkov, 1977), raises serious objections, as the rates of such exchange are so low, even at high temperatures ($\sim 500°C$) (Sackett and Chang, 1979), that at the relatively low temperatures of the sedimentary cover these effects are not visibly manifested even over long periods of geologic time.

The discovery of intramolecular heterogeneity of the distribution of isotopes in biosystems created a new basis for explaining these facts. If one keeps in mind the difference in β_i factors of various structural groups ($\beta_{CH_3} \simeq 1.130$, $\beta_{CH_2} \simeq 1.150$, $\beta_{CH} \simeq 1.168$), then the transfer of the gas-generating role from the CH_3 group in the early stages of the transformation of organic matter to the isotopically heavier structural groups in later stages theoretically explains the observed dependence of the isotopic composition of methane carbon by the depth of its occurrence.

This leads to the geochemically important conclusion that the isotopic composition of methane reflects the extent of transformation of organic matter. The ability of organic matter to produce petroleum hydrocarbons corresponds to a certain degree of its transformation. Therefore, the isotopic composition of methane becomes important as a diagnostic criterion of the process of petroleum formation.

An understanding of the factors determining the isotopic composition of methane has made it possible to formulate a way of establishing the hydrocarbon source of a given gas deposit during its formation. For petroleum geologists, who are concerned with investigating the conditions of formation of deposits, it is always very important to determine whether the hydrocarbons were formed in the immediate vicinity of the area of accumulation of the deposit or have migrated from afar. This is a complicated question, usually requiring an in-depth analysis of the whole geologic setting and is often impossible to answer, although the further geological exploration in a given territory, estimation of prospective reserves, etc., depend on it.

What has been said above concerning the relationship of the isotopic composition of methane to the extent of transformation of organic matter makes it possible to work out criteria for such an estimation. We shall elucidate the criteria with an example of a large gas province in Central Asia where we applied it (Galimov *et al.*, 1973).

In Fig. 11.4 a schematic section of the sedimentary rocks of this territory is represented; the isotopic composition of methane from the gas pools in

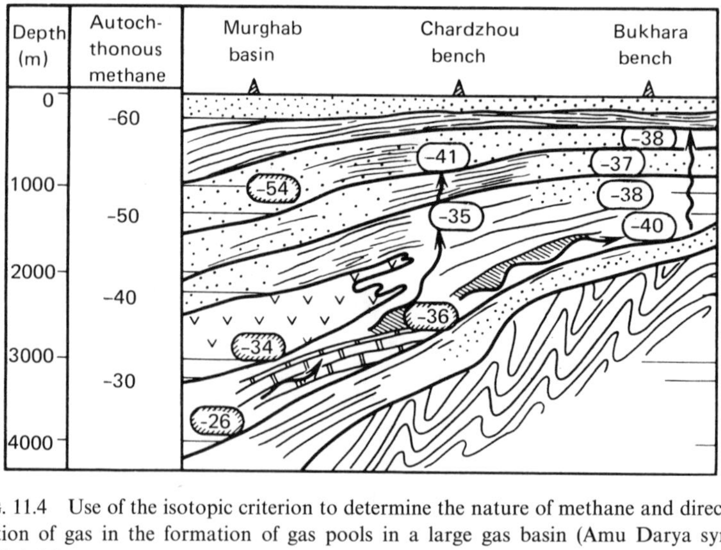

FIG. 11.4 Use of the isotopic criterion to determine the nature of methane and direction of migration of gas in the formation of gas pools in a large gas basin (Amu Darya syneclise, Central Asia).

deposits of different geologic age, measured in a mass spectrometer, is indicated in the ovals. At the left, along the depth scale, the $\delta^{13}C$ values of methane are indicated. The isotopic composition of the methane corresponds theoretically to the extent of transformation of organic matter possible at the corresponding depths.

If the actual isotopic composition of methane of the gas pools falls in the range of $\delta^{13}C$ values corresponding to the extent of transformation of organic matter at this depth, it can be assumed that the methane was formed spatially near the place of its accumulation (autochthonous methane—noted by hatched ovals). If the isotopic composition of the methane does not correspond to the extent of transformation of organic matter in the corresponding deposits, then it must be concluded that it migrated from outside (allochthonous methane—unhatched ovals).

In this case it is obvious that the gas pools of the Bukhara and Chardzhou benches, including one of the richest fields in the Soviet Union—the Gazli on the Bukhara bench—were formed due to an influx of gas from deeper deposits of the Amu Darya syneclise, namely, the Murghab basin. Corresponding recommendations for the direction of geological exploratory work in this territory followed this discovery.

IV. RELATIONSHIP OF THE ISOTOPIC COMPOSITION OF BIOGENIC CARBON TO THE AMBIENT TEMPERATURE

Because biological isotopic effects are of thermodynamic nature and thermodynamic isotopic effects depend on temperature, the isotopic composition of biogenic carbon should be temperature dependent.

In fact, the dependence of the isotopic composition of organisms on the temperature of the environment in which they live has been reported repeatedly.

Apparently Sackett *et al.* (1965) were the first to draw attention to this; they found that plankton collected in the cold, high-latitude waters of the South Atlantic are isotopically lighter ($\delta^{13}C = -27.9\,‰$) than plankton living in warm tropical waters ($\delta^{13}C = -21.7\,‰$).

In our laboratory the isotopic composition of the lipid fraction of plankton from the Indian Ocean was investigated (Galimov *et al.*, 1973a). The $\delta^{13}C$ values of the lipids of plankton were compared with the temperature of the ocean water recorded in the sampling area. The relationship obtained is depicted in Fig. 11.5. A relationship of the isotopic composition of the

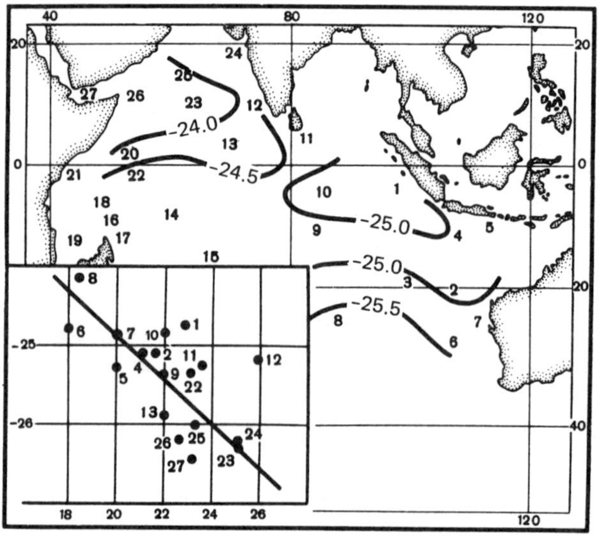

FIG. 11.5 Relationship of the carbon isotopic composition of planktonic lipids to ocean water temperature (Indian Ocean). Numbers on the graph correspond to numbers of sampling sites.

carbon of the biomass to temperature of the environment was observed for organisms grown under laboratory conditions (Degens *et al.*, 1968).

The enrichment of the organisms in the light isotope increases with decreasing temperature, consequently the difference between the isotopic composition of organic carbon and the original carbon dioxide increases. This corresponds to the temperature dependence of thermodynamic isotopic effects (see Chapter 3).

The theoretical dependence of thermodynamic isotopic effects of organic compounds on temperature is 0.2–0.4‰ per 1°C. The natural samples show approximately the same temperature dependence.

It is obvious that, in those cases where for some reason the thermodynamically ordered distribution of isotopes is disturbed, the temperature dependence should disappear. In particular, this would be expected when the organisms develop under extreme conditions, non-steady-state regimes of biosynthesis, etc.

There is another possibility of temperature regulation of isotopic composition, stemming from the formula for enzymatic fractionation. The coefficient κ theoretically can vary with temperature regardless of the β factors, owing to the possible variation with temperature of the values of the constants k_a, k_b, and k_2 and their ratios. This special temperature effect evidently can be associated differently with the temperature dependence of the β factors.

Probably for this reason, investigators in many cases were unable to find a temperature dependence of biological carbon (Smith *et al.*, 1973; Troughton and Card, 1975). In studying isotopic effects in enzymatic reactions *in vitro* a dependence on temperature appeared in some cases (Whelan *et al.*, 1973), in others the effect of temperature was minor (Christeller and Lang, 1976), and sometimes the isotopic effect of temperature had a sign opposite to that expected (DeNiro and Epstein, 1977).

It should be mentioned that there generally are a multitude of factors which mask and complicate the temperature dependence. A whole series of physical and physiological parameters which can affect the isotopic composition or other characteristics of an organism depend on temperature, for instance, the rate of consumption of CO_2 in photosynthesis, the concentration of carbon dioxide in the external environment (water), and the rate of cell growth depend on temperature. In turn, they control the isotopic composition of organisms. In the higher plants this is complicated by the fact that the temperature of the tissue may be different from the air temperature and may vary in different parts of a plant.

Obviously, therefore, the temperature dependence of the carbon isotopic composition for individual organisms is rarely known. Its statistical manifestation for whole biocenoses is more probable; there the effect of many individual factors cancel each other and the dependence on temperature can show up more clearly. This happened, for instance, in the case of the previously mentioned investigations of plankton in ocean waters.

Incidentally, it is this statistical aspect that is important for geochemical applications. The observed cases of correlation of the isotopic composition of organic carbon in sediments with the presumed temperature conditions of the respective geologic epochs find theoretical substantiation. The paleotemperature dependence of the isotopic composition of fossil carbon in sediment was first pointed out by Rogers and Koons (1969). Later Rogers *et al.* (1972), investigating deep-sea sediments of Tertiary and Quaternary age in the North Atlantic, drew attention to the relationship of variations in the $\delta^{13}C$ of organic carbon (kerogen) to the glacial history. Carbon lighter by 1 to 2‰ corresponded to epochs of glaciation. The theory of Rogers and his coauthors concerning the existence of a dependence of the $\delta^{13}C$ of kerogen on paleotemperature was criticized as selective or applicable to very special conditions (Sackett and Rankin, 1970). This criticism was unjustified, in light of what has been said above. The dependence of the $\delta^{13}C$ of kerogen on paleotemperature is a natural reflection of the temperature dependence of biological carbon, for which this property is fundamental in view of the chiefly thermodynamic nature of the isotopic effects. The temperature dependance is inherent despite the fact that the effect of the temperature factor can be masked by the action of other factors.

We have spoken only of the temperature dependence of carbon. It is known that the $\delta^{18}O$ and δD values of plants also correlate with temperature in many cases. However, in most cases this is not really a biological phenomenon but is caused by the dependence of the isotopic composition of hydrogen and oxygen on the rate of evaporation of intracellular water and other temperature-dependent factors that are not related to the mechanism of biosynthesis, proper.

V. VARIATION IN THE ISOTOPIC COMPOSITION OF ORGANIC CARBON OVER GEOLOGIC TIME

Starting from what is presently known about the factors affecting the isotopic composition of biological carbon, the following factors which might produce a change in the isotopic composition of carbon in the course of geologic time can be suggested:

(1) variation in the isotopic composition of carbon of the total carbon reservoir on the earth's surface due to evolution of the process of degassing of the earth's mantle and possible variation in the isotopic composition of the resulting products and due to periodic variations in the scale and rate of recycling of crustal carbon, related to global epochs of magmatism;

(2) variation in the ratio of the organic form of carbon to other forms (especially, between organic and carbonate carbon) as a result of a change in the physicochemical environment on the earth and in the volume of the biosphere;

(3) variation in the character and amount of fractionation of isotopes between organic carbon and the source carbon. (As noted in the preceding chapters the latter can be caused by a change in the mechanism of biosynthesis and consequently in the predominant type of isotopic effects, a change in biochemical composition of organisms, a change in the abundance of carbon dioxide, a change in the chemical form of the original carbon, and variations in temperature of the earth's surface and other factors);

(4) variation in the relationship between the isotopic composition of fossil organic carbon and that of its biological predecessors due to a change in conditions and chemistry of the process of fossilization of organic matter.

The data available at the present time on the isotopic composition of organic carbon in ancient geologic deposits as a rule pertains to the carbon of organic matter as a whole, or at best to the carbon of kerogen and bituminoid. Therefore, the possibilities of using the isotopic criteria formulated above for their analysis are limited. Nevertheless, certain regularities reflecting the evolution of the biosphere during geologic time can be pointed out.

Phanerozoic. Most carbon isotopic data pertain to the Phanerozoic, i.e., the geologic time encompassing the last 600 million years of earth history. In a number of works a tendency toward gradual decrease in enrichment of organic carbon in the ^{12}C isotope from old to young deposits during the Phanerozoic has been pointed out (Hoering, 1967; Welte *et al.*, 1975). In reality this relationship apparently is not monotonic in character. It is better to consider a specific kind of dependence using an example of one large geologic region, as I did with Migdisov and Ronov for the Russian platform. In this work (Galimov *et al.*, 1975a) a representative sample, obtained by mixing several hundred samples from different areas, was prepared for each geologic age of the sedimentary section of the Russian platform. The isotopic compositions of bituminoid and kerogen and of carbonate carbon were determined separately. Figure 11.6 shows the variation in isotopic composition of the investigated forms of organic carbon, and Fig. 11.7, the isotopic composition of carbonates and some parameters characterizing the sedimentary environment during the Phanerozoic on the Russian platform. Comparison of the curves shows no clear-cut dependence of the isotopic composition of organic carbon on the character of the predominant lithofacies setting, rates of sedimentation, concentration of organic carbon, etc.

Two features appearing in the distribution of the isotopic composition of organic carbon during the Phanerozoic on the Russian platform are interesting: sharp depletion of kerogen and bituminoid in the ^{12}C isotope in Carboniferous time and the appearance in the Carboniferous, and continuance through later epochs, of a previously absent difference in isotopic composition between kerogen and bituminoid fractions.

The isotopic composition of carbonate carbon, in contrast to organic carbon, is closely related to other geochemical parameters, as seen from Fig. 11.7. Therefore, carbonate carbon can be a more sensitive indicator of a change in isotopic composition of the surface reservoir (note curves 1 and 2 in Fig. 11.7). The lack of clear-cut correlation between the isotopic composition of carbonate and organic carbon indicates that it was not these factors that produced the observed shifts in isotopic composition of the forms of organic carbon.

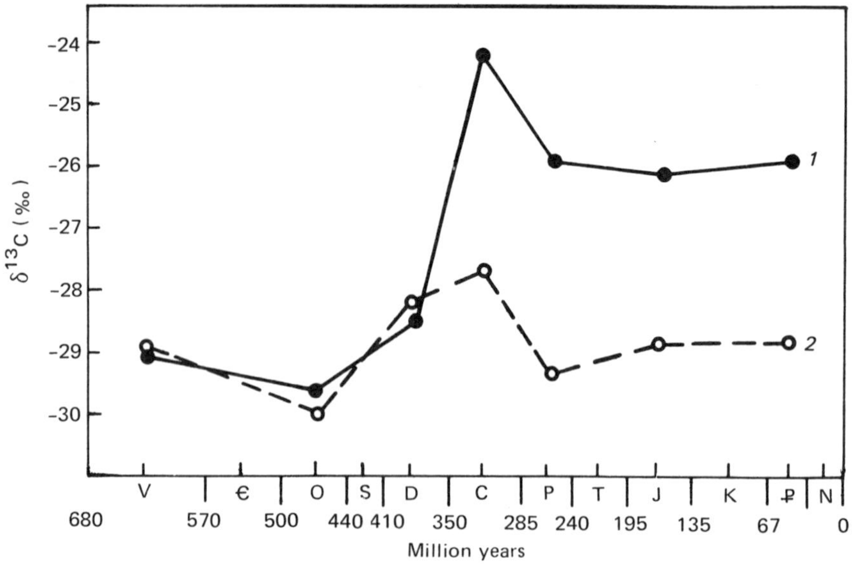

FIG. 11.6 Variation in isotopic composition of organic carbon in the sedimentary deposits of the Russian platform in the Phanerozoic: 1, kerogen; 2, bitumen. (From Galimov *et al.*, 1975a).

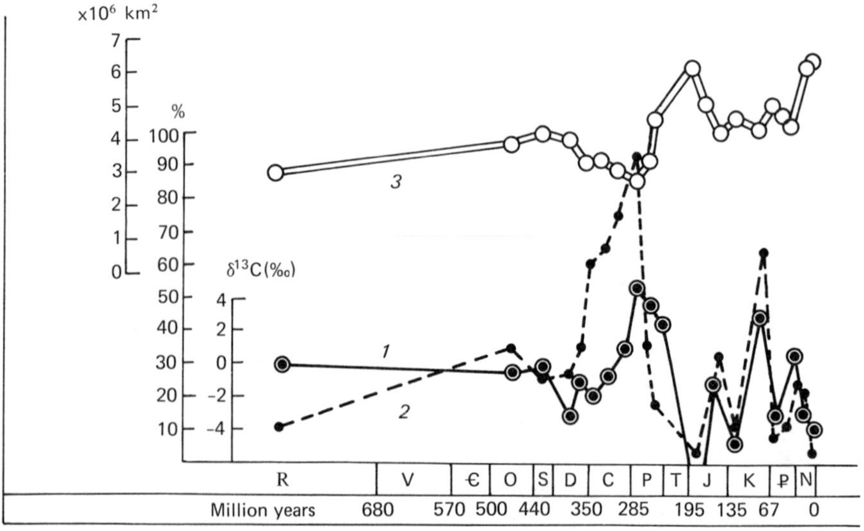

FIG. 11.7 Carbon isotopic composition of carbonates (1), relative abundance of carbonates in percentage of total volume of sediments (2) and area of distribution of continental conditions (3) in different geologic periods during the Phanerozoic on the Russian platform. (From Galimov *et al.*, 1975a).

The characteristic shift in the Carboniferous evidently is related in some way to the appearance at that time of terrestrial vegetation and development of an oxygen atmosphere. However, these facts are not linearly related. Actually, depletion of organic carbon in the ^{12}C isotope in post-Carboniferous time cannot be explained by the contribution of terrestrial vegetation, as its carbon, on the contrary, is enriched in the light isotope relative to the carbon of marine organisms.

The regularities of biological fractionation of isotopes established in this work make it possible to explain the probable cause of the aforementioned shift as a change in the process of transformation of organic matter to the fossil form. It was stated above that the interaction between the amine groups of proteins and the aldehyde groups of carbohydrates, i.e., the melanoidin process, permits stabilization of proteins and carbohydrates, which are easily destroyed in the geologic setting by transforming them into high-molecular-weight polymers, at first in the form of fulvic acids, then humic acids, and finally kerogen. Possibly the melanoidin process was inhibited, either because in pre-Carboniferous time organisms had a different biochemical composition from the present one or because an oxygen atmosphere was absent and the diagenetic setting in the sediment was different. Therefore, predominately chemically and microbiologically resistant lipid components survived in diagenesis. In pre-Carboniferous time they would have constituted the main source of kerogen. As lipids are enriched in the light isotope, this explains the relative enrichment of pre-Carboniferous kerogen in the light isotope. The absence in pre-Carboniferous time of a noticeable difference between the carbon of kerogen and bitumoid is also explained, since the same lipid fraction was the source of the latter.

Precambrian. Study of the isotopic composition of organic carbon in Precambrian deposits (Galimov *et al.*, 1975a; Elchamn and Schidlowski, 1975; Hoefs and Schidlowski, 1967; Hoering, 1967; Ohler *et al.*, 1971) has shown that it was enriched in the light isotope to the usual biological level throughout almost all geologic time accessible to investigation. This, in addition to the great abundance of reduced carbon in the Precambrian (Sidorenko and Sidorenko, 1975), indicates early development of the biosphere.

Carbonate carbon in the Precambrian is characterized by a high amplitude of variation in isotopic composition, with surges of enrichment of carbon in the heavy isotope corresponding to epochs of powerful magmatism (Fig. 11.8). In these epochs, evidently, a substantial part of the previously deposited isotopically heavy carbonate carbon buried in the sedimentary shell was converted to a surface fund of the atmosphere–hydrosphere system (Galimov *et al.*, 1975a, 1976).

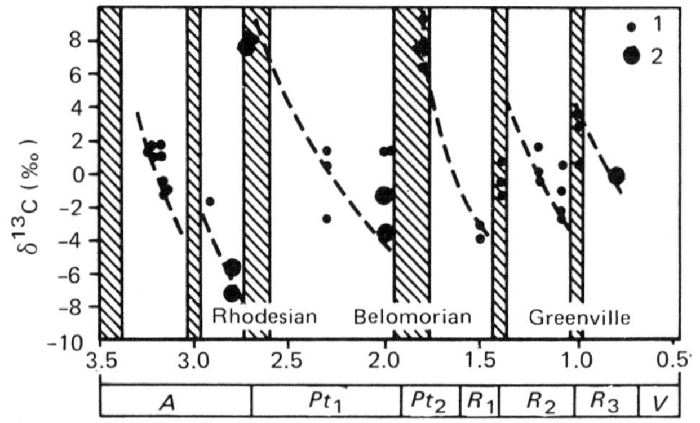

FIG. 11.8 Carbon isotopic composition of Precambrian carbonates and their relationship to epochs of magmatism: 1, single determinations; 2, mixed sample characterizing a multitude of rock specimens of corresponding geologic age (From Galimov *et al.*, 1975a).

For the Phanerozoic, organic carbon shows no close relationship to the isotopic composition of carbonate carbon in the corresponding geologic periods. Figure 11.9 depicts data on the carbon isotopic composition of kerogen from Precambrian deposits obtained by different authors. The abundance of isotopes in organic carbon in the Precambrian is characterized by two characteristics.

The first consists of the relative enrichment of kerogen in the ^{12}C isotope in the Early Precambrian ($\delta^{13}C$ from -28 to $-38\,‰$, whereas for the Phanerozoic, values from -23 to $-30\,‰$ are characteristic). This condition was first noted by Hoering (1967).

The second characteristic is the anomalous depletion of organic carbon in the ^{12}C isotope in the oldest sedimentary deposits; Schopf *et al.* (1971) found that kerogen from the Theespruit formation (deposits of the Onverwacht group of the Swaziland series, South Africa), which is 3.5 billion years old, is characterized by $\delta^{13}C$ values averaging about $-16\,‰$. This isotopic shift, in the opinion of these authors, possibly marks the transition from abiogenic to biogenic carbon, or the time when a mechanism of biological fractionation of isotopes similar to the present one appeared.

Several reasons have been offered for the enrichment of Early Proterozoic organic matter in the ^{12}C isotope. Epstein (1969) suggested that it is related to the low abundance of biogenic carbon at that time and the corresponding shift of the isotopic balance. Degens (1969) proposed that isotopically light organic carbon in the Precambrian possibly reflects conditions of a lower pH and higher concentration of molecular CO_2 in the Precambrian sea. Pardue *et al.* (1976), who experimented with a culture of

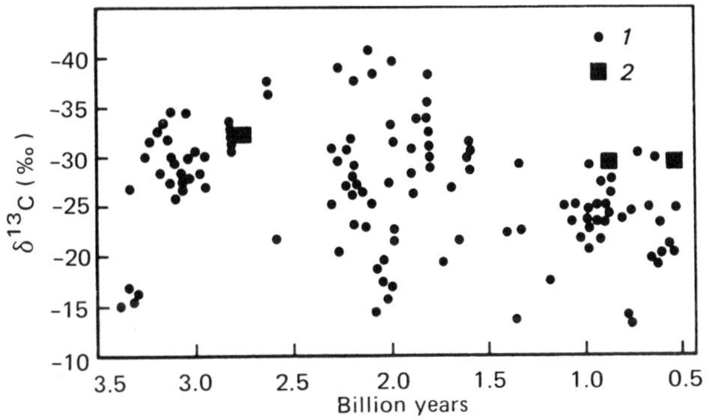

FIG. 11.9. Data on the isotopic composition of organic carbon in the Precambrian: 1, single determinations; 2, mixed sample characterizing a multitude of rock specimens of corresponding geologic age (From Galimov, 1976).

blue–green algae, showed that, under certain conditions, in particular in the case of a reduction in concentration of cells in the medium, isotopic fractionation increases. They suggested that this might explain the isotopically light carbon in the Precambrian.

Schidlowski and coauthors (Eichmann and Schidlowski, 1975; Junge *et al.*, 1975) believe that variations in $\delta^{13}C$ in Precambrian organic carbon do not generally extend beyond the limts of the range characteristic of the Phanerozoic and are explained by variations in the proportion of carbonate and organic carbon during geologic time. We note that in this case synchronous variations of $\delta^{13}C$ of carbonate and organic carbon would be expected.

In summary, the general tendency consists of a change in the isotopic composition of organic matter at three boundaries: the geologic times of 3.5, 2–1.8, and 0.4 billion years ago. The last corresponds to the appearance of terrestrial plants and an increase in concentration of atmospheric O_2 to the present level. In the 1.8–0.4 billion year interval the isotopic composition of organic carbon remained at the level typical of pre-Carboniferous kerogen. The presumed reasons for its relative enrichment in the light isotope relative to kerogen of younger age have been considered above. Although the still greater enrichment of organic carbon in the ^{12}C isotope in rocks older than 2 billion years is possibly explained in part by the factors indicated by the authors cited above, it may be related to a feature of the biochemical composition of ancient organisms. The 3.5 billion year boundary has not yet been sufficiently established. There are very few data for this period of the earth's history. We note only that magmatic rocks which might have been melts from ancient sedimentary deposits that have not been preserved almost always contain isotopically light disseminated carbon. This may be explanable by the characteristics of mantle carbon, but it is not inconsistent with the idea that biological fractionation of isotopes—and therefore organisms—appeared on the earth much earlier than 3.5 billion years ago.

REFERENCES

Berman, Ye. L., Sakharov, A. M., and Galimov, E. M. *et al.* (1977). Study of the structure of the active center in polymerization of 1,3-dioxalane. *Doklady AN SSSR* **234**(4).

Berman, Ye. L., Sakharov, A. M., and Galimov, E. M. *et al.* (1978). Kinetic isotopic effect ^{13}C in ionic polymerization of ethylene oxide. *Doklady AN SSSR* **239**(5).

Bogacheva, M. P., Kodina, L. A., and Galimov, E. M. (1980). Intramolecular carbon isotope distribution in chlorophyll and in geochemical derivatives. *In* Advances in Organic Geochemistry 1979. "Proceed. 9th Int. Meet. Org. Geoch." Pergamon, Newcastle upon Tyne.

Brown, F. S., Baedecker, M. J., Nissenbaum, A., and Kaplan, I. R. (1972). Early diagenesis in Saanich Inlet, a reducing fiord. 3. *Geochim. Cosmochim. Acta.* **36**, 1185.

Christeller, J. T., Lang, W. A., and Troughton J. H. (1976). Isotope discrimination by ribulose-1,5-diphosphate carboxylase. *Plant Physiol.* **57**, 580–582.

Degens, E. T. (1969). Biogeochemistry of stable carbon isotopes. *In* "Organic Geochemistry" (G. Eglinton and M. T. J. Murphy, eds.). Springer-Verlag, Berlin and New York.

Degens, E. T., Guillard, R. R. L., Sackett, W. M., and Hellebust, J. A. (1968). Metabolic fractionation of carbon isotopes in marine plankton I. Temperature and respiration experiments. *Deep-Sea Res.* **15**, 1–9.

DeNiro, M., and Epstein, S. (1977). Mechanism of carbon isotope fractionation associated with lipid synthesis. *Science* **197**, 261.

Eichmann, R., and Schidlowski, M. (1975). Isotopic fractionation between coexisting organic carbon–carbonate pairs in Precambrian sediments. *Geochim. Cosmochim. Acta* **39**, 585.

Epstein, S. (1969). Distribution of carbon isotopes and their biochemical and geochemical significance. *In* "Proc. Symp. on CO_2: Chemical, Biochemical and Physiological Aspects" (Forster *et al.*, ed.), pp. 5–14. (NASA SP-188), Harerford.

Galimov, E. M. (1968a). "Geochemistry of Stable Carbon Isotopes." Nedra, Moscow.

Galimov, E. M. (1968b). Isotopic composition of carbon in gases of the earth's crust. *Izv. AN SSSR, Ser. Geol.* (5), 29–44.

Galimov, E. M. (1973). "Carbon isotopes in oil and gas geology." Nauka, Moscow. English translation: NASA TT-682, Washington, D.C., 1975, 395 pp.

Galimov, E. M. (1976). Variations of the carbon cycle at present and in the geological past. *In* "Environmental Biogeochemistry" (J. Nriagu, ed.), Vol. 1, p. 3.

Galimov, E. M. (1980). $^{13}C/^{12}C$ in kerogen. *In* "Kerogen" (B. Durand, ed.), pp. 271–300. Editions Technip, Paris.

Galimov, E. M., *et al.* (1973). Conditions of formation of gas pools in the eastern part of the Turan plate according to data on the carbon isotopic composition of the gases. *Geokhimiya* (11), 1689–1703.

Galimov, E. M., Kodina, L. A., Generalova, V. N., Bogacheva, M. P., and Shirinskiy, V. G. (1977). Investigation of the distribution of carbon isotopes in biogenic compounds. *In* "Proc. VIII Int. Congress on Organic Geochemistry," pp. 189–203. Nauka, Moscow.

Galimov, E. M., Migdisov, A. A., and Ronov, A. B. (1975a). Variations in the isotopic composition of carbonate and organic carbon of sedimentary rocks in Earth history. *Geokhimiya* (3), 323–342.

Galimov, E. M., Shirinskiy, V. G., Bordovskiy, O. K., Zaikin, V. G. (1975b). Isotopic composition of the carbon of lipids in Indian Ocean plankton. *Geokhimiya* (6), 895–904.

Galimov, E. V., Kodina, L. A., Bogacheva, M. P., and Shirinsky, V. G. (1980). Organic geochemistry study of samples from DSDP Sites 474, 477, 478, 479, 481 in Gulf of California, Leg 64. *In* "Initial Reports of Deep Sea Drilling Project," Vol. 64. U.S. Government Printing Office, Washington, D.C., 819–836.

Hayes, J. M. (1970). Organic constituents in meteorites: a review. *Geochim. Cosmochim. Acta* **31**(12), 1395.

Hoefs, J. M., and Schidlowski, M. (1967). Carbon isotope composition of carbonaceous matter from the Precambrian of the Witwatersrand system. *Science* **155**, 1096.

Hoering, T. C. (1967). *In* "Researches in Geochemistry" (P. Abelson, ed.), p. 2. Wiley, New York.

Junge, C. E., Schidlowski, M., and Eichman, R. Pietreck. (1975). Model calculations for the terrestrial carbon cycle: carbon isotope geochemistry and evolution of photosynthetic oxygen. *J. Geophys. Res.* **80**, 4542.

Manskaya, S. M., and Drozdova, T. V. (1964). "Geochemistry of Organic Matter." Nauka, Moscow.

Melander, L. (1964). "Isotopic effects in rate reactions." Mir. Moscow (Russian translation).

Nissenbaum, A. (1974). "The organic geochemistry of marine and terrestrial humic substances: implications of carbon and hydrogen isotope studies" (Adv. Org. Geochem.). Editions Technip, Paris.

Nissenbaum, A., and Kaplan, I. R. (1972). Chemical and isotopic evidence for the *in situ* origin of marine humic substances. *Limnol. Oceanogr.* **17**(4), 570–582.

Nissenbaum, A., Baedecker, M. J., and Kaplan, I. R. (1971). "Studies on dissolved organic matter from interstitial water of a reducing marine fjord" (Adv. Org. Geochem.), pp. 427–440. Pergamon Press, Oxford.

Nissenbaum, A., Presley, B. J., and Kaplan, I. R. (1972). Early diagenesis in Saanich Inlet, a reducing fjord. 1. *Geochim, Cosmochim. Acta* **36**, 1007.

Ohler, D. Z., Schopf, J. M., and Kvenolden, K. A. (1971). Carbon isotopic studies of organic matter in Precambrian rocks. *Science* **173**, 1246.

Pardue, J. W., Scalan, E. S., van Baalen, C., and Parker P. L. (1976). Maximum carbon isotope fractionation in photosynthesis by blue-green algae and green algae. *Geochim. Cosmochim. Acta.* **40**, 309.

Philp, R. P., and Calvin, M. (1976). Kerogen structures in recently deposited algal mats in Laguna Mormona, Baja California: a model system for the determination of kerogen structures in ancient sediments. *In* "Environmental Geochemistry" (Nriagu, ed.) *Ann Arbor Sci.* **1**, 131.

Prasolov, E. M., and Lobkov, V. A. (1977). On the conditions of formation and migration of methane (according to the carbon isotopic composition). *Geokhimiya* (1).

Raschid, M. A., and King, L. H. (1970). Major oxygen-containing functional groups present in humic acid fractions isolated from contrasting marine environments. *Geochim. Cosmochim. Acta.* **34**, 193–202.

Rogers, M. A., and Koons, C. B. (1969). Organic carbon $\delta^{13}C$ values from Quaternary marine sequences in the Gulf of Mexico: a reflection of paleotemperature changes. *Trans. Gulf Coast Assoc. Geol. Soc.* **19**, 529–534.

Rogers, M. A., and Hinte, J. E., and van Sugden, J. G. (1972). Organic carbon $\delta^{13}C$ values from Cretaceous, Tertiary and Quaternary marine sequences in the North Atlantic. *In* "Initial Reports of Deep Sea Drilling Project," Vol. 12, p. 1115. U.S. Government Printing Office, Washington, D.C.

Sackett, W. M. (1968). Carbon isotope composition of natural methane occurrences. *Bull. Am. Assoc. Petrol. Geol.* **52**(5), 853.

Sackett, W. M., and Chang, H. M. (1979). Experimental confirmation of the lack of carbon isotope exchange between methane and carbon oxides at high temperature. *Geochim. Cosmochim. Acta* **43**, 273–276.

Sackett, W. M., Eckelman, W. R., Bender, M. L., and Be, A. W. H. (1965). Temperature dependence of carbon isotope composition in marine plankton and sediments. *Science* **148**, 235–237.

Sackett, W. M., and Rankin, J. G. (1970). *J. Geophys. Res.* **75**, 4557.

Schitzer, M., and Neyroud, J. A. (1975). Alkanes and fatty acids in humic substances. *Fuel* **54**, 17–19.

Shirinskiy, V. G. (1978). "Regularities of Fractionation of Carbon Isotopes in Natural Lipids." Dissertation abstract, GEOKhI AN SSSR, Moscow.

Sidorenko, S. A., and Sidorenko, A. B. (1975). "Organic Matter in Precambrian Metasedimentary Rocks." Nauka, Moscow.

Smith, B. N., Hearith, H. M. W., and Chase, J. B. (1973). Effects of growth temperature on carbon isotope ratios in barley, pea and rape. *Plant. Cell. Physiol* **14**, 177–182.

Stevenson, F. J., and Goh, K. M. (1971). Infrared spectra of humic acids and related substances. *Geochim. Cosmochim. Acta* **35**(5), 471.

Troughton, J. H., and Card, K. A. (1975). Temperature effects on carbon isotope ration of C_3, C_4 and crassulacean–acid-metabolism (CAM) plants. *Planta* **123**, 185–190.

Stahl, W. (1968). On the origin of natural gases of northwest Germany. *Erdölund Kohle, Erdgase, Petrochem.* **21**(9).

Teichmüller, R., Teichmüller, M., and Colombo, U., *et al.* (1970). The carbon isotope ratio in methane from crude gas and refined gas and its dependence on the geologic relationships—geologic communications. *Ztschr. Allg. Region. Angew. Geol.* (9), 181–206.

Welte, D. H., Hageman, H. W., and Hollerbach, A., *et al.* (1975). World Petr. Congr. Tokyo, PD3, Topic 5.

Whelan, T. W., Sackett, M., and Benedict, C. R. (1973). Enzymatic fractionation of carbon isotopes by phosphoenolpyruvate carboxylase from C_4 plants. *Plant Physiol.* **51**, 1051–1054.

CHAPTER 12

The Thermodynamic State of Life

Is a thermodynamically ordered distribution of isotopes in biosystems compatible with the idea of their essentially nonequilibrium composition and structure?

Some investigators [see Galimov (1978)] have expressed doubts as to the equilibrium nature of the observed isotopic effects because of the following consideration: Isotopic equilibrium is established as a result of chemical equilibrium. Therefore, they feel that thermodynamic isotopic effects can take place only in chemically equilibrated systems, and the supposition of chemical equilibrium is absurd when applied to living systems.

That reasoning, however, contains an essential inaccuracy. The rule according to which isotopic equilibrium occurs after chemical equilibrium pertains only to a chemical reaction in which isotope exchange is inherent but not to equilibrium of the system in general. For example, let us take a system consisting of acetone and water. Under normal conditions these substances do not react with each other to form other more stable products. Nevertheless, if one of them is deuterated, after a certain time equilibrium distribution of deuterium is established in both compounds. The reactions providing isotopic equilibrium in this case are the exchange between the hydrogen of the water and the hydroxyl of the enol form of acetone, accompanied by keto-enol tautomerization of acetone.

Only the equilibrium of this last reaction, not the elementary equilibrium in the acetone–water system, is the isotopic exchange equilibrium condition. Applied to biosystems this means that equilibrium of the process of selection of isotopic forms in the enzyme–substrate complex does not require elemental equilibrium in the biomolecules.

This determines the answer to the question raised at the beginning of the chapter. But the root of the question is deeper. It seems that manifestation of thermodynamically ordered distribution of isotopes in biosystems emphasizes the fact that the accepted interpretation of biological systems as exclusively nonequilibrium systems does not answer the questions concerning the thermodynamic state of living matter. Here it is important to introduce some related considerations.

An inherent property of biological systems is the fact that biochemical reactions are governed by enzyme catalysts. The catalyst affects the rate of reactions only secondarily—accelerating one reaction and retarding another. The primary importance of the catalyst is that by interacting with the force field of the atoms of the reacting molecules, it modifies the pathway of conversion of the system to a more stable state. The presence of the catalyst changes the topography of the potential surface which characterizes the interaction in the system. In this sense catalysis is a special case of the manifestation of the force fields in the system. These are diverse, beginning with the external fields and ending with interaction between atoms of the compounds which form the system. The orienting action of a seed in the process of crystallization or the role of the matrix in the reproduction of copies can also be regarded as catalysis. The presence of a catalyst means the presence of some specific (determined by the properties of the catalyst) interaction in the system. Any interaction, in essence, is an instruction as to how the process of conversion from one energy state to another should be accomplished.

The existence of interactions predetermines the ordering of the system to a certain extent. The idea that a completely disordered, chaotic distribution of particles is most likely (maximum entropy) is valid only in the absence of force fields. The state of a system is determined by the competition of interactions and thermal movement. The lower the temperature, the more important the role of interactions. At the temperature of absolute zero, the entropy is zero. The system is completely ordered in the sense that its state is determined only by the existence of interactions.

Ordered, including highly ordered, systems may be equilibrium systems. The distribution of metallic particles in a magnetic field can form an intricate pattern in correspondence with the directions of the lines of force. This distribution will be in equilibrium for the given conditions, although also fairly highly ordered. The same is true for chemical systems. Therefore, in a state of equilibrium it is not necessary that the elements constituting living matter must form simple compounds such as CH_4, NH_3, CO_2, and H_2S. Complex polyatomic compounds typical of living matter can, in principle, represent equilibrium compositions in the structure of interactions, which is due to the set of catalysts operating in biological systems.

It must be stressed that a feature of thermodynamics is that it is abstracted from the idea of interactions and considers a process as if no interactions existed, i.e., the subsystems forming a given system behave independently. Sometimes it is assumed that the independence of the particles forming the system is analogous to the equation of state of an ideal gas and that the limitations connected with it can be eliminated with the passage to more complex equations of state of a real substance. But this is not so. The idea of the independence of subsystems and thus of the additivity of energies is much more fundamental. In addition to the requirement of maximum thermodynamic probability, it is the starting point for obtaining the expression for the partition functions through which all thermodynamic functions are represented. If particles having energies ε_1 and ε_2 interact, their total energy is not equal simply to the sum $\varepsilon_1 + \varepsilon_2$. It differs from that by some value ε_{12}, which generally depends on the coordinates and particle velocities. In statistical mechanics courses, it is stressed that "actually obtaining valuable results is possible only in the case where the states and movements of a complex material system can be broken down into states and movements of almost independent systems" (Fowler and Guggenheim, 1949, p. 16).

If thermodynamics nevertheless considers real interacting systems, the boundaries of the assumptions involved must be clearly recognized. The energy of an interaction with conservative fields can be represented as a potential function depending only on the coordinates and added to the energy of each subsystem. Then energies ε_i can be summarized. The so-called configuration integral, which contains the corresponding potential function, enters into the expression for the partition function. As a result, the equilibrium distribution of particles can be described (in a magnetic or gravitational field, for instance). By choosing the potential function in a certain way, the state of real gases and liquids can be described in "conservative" approximation. In a chemical reaction the interactions of atoms with each other are similar to interactions with a nonconservative field, i.e., a field depending on time. However, if an atom is observed for a long enough period of time that it can enter into different reactions, then the energies of its interactions can be averaged. Then the value ε_i will contain a term determined by the interaction but not dependent on time (time-averaged), i.e., the value ε_i acquires the property of additivity. The duration of the time during which ε_i can be summarized in conservative approximation depends on the nature of the interactions. This time is the actual time of establishment of equilibrium in a system of interacting particles.

In the time scale of the establishment of equilibrium, the respective chemical compounds can be considered to be freely interconvertible, and the atoms in the direction of the reaction can be considered to be independent particles. In some cases regrouping of atoms can be accomplished in some certain direction more freely than in others, corresponding to the equilibrium of a certain chemical reaction.

Generally, a catalyst is the source of interactions which provide equilibrium in specific reactions. For instance, if the original compound is ethyl alcohol, using various catalysts one can obtain the following equilibria†:

$$(1)\quad C_2H_5OH \xrightleftharpoons{(Na)} \tfrac{1}{2}C_4H_9OH + \tfrac{1}{2}H_2O,$$

$$(2)\quad C_2H_5OH \xrightleftharpoons{(Cu)} CH_3CHO + H_2,$$

$$(3)\quad C_2H_5OH \xrightleftharpoons{(Al_2O_3)} C_2H_4 + H_2O,$$

$$(4)\quad C_2H_5OH \xrightleftharpoons{(Cu\ active)} \tfrac{1}{2}CH_3COCH_3 + \tfrac{3}{2}H_2 + \tfrac{1}{2}CO.$$

If the condition of free interconversion (independence of subsystems) is fulfilled, the reaction can be investigated by the method of equilibrium thermodynamics (the equilibrium constants and ratio of the components are determined), and the presence or absence of a catalyst has no effect on the result.

Thus to make practical use of the thermodynamic method, it is necessary to isolate in a real system that part of it, or that reaction in it, for which the assumption of independence of subsystems would be maximally applicable. Systems in which the aforesaid condition is not fulfilled cannot feasibly be examined as a whole, any more than it would be feasible to consider reactions between reagents contained in different vessels on a laboratory table. Biological systems are precisely such systems. While they are ordinary physiochemical systems, they are distinguished by the fact that the interactions in them are so essential and so specific that it is nearly impossible to idealize them to a sense convenient for application of the thermodynamic method.

The secret of highly ordered biological systems and of the remarkable efficiency of the functioning of living organisms is not a consequence of their being physicochemical systems but rather of the specific structure of the interactions in the system, which was developed by evolution. It is extremely important to note that this structure is not the product of development of a given physiochemical system. An information system, including DNA and transcribing enzymes, transmitted to the organism in the zygote at conception, bears the instruction as to which enzymes should arise and which reactions they should govern. The chemistry of development of an organism in ontogeny is the process of development of equilibrium in the "force field" given the instructions.

During the life of an organism, a mass of substance (from the surrounding environment) which previously was at a lower level of organization is

† In practice these reactions are carried out at somewhat different temperatures, but in this context that is of no importance in principle.

involved in a high degree of ordering—the entropy of the system is decreased. A point of view exists that this occurs due to extraction of low-entropy material from the environment. Schrödinger wrote:

> A living organism continually increases its entropy, or else produces positive entropy and thus approaches the dangerous state of maximum entropy which is its death. It can avoid this state, i.e. remain living, only by constantly extracting negative entropy from its surrounding environment.... In fact, in the higher animals we know fairly well the kind of ordering by which they are nourished, namely, an extremely well ordered state of material of more or less complex organic compounds is their food. After use, the animals return these substances in a very degraded form, but not completely degraded because they can still be used by plants (Schrödinger, p. 121).

Lehninger (1974) in his "Biochemistry" develops this approach with respect not only to the organism as a whole, but also to individual biomolecules. He writes:

> The conclusion according to which the native biologically active conformation of the polypeptide chain apparently is the most stable conformation can be shown to be paradoxical and incompatible with the laws of thermodynamics operating in "nonliving" systems. Systems with a high degree of ordering are by definition characterized by relatively low entropy. Therefore it might seem strange that at pH = 7.0 an unwound polypeptide chain (unraveled into a disordered tangle), having no biological activity and characterized by a relatively high entropy, can spontaneously form a more ordered structure with specific biological activity, i.e. a system with less entropy. But this is only a seeming paradox. We can resolve it by remembering that according to the second law of thermodynamics, in the case of equilibrium the sum of the entropies of a system (polypeptide chain) and of the surrounding environment (aqueous solution) becomes maximal. This law completely allows spontaneous decrease in entropy of a nonisolated system, which the polypeptide chain is, under the condition that the entropy of the surrounding environment, i.e. of the aqueous solution, increases at the same time... (p. 852).

This is valid. But the nonisolated nature of the system is only a condition, not a cause. The question, Because of what properties of biological systems is negative entropy directed precisely toward the organism?, remains open.

A decrease in entropy in a certain part of a system is possible when it corresponds to a lower level of its free energy. An example is the process of crystallization. The entropy of the crystalline phase is lower than that of the original liquid. The process of crystallization is accompanied by an increase in ordering and a decrease in entropy. This process is possible because the crystalline state corresponds to a lower level of free energy of the system. For that process to be realized, the system must transfer the excess internal energy to the surrounding environment (constant temperature), i.e., the system must be open (in an insulated system, entropy cannot be decreased). In our view, a decrease in entropy in the transformation of nonliving into living matter, as in other natural processes, is accompanied by minimization of free energy in the system. However, it is important to point out that such a process, related to a decrease in entropy of an open system, leads to equilibrium, for the minimum free energy corresponds to a state of equilibrium. When we say that an organism "lives on negative entropy," we mean, on the contrary, that the system is far from being in equilibrium.

Equilibrium actually is not achieved during the life of an organism because the enzymes governing the synthesis of biomolecules are in turn the product of biosynthesis.

Here it is necessary to point out that a unique test of the thermodynamic state of biological systems is the ratio of the isotopic compositions of biomolecules. The fact is that the isotopic composition primarily is the ratio of the contents of the isotopic forms. Correspondingly, the thermodynamic value characterizing the isotopic properties of a compound is the ratio of the partition functions of the corresponding isotopic forms (β factors). It is important that the force field of the molecules does not change upon isotopic substitution. The various isotopic forms of the molecules are all characterized by the same set of force constants. Therefore, the force field in which the transformation of isotopic molecules take place in the process of catalysis is indistinguishable. Fractionation of isotopes occurs not because the force fields of the molecules are different but because the behavior of particles (atoms) of different masses in the same force field is different. This means that the configuration integrals characterizing the interaction in which the isotopic forms participate (at least according to conservative approximation) are the same and cancel out in the expression for the β factor. Therefore, no instructions can be imparted to the instructional molecules concerning the choice of isotopic forms of molecules. The isotopic composition of the biomolecules cannot be coded. There are no isotopically specific enzymes.

Consequently, the specific structure of the reactions, which instructs the elemental distribution in biological systems and makes the principle of independence of subsystems inapplicable with respect to them, does not extend to isotopic distribution. Therefore, if a tendency toward equilibrium is an essential feature of a biological process, it should be manifested in a tendency toward equilibrium distribution of isotopes. The observed thermodynamically ordered distribution of isotopes once more reflects this tendency.

Thus biological systems do not belong to any special class of systems which by-pass equilibrium or are on a special entropy diet in contrast to other natural systems ("live on negative entropy"). Ordering of biological systems is not the result of the biological process itself but is produced by interactions, the instructions for which are transmitted to the organism at conception. The biological process proceeds in the direction of a decrease in free energy in the system. Theoretically, the development of an organism in ontogeny is no different from other natural processes, for instance, the process of crystallization, although in the biological process the "seed" is incomparably more complex.

The entropy problem mentioned above has two aspects. One is to reconcile the high ordering of the organism as a chemical system with the second law of thermodynamics. The other is to reconcile the biological evolutionary principle with the thermodynamic evolutionary principle, i.e., to reconcile the appearance of more and more highly organized forms of life in the course of evolution with the assertion of thermodynamics concerning the general direction of the processes toward disordered systems.

While the low-entropy state of an organism, acquired in ontogeny, is caused by an element of order introduced from without (DNA), evolution of the information molecule itself from the conception of life through increasingly higher degrees of phylogenetic organization is truly a spontaneous process. Thus the problem of entropy has a different meaning in these two cases.

"It is generally known that it is difficult to give a definition of life." Quoting these words of Crick, Eigen comments: "because the transition between the living and nonliving is not sharp." And later, "returning to this stage [the moment of transition from nonliving to living] we must begin with systems which formerly were not "living" (for instance, minerals) and end with those from which at least as a result of evolution there results what we definitely call "life": bacterial cells, plants, animals" (Eigen, 1973, p. 26). Eigen thus speaks of the intuitive ability to distinguish the living from the nonliving, precisely as did Schrödinger, who wrote:

> When do we think a material is living? When it continues to "do something," moves, takes part in exchange of matter with the surrounding environment, etc.—all this in the course of a longer time than we would expect inanimate matter to do under similar conditions (Schrödinger, 1973).

One nearly always tries to define life in terms of properties of the living. But it is clear that if the physicochemical content of biological development of a cell or organism is a trivial process similar to crystallization, then the definition of life should be sought in ideas of a different category.

An organism must belong to a class of entities constructed according to instructions. Also included in this class are human-made structures and mechanisms. There is no difference between them in principle. In both cases there is ordering of the material. In both cases this ordering is produced according to a plan prescribed from without. An organism is a material formation created from elements of the external environment by instructions carried in the information molecule. In a technological system the plan of the person who conceives it and directs its creation is realized precisely this way. The information molecule organizes space at the atomic level; man organizes material at the macroscopic level. The circumstance that creates the obvious difference between organisms and artificial systems and permits us to distinguish them intuitively is that we always perceive an organism in harmony with its molecular informational mechanism, while artificial systems are spatially separated from man, who plays the role of carrier of the instructions with respect to them.

If the essence of life is impossible to define adequately through the properties of the organism, then evidently its main content is somehow related to the evolution of the instruction itself, or the type of evolution of the material. But a single law of evolution of material is known, expressed by the second law of thermodynamics. According to it the evolution of material consists of its disordering and increase in entropy. But it is essential that the second law of thermodynamics be statistical in nature. It indicates the direction of the process with great probability but is not determinate. There always remains some probability that a process will be accomplished against the general stream of evolution—negative fluctuation of entropy. In individual parts of a system or in individual subsystems, negative fluctuations of entropy are constantly being realized. However, not being interconnected, they flash like a spark, only to be extinguished immediately. But in principle, accumulation of fluctuations of entropy is possible. For this, the probability that negative fluctuations of entropy will arise has to be higher in the subsystem which underwent the last negative fluctuation of entropy.

The mechanism of such accumulation is known—it is selection. The general idea of selection as the principle of biological evolution belongs to Darwin. In Oparin's 1924 book, we first find the idea of natural selection at the molecular level. It was expanded by Eigen (1973).

It is obvious that the evolution of instructions would be accomplished by organization of the elements of the environment. Selection includes the production of copies and testing the stability of a subsystem in given conditions. The total of physiochemical processes carried out in this case, i.e., processes accompanying the development of living entities, is in complete agreement with the second law of thermodynamics, as has already been stressed. The meaning of negative fluctuations of entropy comes down to the spontaneous change in the instructing molecule which appears in the selective advantages of its derivatives. The latter may be on the level of molecules, cells, organisms, or societies. Their displacement of the derivatives of a former version of the information molecule predetermines the emergence of a new negative fluctuation of entropy in the subsystem which was strengthened as a result of a previous fluctuation.

Thus life, defined as a type of evolution of matter consisting of the accumulation of negative fluctuations of entropy, falls in a special relationship to the second law of thermodynamics. It is not inconsistent with it, but is not identical to it. Life can be regarded as a type of evolution of matter which in its essence is complementary to the second law of thermodynamics. Life is the result of its nonabsolute determinism (statistical character), which allows for very unlikely departures from the main trend of the evolution of material. This main trend can accumulate by means of selection, producing a line of evolution of material opposite to the main direction of material movement, which is toward disorder.

Hence the concept of life should occupy the following place in the hierarchy of fundamental concepts: It is not a way of existing, nor a form of organization; it is not a property and not an ability. A much more general concept is that life is an evolution of material. The possibility of life is intrinsic to all spaces and times.

REFERENCES

Eigen, M. (1973). "Self-Organization of Matter and Evolution of Biological Macromolecules." Mir. Moscow (Russian translation).

Fowler, R., and Guggenheim, E. (1949). "Statistical Thermodynamics." Izd-vo Inostr. Lit., Moscow (Russian translation).

Galimov, E. M. (1978). On the concept of thermodynamically ordered distribution of isotopes in biological systems and errors related to its understanding. *Geokhimiya* (10), 1570–1586.

Lehninger, A. (1974). "Biochemistry." Mir. Moscow (Russian translation).

Oparin, A. I. (1924). "The Origin of Life." Moskovskiy Rabochiy, Moscow.

Oparin, A. I. (1968). "Life, Its Nature, Origin and Evolution." Nauka. Moscow.

Schrödinger, E. (1973). "What Life Is from the Point of View of Physics." Atomizdat, Moscow.